Synergetics Far from Equilibrium

Structural Stability in Physics
Editors: W. Güttinger and H. Eikemeier

Pattern Formation by Dynamic Systems and
Pattern Recognition
Editor: H. Haken

Volume 6 Dynamics of Synergetic Systems
Editor: H. Haken

Springer Series in Synergetics

Synergetics, an interdisciplinary field of research, is concerned with the cooperation of individual parts of a system that produces macroscopic spatial, temporal or functional structures. It deals with deterministic as well as stochastic processes.

Volume 1 **Synergetics** An Introduction 2nd Edition
By H. Haken

Volume 2 **Synergetics** A Workshop 3rd Printing
Editor: H. Haken

Volume 3 **Synergetics** Far from Equilibrium
Editors: A. Pacault and C. Vidal

Volume 4 **Structural Stability in Physics**
Editors: W. Güttinger and H. Eikemeier

Volume 5 **Pattern Formation** by Dynamic Systems and **Pattern Recognition**
Editor: H. Haken

Volume 6 Dynamics of **Synergetic Systems**
Editor: H. Haken

Dynamics of
Synergetic Systems

Proceedings of the
International Symposium on Synergetics,
Bielefeld, Fed. Rep. of Germany,
September 24–29, 1979

Editor: H. Haken

With 146 Figures, Some in Color

Springer-Verlag Berlin Heidelberg New York 1980

Professor Dr. Hermann Haken

Institut für Theoretische Physik der Universität Stuttgart
Pfaffenwaldring 57/IV, D-7000 Stuttgart 80, Fed. Rep. of Germany

ISBN-13: 978-3-642-67594-2 e-ISBN-13: 978-3-642-67592-8
DOI: 10.1007/978-3-642-67592-8

Softcover reprint of the hardcover 1st edition 1980

2153/3130-543210

Preface

This book contains the invited papers of an international symposium on Synergetics which was held at ZIF (Center for interdisciplinary research) at Bielefeld, Fed. Rep. of Germany, Sept. 24.-29., 1979. In keeping with our previous meetings, this one was truly interdisciplinary.

Synergetic systems are those that can produce macroscopic spatial, temporal or functional structures in a self-organized way. I think that these proceedings draw a rather coherent picture of the present status of Synergetics, emphasizing this time theoretical aspects, although the proceedings contain also important contributions from the experimental side.

Synergetics has ties to many quite different disciplines as is clearly mirrored by the following articles. Out of the many ties I pick here only one example which is alluded to in the title of this book. Indeed, there is an important branch of mathematics called dynamic systems theory for which the problems of Synergetics might become an eldorado. While, undoubtedly, a good deal of dynamic systems theory had been motivated by mechanics, such as celestial and fluid dynamics, Synergetics provides us with a wealth of related problems of quite different fields, e.g., lasers or chemical reaction processes. In order to become adequately applicable, in quite a number of realistic cases dynamic systems theory must be developed further. This is equally true for a number of other approaches.

I hope the reader will agree with the opinion of the participants at the meeting, namely, that Synergetics is a fascinating new field with many challenging problems waiting for an experimental or theoretical solution.

It is a pleasure to thank ZIF, Bielefeld, and especially its present director, Prof. Horn, for inviting us to ZIF. I wish to thank the staff members of ZIF for the perfect organization and their assistance at every stage of the conference. My thanks go further to Prof. Dress of the University of Bielefeld for his initiative and support. Last but not least I express my cordial thanks to my secretary, Mrs. U. Funke, for her valuable assistance in the preparation of that meeting and in editing these proceedings.

Stuttgart, December 1979 *H. Haken*

Contents

Part VI. *Self-organization of Biological Macromolecules*

Part VII. *Dynamics of Multi-unit Systems*

Part VIII. *Models of Psychological and Social Behavior*

Part IX. *Mathematical Concepts and Methods*

Part I

Introduction

Lines of Developments of Synergetics

H. Haken

Institut für Theoretische Physik der Universität Stuttgart
D-7000 Stuttgart 8, Fed. Rep. of Germany

1) Introduction and survey

According to its definition Synergetics is concerned with the cooperation of the individual parts of a system that produces macroscopic spatial, temporal or functional structures. Quite in the spirit of our previous meetings the symposium at Bielefeld brought together scientists from various disciplines. The topics treated at this symposium and being dealt with by the articles of the present volume mirror quite well the present status of synergetics, in particular with respect to theoretical developments. A survey is given in fig.1.

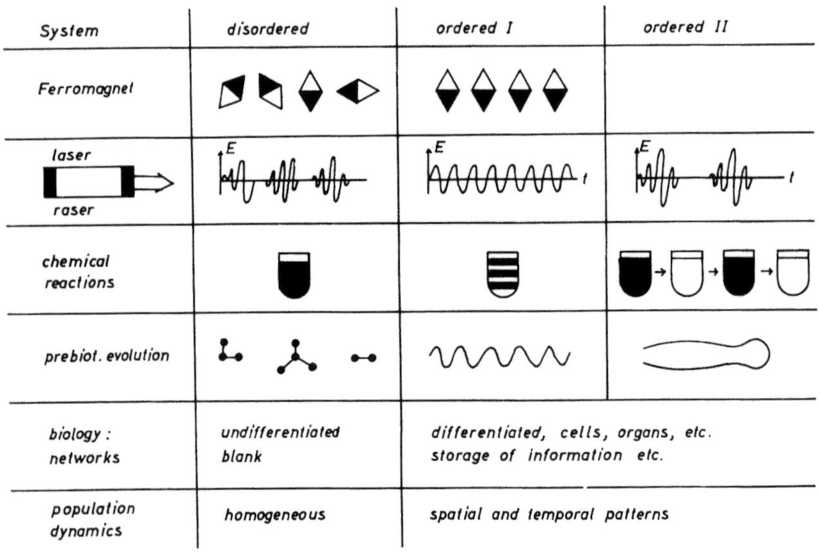

Fig. 1 Compare text

In physics quite a number of systems are known which, being composed of many subsystems, change their macroscopic behaviour dramatically when certain external parameters are varied. An important class of such phenomena are phase transitions of systems in thermal equilibrium and this field is still treated intensely by statistical physics as well as experimentally. Though it is not quite in the focus of synergetics we can learn quite a lot of the basic concepts and methods developed in this field. I am therefore particularly glad that Stanley [1] gave us a detailed review on its status. Miller communicated recent results on phase transitions in the relativistic case. As is known, Synergetics deals mainly with physical systems far from thermal equilibrium or with cooperative effects shown by nonphysical systems. An example which I am presenting again and again is the laser which can show a number of ordering effects. Some of them can be characterized as nonequilibrium phase transitions. Over the past years Brun [1] has performed beautiful experiments on this kind of transitions in the radio frequency range and he gives us a detailed account of his results. The contribution by Haug and Koch shows how the concept of nonequilibrium phase transitions applies to current problems of solid state physics. As we know, when disorder-order transitions occur, fluctuations play an important role. A new class of transition phenomena has been found by Horsthemke, Mansour, Arnold and Lefêver, who show that nonequilibrium phase transitions can be even caused by fluctuations, and we can find more about this in the paper by Horsthemke. A broad class of spatio-temporal organization on macroscopic scales is provided by chemical reactions. A number of speakers, namely Smoes, Arnold, Fife, Kuramoto and Nitzan illuminated this fascinating and extremely rapidly growing field from various points of view. This leads us to the question of selforganization in the domain of biology. An important fundamental role is played here by problems of prebiotic evolution or, in other words, of evolution processes on the molecular level. These topics are covered by two experts in this field, namely Schuster and Sigmund.

We then proceed to questions of mathematical biology dealing with the behaviour of systems composed of many units. Questions treated here are how undifferentiated cells become differentiated within morphogenetic fields so to produce patterns or even organs. Problems of this and related kinds are treated by Babloyantz and Othmer. A rather complex biological system is the immune system which is dealt with theoretically by Lefêver. When we try to deal with synergetic systems on the next level we are led into the problems of the behaviour of groups of individuals which is studied by population dynamics. An interesting aspect, namely the motion and settlement of different social groups in a city is treated by Weidlich. An exciting approach in the field of behavioural science is presented by J. Nicolis combining game theory with bifurcation theory. On the other hand, concomitant with the exposition of various explicit examples, other papers deal with general mathematical methods to cope with these problems. Two papers are devoted to the further development of methods with stochastic processes. As is known, the master equation is a very useful tool and Gardiner and Turner report on some recent developments how to solve master equations. The relation between deterministic and stochastic approaches is studied by Arnold, while Rössler presents recent results on chaotic processes. The paper of Poston gives us a general overview starting from catastrophe theory and notions such as transversality and generic situations. In my present paper I should like to follow up again the traditional line of the synergetics enterprise. I wish to analyse a specific example further and I shall indicate on the other hand a general method of approach. As the reader will notice, neither the analysis of my example nor my suggestions of a general approach are complete. Quite on the contrary synergetics is an open field of study with many challenging problems. In the following I shall describe a new type of synergetic phenomenon, and I shall show how the laser is a beautiful example of a dynamic system showing a bifurcation scheme including nodes, limit cycles, tori and strange attractors. Then I shall sketch a general basis of a theory of qualitative changes of systems and shall conclude with some epistemological remarks.

2) A new type of synergetic phenomenon: selforganization through increase of number of components

Whenever I have attended a meeting dealing with problems of synergetics and I had presented the laser as an example I told myself that next time I should not bore my audience any more by that example. However, each time I discovered a new facet of laser theory which seems to illustrate general aspects of synergetics quite well. So it happens again this time.

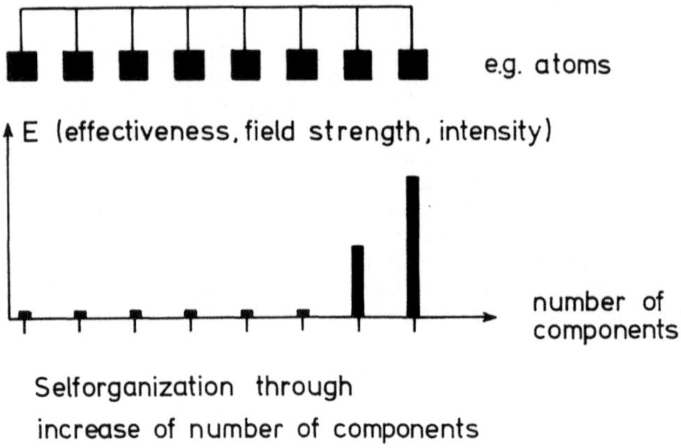

Fig. 2 The increase of number of components can lead to qualitatively new properties of a system (schematic)

As is well known the laser consists essentially of two main physical parts, namely the laser atoms and the field produced by the atoms [2]. So far our picture was the following. When the laser atoms are energetically pumped weakly, the light field produced consists of incoherent wave tracks representing the incoherent light of lamps. On the other hand, when the pump strength exceeds a certain critical value, laser action sets in and a practically infinitely long light wave is emitted. We can, however, induce the transition from "lamp" to "laser" not only by increasing the pump, i.e. the inversion of the atoms, but also in the following way. Let us assume a laser cavity into which we add one laser atom after the other each being pumped at a certain strength. Then first no laser action occurs. But suddenly, at a critical number of atoms, laser action sets in. This is a new type of synergetic phenomenon. The laser atoms start coherent oscillation in a selforganized way when their number exceeds a certain critical value. It is not too hard to cast this phenomenon into a rigorous mathematical form not restricted to laser physics. We then are led to some kind of extended bifurcation theory in which the control parameter is provided by the number of components. I think that this new approach has a number of promising applications. Indeed we see that just by mere increase of the number of components a system can quite suddenly act in a qualitatively entirely different manner from what it could do before. It is remarkable that all components of the total system have exactly the same structure so that the newly added components are not distinguished in any way from the former ones. I shall publish the details of this approach elsewhere.

3) The laser as an example of a dynamic system: bifurcation scheme including nodes, limit cycles, tori and strange attractors

In the course of its developments laser theory has dealt with various aspects. Quite naturally first the properties of laser light, such as its coherence and other statistical properties, stood in the foreground. Though my first paper on laser noise and coherence showed analogies between the laser transition and phase transitions [3] the most explicit demonstration of this analogy was provided in the next step [4],[5] showing, for instance, the analogy between lasers and one-dimensional superconductors [6] . In this section, I should like to show that semi-classical laser theory provides us with a rather complete picture of a dynamic system showing various phenomena envisaged by dynamic systems theory. For sake of completeness let me first describe the laser equations. In the following we shall treat the laser as a one-dimensional device with periodic boundary conditions which can be achieved for instance by a triangular arrangement of mirrors. The three quantities we are dealing with are

$E(x,t)$ electric field strength

$P(x,t)$ atomic polarization of the medium

$D(x,t)$ the atomic inversion

x is the spatial coordinate along the laser axis and t is the time-coordinate.

As is shown in laser theory [2] these quantities obey the following partial differential equations

$$-\frac{\partial^2 E(x,t)}{\partial x^2} + \frac{1}{c^2}\frac{\partial^2}{\partial t^2}E(x,t) + \frac{4\pi\sigma}{c^2}\frac{\partial}{\partial t}E(x,t) =$$

$$= -\frac{4\pi}{c^2}\frac{\partial^2}{\partial t^2}P(x,t) \qquad (3.1)$$

$$\frac{\partial^2}{\partial t^2}P(x,t) + 2\gamma\frac{\partial}{\partial t}P(x,t) + \omega_0^2 P(x,t) =$$

$$= -2\omega_0\left(\left(|\vartheta_{12}|^2\right)/\hbar\right)E(x,t)D(x,t) \qquad (3.2)$$

$$\frac{\partial}{\partial t}D(x,t) = \frac{D_0 - D(x,t)}{T} + \frac{2}{\hbar\omega_0}E(x,t)\frac{\partial}{\partial t}P(x,t) \qquad (3.3)$$

c is the velocity of light, σ the electric conductivity, γ the damping constant of the atomic polarization, ω_0 the eigenfrequency of the polarization, ϑ_{12} the optical dipole matrix element, $\hbar$ the Planck's constant. D_0 serves as a control parameter which can be manipulated from the outside by the pump strength. T is the relaxation time of the inversion. For what follows the meaning of these different constants is not important except for the fact that D_0 serves as control parameter. It is useful to describe the solutions of the systems (3.1) - (3.3) using the decomposition of the electric field strenth according to Fig. 3, namely by

$$E(x,t) = e^{ik_o x - i\omega_o t} E^{(+)}(x,t) + c.c. \tag{3.4}$$

$$P(x,t) = e^{ik_o x - i\omega_o t} P^{(+)}(x,t) + c.c. \tag{3.5}$$

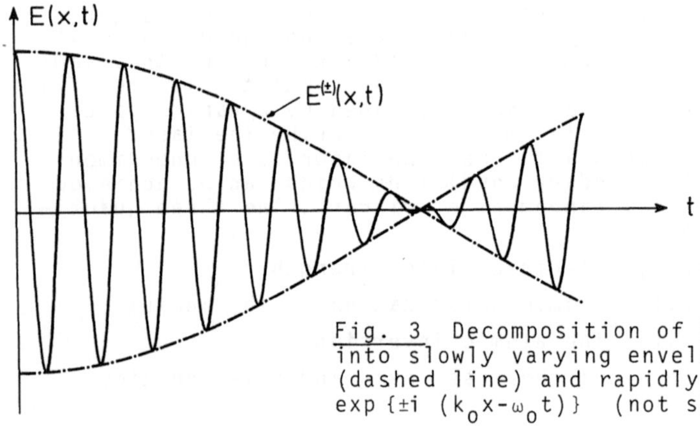

Fig. 3 Decomposition of $E(x,t)$ (solid line) into slowly varying envelope $E^{(\pm)}(x,t)$ (dashed line) and rapidly oscillating functions $\exp\{\pm i(k_o x - \omega_o t)\}$ (not shown)

In it we decompose the field strength and the polarization into a rapidly oscillating exponential function, which contains the eigenfrequency ω_o of the polarization and the corresponding wave vector k_o, and into slowly varying amplitudes.

In order to understand the following tables I must make a few remarks about the usual approach in laser theory. Adopting the hypotheses (3.4), (3.5) one applies the "rotation wave approximation" and the "slowly varying amplitude approximation". These approximations, which I shall not discuss here, allow one to reduce the system of differential equations (3.1) - (3.3) to the following equations:

$$\frac{\partial}{\partial t} E^{(+)}(x,t) + c \frac{\partial}{\partial x} E^{(+)} + 2\pi\sigma E^{(+)} + \delta_1 = +2\pi i\omega_o P^{(+)} + \delta_2 \tag{3.6}$$

$$\frac{\partial}{\partial t} P^{(+)} + \gamma P^{(+)} + \delta_3 = -\frac{i}{\hbar} |\vartheta_{12}|^2 E^{(+)} D \tag{3.7}$$

$$\frac{\partial}{\partial t} D(x,t) = \frac{D_o - D(x,t)}{T} + \frac{2i}{\hbar}(E^{(+)}P^{(-)} - E^{(-)}P^{(+)}) + \delta_4 \tag{3.8}$$

δ_1, δ_2, δ_3 and δ_4 are small quantities (provided ω_o is sufficiently big) which are neglected by the above mentioned approximations. I shall call the set of equations (3.6), (3.7), (3.8) the reduced equations, or "R" - equations. We have analysed these equations with increasing control parameter D_o and found the following bifurcation scheme (compare Fig. 4a - d. For small D_o, we find a stable node corresponding to a state without laser action.

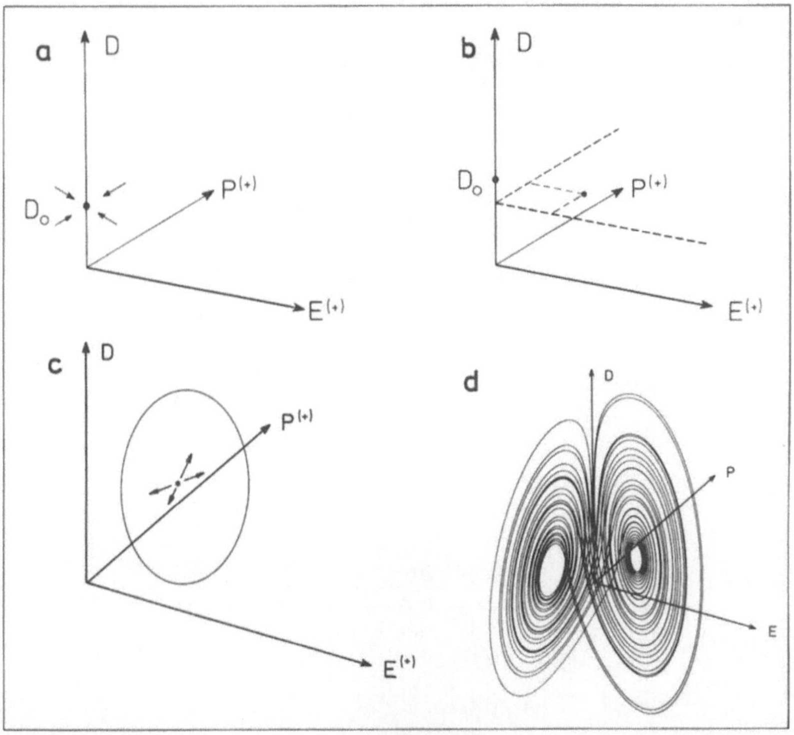

Fig. 4a-d Bifurcation hierarchy of the laser in the frame of the R-equations. (a) Stable node, no laser action; (b) stable focus, normal laser action; (c) limit cycle, laser pulses; (d) Lorenz attractor, laser turbulence

In it $E^{(+)} = P^{(+)} = 0$, $D = D_o$. Above a critical value of D_o laser action occurs which is characterized by a stable focus with $E^{(+)} = \text{const}$, $P^{(+)} = \text{const.}$, $D < D_o$ [7]. With further increase of D_o the stable focus becomes unstable and is replaced by a limit cycle which corresponds to ultrashort laser pulses [8]. For other parameter values of γ and σ and T another kind of instability occurs, namely the laser equations are exactly identical with those of the Lorenz attractor [9] describing laser turbulence [1o].

A still more interesting bifurcation scheme for laser occurs when we go back to the original equations dropping the rotating wave approximation or the slowly varying amplitude approximation or both. In this case we deal with the full expressions E and P rather than with $E^{(+)}$ and $P^{(+)}$, compare eqs. (3.4) and (3.5). Then we obtain the following scheme. First for small D_o the system has a stable focus, i.e. no laser action occurs. Above a critical control parameter D_o the stable focus becomes unstable and is replaced by a limit cycle i.e. a coherently oscillating electric field. This is normal laser action. When we increase D_o further, again this limit cycle becomes unstable.

A new frequency occurs and the motion proceeds now on a torus, representing ultrashort laser pulses. They are characterized by two frequencies, the carrier frequency and the pulse repetition frequency (see Fig. 5a-d). I have treated this case treating the small quantities δ_1, δ_2, δ_3 as perturbations and neglecting δ_4.

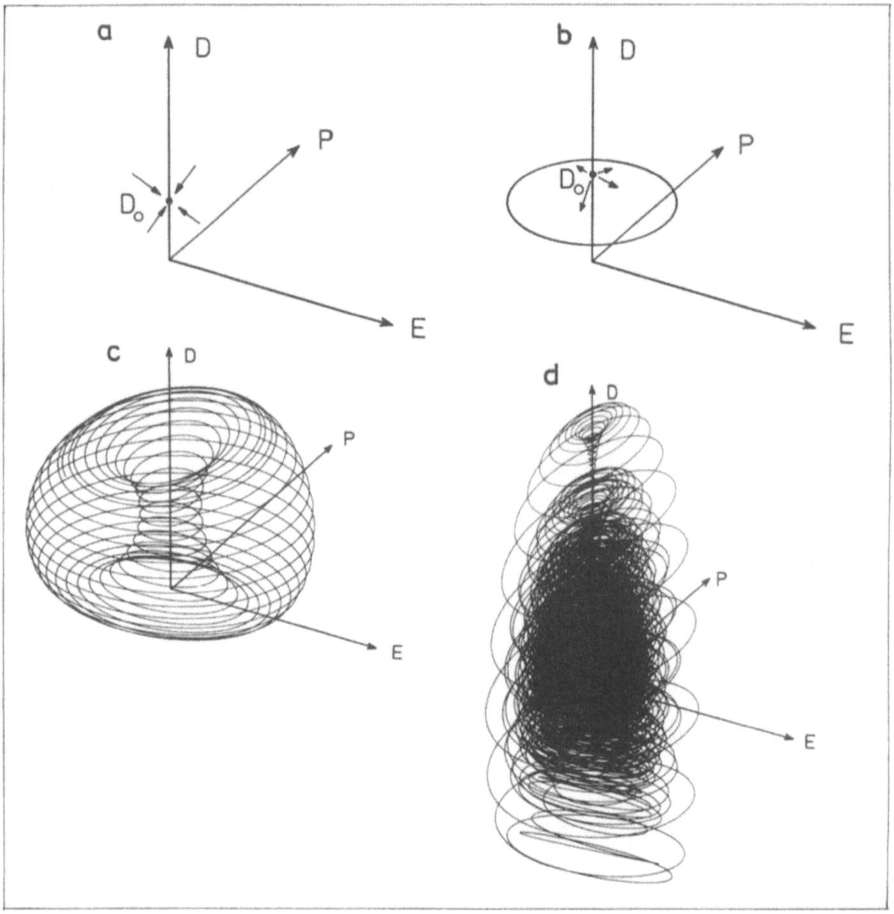

Fig. 5a-d Bifurcation hierarchy of the laser in the frame of the unreduced equations (3.1-3). (a) Stable focus, no laser action; (b) limit cycle, laser action; (c) torus, laser pulses; (d) chaos, laser turbulence

I suggested this problem to a young mathematician, M. Renardy, who is presently working at my institute, to apply to it bifurcation theory. He showed that due to the rotation symmetry the problem can be reduced to that of Hopf bifurcation which allowed him to incorporate δ_1, δ_2 and δ_3 in the linear operator. At present H. Ohno and myself are treating the additional term δ_4 which contains nonlinearities. For different parameter values for σ and γ we obtain the bifurcation of the limit cycle into chaos instead of its bifurcation into a torus. I should mention that laser theory supplies us with many further examples. For instance, in inhomogeneously broadened atomic lines or using spatial hole burning one can obtain the independent operation of n laser modes ("free running laser") at frequencies $\omega_1, \omega_2 \ldots \omega_n$. This "motion" can be represented by an n-dimensional torus T_n. However, it is known that under certain conditions frequency locking occurs so that the n-dimensional torus collapses into a limit cycle. We may have also the transition of a single limit cycle into two limit cycles, for instance in optical bistability [11]. Furthermore, in somewhat more complicated laser configurations using saturable absorbers we may have the transition from a focus directly to a torus. I hope these remarks show that the study of the laser provides us with a beautiful example of a dynamic system. However, we should bear in mind that laser physics is still richer because we have to deal with fluctuations. I shall make a few comments in the next section.

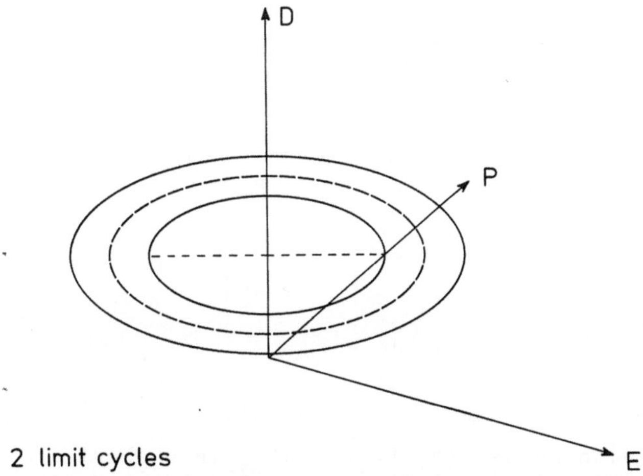

Fig. 6 Bifurcation of a limit cycle into two limit cycles, realized by optical bistability

4) The role of noise

Within the frame of semiclassical laser theory, noise is introduced by adding fluctuating forces to the right hand sides of the eqs.(3.1) to (3.3) or (3.6) to (3.8). Then the bifurcation scheme outlined above must be extended. Now we find that the former node is replaced by a fuzzy ball within which the trajectories perform random motion. An adequate way to deal with this problem is the use of probability distributions and correlation functions. When laser action sets in, the ball is replaced by a fuzzy limit cycle or, in other words, the trajectory fills a fuzzy ring tube. In radial direction of the tube we find a probability distribution of the electric field strength as indicated in Fig. 7d. In addition, the representative point of the trajectories diffuses along the axis of the tube (phase diffusion).

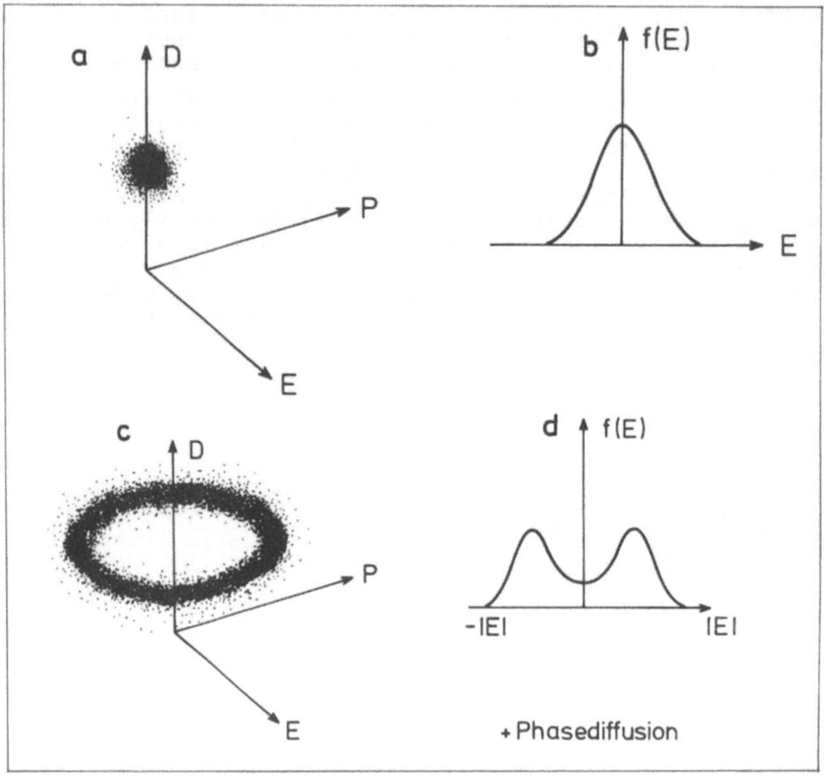

Fig. 7a-d The laser problem with noise. (a) Fuzzy ball, incoherent light emission; (b) distribution function f of light field amplitudes, incoherent emission; (c) fuzzy limit cycle, laser light emission with amplitude fluctuations and phase diffusion; (d) distribution function f of light field amplitude, laser light emission

5) Need for new mathematical approaches

In my opinion the laser example is so important because the laser is a real system which allows us to check all theoretical details experimentally. As we know, in it fluctuations play an important role and from physics we know that this must be so in other systems, too. Thus the laser provides us with powerful criteria on which kind of approach is adequate and which needs further development. In the light of these facts let me make a comparison between bifurcation theory in its traditional form and some new approaches. We have just mentioned that at critical points fluctuations are very important. Furthermore, we wish to know the stability of the new evolving solutions. Bifurcation theory is based to a large extent on the Ljapunov-Schmidt method or on methods using the center manifold. A typical equation is

$$L(\alpha)\, \underset{\sim}{q} + N(\underset{\sim}{q}) = 0 \qquad (5.1)$$

where L is a linear operator depending on a control parameter α and N is a nonlinear operator. The solutions q are usually assumed to belong to a Banach space. I shall not dwell on the mathematical details, however, but illuminate the essential features by a most simple example, namely an equation of the form

$$0 = \alpha q - \beta q^3 \qquad (5.2)$$

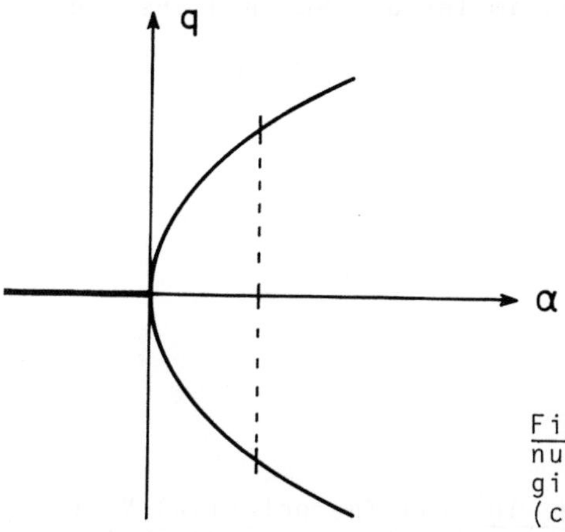

Fig. 8 A discrete set of values of the null-space variable q is attached to a given value α of the control parameter (conventional bifurcation theory)

In bifurcation theory, the solutions belonging to (5.2) are as follows: for each α one or several fixed values q are determined. In the very moment we have to take into account fluctuations, this scheme is not sufficient, however, because fluctuations push the values of q around and render the equations (5.1) or (5.2) into time-dependent equations. Even if the fluctuations are very small, all the time they test the stability of the newly bifurcating solutions. Thus we have to deal with equations of the form

$$L(\alpha)\, \underset{\sim}{q} + N(\underset{\sim}{q}) = \frac{d}{dt}\, \dot{\underset{\sim}{q}} \quad (+F(t)) \qquad (5.3)$$

or, to take up again our specific example

$$\frac{d}{dt} q = \alpha q - \beta q^3 = \frac{\partial V}{\partial q} \quad (+F(t)) \qquad (5.4)$$

As I have shown at various occasions (e.g. [3]), the behaviour of q can be best studied by means of the potential function V which is plotted in Fig. 9 for two typical parameter values $\alpha<0$ and $\alpha>0$.

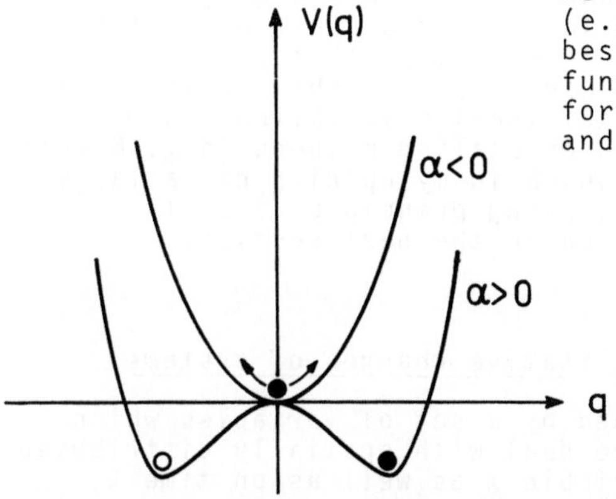

Fig. 9 A potential curve (or, more generally, a differential equation) for the order parameter (null-space variable) q is attached to a given value α of the control parameter

Thus in this approach we attach to each value of α a differential equation, and in the special case treated here a potential curve. From it we can immediately read off the stability of the solution and the impact of fluctuations. It is sometimes useful to plot V as a function of both q and of the control parameter α . We then are led to the Fig.1o.

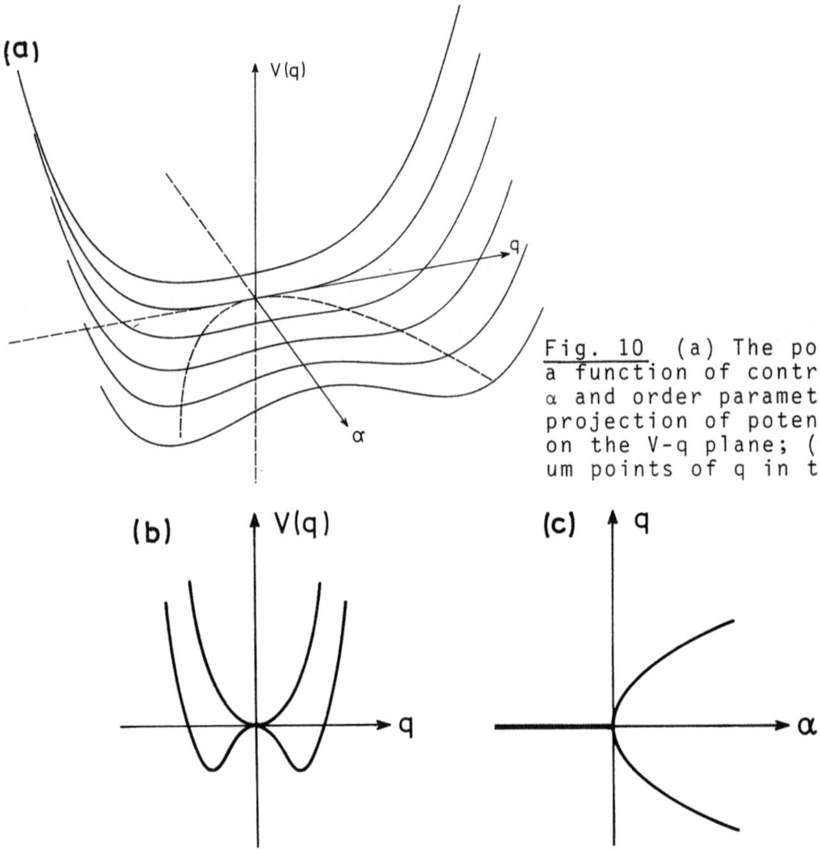

Fig. 10 (a) The potential V as a function of control parameter α and order parameter q; (b) projection of potential curves on the V-q plane; (c) equilibrium points of q in the q-α plane

The curve in the α - q - plane corresponds to the points $q(\alpha)$ of the minimum of the potential (compare Fig. 1oc). On the other hand, the projections on the V-q-plane are shown in Fig. 1ob. This example which stems from a concrete example (namely laser physics) shows quite clearly the need for an extension of the traditional bifurcation schemes. Some of such extended schemes are based on scaling methods (e.g. Newell and Whitehead [12]). Another method which in my opinion has a large scope of applications is based on the slaving principle. I shall sketch the main features of this approach in the next section.

6) A general basis for a theory of qualitative changes of systems

Let us assume that a system is described by a set of variables which I shall describe by a vector $\underset{\sim}{q}$. When we deal with spatially distributed systems, $\underset{\sim}{q}$ may depend on the space variable $\underset{\sim}{x}$ as well as on time t.

A good deal of processes in physics, chemistry and many other disciplines can be described by evolution equations of the form

$$\dot{\underset{\sim}{q}} = \underset{\sim}{N}(\underset{\sim}{q},\alpha,t) \qquad (6.1)$$

$\underset{\sim}{N}$ is in general a nonlinear function of $\underset{\sim}{q}$ and it depends on one or a set of control parameters α . It may or may not depend explicitly on time. In general a time dependence is introduced via fluctuations. When $\underset{\sim}{q}$ depends on $\underset{\sim}{x}$, $\underset{\sim}{N}$ contains in general partial derivatives with respect to space. A general solution of these nonlinear stochastic partial equations is probably hopeless. However, there are classes of interesting phenomena in which these equations can be considerably simplified and reduced. As I have pointed out at various occasions, such a reduction becomes possible at critical points of control parameters α where the macroscopic behaviour changes qualitatively. In the following I shall assume that the fluctuations do not shift the critical points appreciably. Therefore I shall assume that we can decompose $\underset{\sim}{N}$ according to

$$\underset{\sim}{N} = \underset{\sim}{N}_0 + \underset{\sim}{F} \qquad (6.2)$$

into a deterministic part, $\underset{\sim}{N}_0$, and a stochastic force, $\underset{\sim}{F}$, which I shall neglect in the first step of our following analysis but which can be taken into account in the following steps. Our procedure consists then of the following steps:

1)Let us assume that for the control parameter value $\alpha = \alpha_0$ an old state is realized which obeys the equation

$$\dot{\underset{\sim}{q}}^0 = \underset{\sim}{N}(\underset{\sim}{q}^0, \alpha_0, t) \qquad (6.3)$$

For this α_0, $\underset{\sim}{q}^0$ is assumed stable. We now study what happens when we change the control parameter α .

2) Stability analysis
We assume that $\underset{\sim}{q}^0$ is still a solution of (6.1) but for a new value α. To check the stability of that solution we put

$$\underset{\sim}{q}(t) = \underset{\sim}{q}^0 + \underset{\sim}{u}(t) \qquad (6.4)$$

and make a linear analysis by inserting (6.4) into (6.1) and linearizing $\underset{\sim}{N}$. Provided the system is autonomous, i.e. $\underset{\sim}{N} = 0$ is independent of time t, the linearized equations read

$$\dot{\underset{\sim}{u}} = L(\underset{\sim}{q}^0)\, \underset{\sim}{u} \qquad (6.5)$$

Let us discuss the following special cases:

a) $\underset{\sim}{q}^0$ is time independent. Then L is also time independent. The solutions of (6.5) read

$$\underset{\sim}{u}(t) = \exp(\lambda t)\, \underset{\sim}{v} \qquad (6.6)$$

provided the characteristic exponents λ are nondegenerate. We shall call modes $\underset{\sim}{u}$ for which

$$\mathrm{Re}\lambda \geq 0 \qquad (6.7)$$

unstable and we shall write

$$\underset{\sim}{u}_u(t) = \exp(\lambda_u t)\underset{\sim}{v}_u \tag{6.8}$$

Modes with

$$\mathrm{Re}\lambda < 0 \tag{6.9}$$

are called stable and will be denoted by

$$\underset{\sim}{u}_s(t) = \exp(\lambda_s t)\underset{\sim}{v}_s \tag{6.1o}$$

b) the solution $\underset{\sim}{q}^o$ is periodic

So is L. By virtue of Floquet's theorem the solution of the nondegenerate case reads

$$\underset{\sim}{u}(t) = \exp(\lambda t)\underset{\sim}{v}(t) \tag{6.11}$$

where $\underset{\sim}{v}(t)$ is again periodic.

c) $\underset{\sim}{q}^o$ is quasi-periodic i.e. it can be described as a motion with basic frequencies $\omega_1, \omega_2, \ldots, \omega_M$, for instance as the multiple Fourier series

$$\underset{\sim}{q}^o(t) = \sum_{\underset{\sim}{m}} \underset{\sim}{M}_m \exp\{i(m_1\omega_1 + m_2\omega_2 + \ldots)t\} \tag{6.12}$$

In this case, also L is quasi-periodic.

Since I could not find references for the structure of the solutions of the linear equations (6.5) for this case and a number of mathematicians told me that no theorem is known I treated this case by my own. I will give a sketch of the results in the appendix.

It turns out that there are large classes of linear operators L for which the solution has the form

$$\underset{\sim}{u} = \exp(\lambda t)\ \underset{\sim}{v}(t) \tag{6.13}$$

$\mathrm{Re}\lambda$ is as usual called generalized characteristic exponent. It can be shown that $\underset{\sim}{v}(t)$ is again quasi-periodic.

3) The linear analysis of 2) gives us a set of solutions $\underset{\sim}{u}$ or, when we split off the exponential factor, $\underset{\sim}{v}$. We now extend the wanted solution $\underset{\sim}{q}(t)$ of the total nonlinear equations into the superposition of the form

$$\underset{\sim}{q}(t) = \underset{\sim}{q}^o + \sum_j \xi_j(t)\ \underset{\sim}{v}_j(t) \tag{6.14}$$

When we deal with the bifurcation of limit cycles or tori some care must be exercised with this hypothesis because of shifts of phase angles which in the course of time can become arbitrarily large. These shifts can be taken care of by introducing adequate phase angles into

q^o and v as we have shown elsewhere [13]. I shall not enter into the discussion because it is rather lengthy and I want just to give a sketch. In the next step of the analysis the hypothesis (6.14) is inserted into (6.1) and the equations are projected on the eigenvectors v. This leads us to equations for ξ which must be put into two classes namely the unstable modes ξ_u and the phases ϕ on the one hand and the stable mode amplitudes ξ_s on the other hand.

4) Slaving principle

As can be shown the resulting equations for ξ_s are of such a structure that for big enough damping i.e. big enough $|Re\lambda_s|$ and small enough amplitudes ξ_u the equations of ξ_s can be solved uniquely by explicit rules. This allows us to express ξ_s by means of

$$\xi_s(t) = F(\xi_u) \tag{6.15}$$

Since I have discussed the slaving principle at various occasions I shall not dwell on it but just repeat the following feature. In many dissipative systems we have studied, the number of ξ_u is much smaller than that of ξ_s. Therefore we obtain an enormous reduction of the degrees of freedom. Inserting (6.15) into the equations for $\dot{\xi}_u$ we end up with a closed set of equations for ξ_u.

$$\dot{\xi}_u(t) = \tilde{N}(\xi_u(t')) \tag{6.16}$$

A study of these reduced equations has just begun, for instance they contain a class of equations which I derived earlier and which I called generalized Ginzburg-Landau equations. They also are capable of describing the bifurcation of limit cycles or of tori.

7) Some epistemological remarks

This meeting, as previous ones, has shown that there exist striking analogies between the behaviour of quite different systems belonging to entirely different disciplines. Thus from this point of view this meeting is just another example for the existence of the new field of Synergetics. But, of course, Synergetics is related to other disciplines, at least in two ways. First of all, the systems it studies belong to quite different disciplines. Second, the other disciplines contribute with their own thoughts. Thus it is quite natural that scientists try to enter this new field by seeing it as an extension of their own field. This remark can be substantiated just by looking at titles of some recent papers or conferences [14] - [17]. For instance, I myself cannot deny to have started from laser physics. So a typical title of a talk of mine is: "The laser - trailblazer of synergetics". Mathematicians having dealt with bifurcation theory choose the title "bifurcation theory and its applications". Physicists having worked in the field of phase transitions are inclined to talk about "non-equilibrium phase transitions" and scientists who are experts in the field of statistical mechanics call this approach "nonequilibrium nonlinear statistical mechanics". Others see it as an extension of irreversible thermodynamics, and again others apply catastrophe theory (sometimes calling problems not yet accessible "generalized catastrophes"). In particular mathematicians look at these problems from the view point of structural stability.

All of these disciplines are extremely important for an understanding of the selforganized formation of macroscopic structures. But at the same time each of these disciplines misses equally important points. Let me list some: the world is not a laser. At bifurcation points, fluctuations, i.e. stochastic processes are decisive. Nonequilibrium phase transitions have some important features different from ordinary phase transitions, for instance finite size effects, impact of shape of boundaries etc. In equilibrium statistical mechanics selfsustained oscillations do not occur. Irreversible thermodynamics is largely using concepts such as entropy, entropy production etc. which are no more adequate to deal with nonequilibrium phase transitions. Catastrophe theory is based on the existence of certain potential functions which do not exist in systems driven away from thermal equilibrium. The intention of my remarks is definitely not to critizise the individual disciplines but to underline my point: There is the undeniable urgent need for a discipline which unifies all these different aspects. Whether one calls this discipline "Synergetics" or not is quite unimportant for science. But it exists.

Appendix

On the structure of the solutions of coupled linear differential equations with quasi-periodic coefficients

In my theory of bifurcation of tori (Z.Physik,B3o, 423(1978) and unpublished work) I have used a specific form of solutions (cf. section 6.2c) of coupled linear differential equations with quasi-periodic coefficients. Since, to my knowledge, no theorems of this kind seem to exist in the literature, I give a brief sketch of some of the main results.

We treat an equation of the form

$$\frac{dq}{dt} = M(t)q \tag{A1}$$

where q is a vector with N components and M is a NxN matrix which can be represented as multiple Fourier series,

$$M(t) = \sum_{\underset{\sim}{m}} M_{\underset{\sim}{m}} \exp\{i(m_1\omega_1+m_2\omega_2+\ldots+m_n\omega_n)t\} \tag{A2}$$

We embed the solution of (A1) into the solution of the following problem:

$$\frac{d}{dt}\, q(t,\phi_1,\phi_2,\ldots,\phi_n) = M(t,\phi_1,\phi_2,\ldots,\phi_n)q \tag{A3}$$

where

$$M(t,\phi_1,\phi_2,\ldots) = \sum_{\underset{\sim}{m}} M_{\underset{\sim}{m}} \exp\{-i(m_1\omega_1\phi_1+m_2\omega_2\phi_2+\ldots)\} \times$$

$$\times \exp\{i(m_1\omega_1+\ldots)t\} \tag{A4}$$

We introduce the translation operator

$$T_\tau : t \to t + \tau, \quad \phi_j \longrightarrow \phi_j + \tau, \quad j = 1,\ldots,n; \; -\infty < \tau < \infty \quad .$$

Evidently, the T_τ's form an Abelian group. Furthermore

$$T_\tau \, M = M \, T_\tau \tag{A5}$$

and consequently

$$T_\tau Q = Q \, A(\tau,\phi_1,\ldots,\phi_n) \tag{A6}$$

where Q is the fundamental solution matrix of (A3).
We now assume

M is C^k, $k \geq 0$, with respect to the ϕ s.

Thus M is bounded and C^k with respect to t. Consequently we can introduce generalized characteristic exponents λ_k:

$$\lambda_k = \limsup_{t \to \infty} (1/t) \, \ln \| q_k(t) \| \tag{A7}$$

In the realm of this appendix I make the stronger assumption

$$\lambda_k = \lim_{t \longrightarrow \infty} (1/t) \, \ln \| q_k(t) \| \qquad \text{exists.} \tag{A8}$$

I further assume that the trajectory $(\tilde{\Phi}) = (\tau,\tau, \ldots,\tau)$, $0 \leq \tau \leq \infty$ (or $(\phi_1^0 + \tau, \phi_2^0 + \tau,\ldots))$ lies dense on the torus

$$(\phi_1, \phi_2,\ldots,\phi_n); \quad 0 \leq \phi_j \leq 2\pi/\omega_j \quad . \tag{A9}$$

The following Lemmata hold:

1) $$q(t,\phi) = \tilde{T} \exp \left\{ \int_0^t M(t',\phi)dt' \right\} q(0,\phi^0), \tag{A1o}$$

$\tilde{T}$ time-ordering operator,

is C^k with respect to ϕ's for all finite t and it is invariant under

$$\phi_j \longrightarrow \phi_j + 2\pi n_j/\omega_j; \; n_j \text{ integer, for each } j = 1,\ldots,n.$$

2) The set λ_k is invariant against T_τ

3) If the λ_k's are different from each other, a (time-independent) Matrix $B(\phi_1, \phi_2, .., \phi_n)$ can be constructed, which transforms Q and A in (A6) so that A becomes half-diagonal, i.e. of the form

$\tilde{A}$ =  (A11)

B is periodic in the ϕ's and continuous.
Thus (A6) is replaced by

$$T_\tau \tilde{Q} = \tilde{Q} \tilde{A}_\tau (\phi_1, ..., \phi_n), \tag{A12}$$

4) If M is C^0 with respect to t and the ϕ's, then $\tilde{Q}$ is C^1 with respect to t and C^0 with respect to the ϕ's. $\tilde{Q}$ is periodic in the ϕ's.

We are now concerned with a further reduction of (A12).

5) When the exponents λ exist also for $t \longrightarrow -\infty$, $\tilde{A}$ can be diagonalized using similar steps as above.

6) If $\tilde{M}(\tau, \phi)$ is half-diagonal, C^0 with respect to ϕ,

$$\int_0^t \tilde{M}_{jj}(\tau, \phi) d\tau = \lambda_j t + \bar{M}_{jj}(t, \phi) \tag{A13}$$

with

$$T_\tau \bar{M}_{jj} = \bar{M}_{jj} T_\tau , \quad |\bar{M}| < \infty \tag{A14}$$

then $\frac{dq}{dt} = \tilde{M}q$ can be transformed by matrix C, which is C^1 with respect to t and C^0 with respect to ϕ, into

$$\frac{d\tilde{q}}{dt} = \hat{\tilde{M}}q \quad , \tag{A15}$$

where $\hat{M}$ is half-diagonal, quasi-periodic, C^0w.r. to t and ϕ and $\hat{M}_{jj} = \lambda_j$.
$\tilde{q}$ can be chosen so that

$$T_\tau \tilde{q} = e^{\lambda\tau}\tilde{q} \quad , \tag{A16}$$

or equivalently

$$\tilde{q} = \exp \lambda_j t \, u(t, \phi_1, ..., \phi_n) \tag{A17}$$

where T_τ u = u; u periodic in ϕ's.

If follows that $u(t,0,0,...,0)$ is quasi-periodic.

7) A sufficient condition for $\tilde{M}$ so that (A13) holds is the following: $\tilde{M}_{jj}$ is C^2 with respect to the ϕ's and the ω_j's obey a Kolmogorov-Arnold - Moser condition.

8) The above results can be extended to almost periodic M's by formally replacing $(\omega_1,\omega_2,...,\omega_n)$ and $(\phi_1,\phi_2,...,\phi_n)$ by corresponding infinite sets.

9) If all λ_k's are equal and $e^{-\lambda n\tau} \| T_\tau^n q \|$, $e^{\lambda n\tau} | T_\tau^{-n} q |$ bounded for $n \longrightarrow \infty$, we may invoke theorems on linear operators. The (group) representations belonging to T are equivalent to unitary transformations and we can diagonalize, so that (A16) results (λ imaginary) provided the spectrum is a point spectrum. The case of a continuous spectrum requires some additional discussion which I shall not present here.

1o) A third class of solution can be identified when $\| T_\tau^{\pm n} q \|$ goes with n^m. The solutions are then polynomials in t (up to order m) with quasi-periodic coefficients.
I shall publish the details, including the treatment of some "mixed" cases (some of the λ's coincide) and further generalizations elsewhere.

References

[1] For references consult the individual articles
[2] H.Haken, "Laser Theory", Encyclopedia of Physics, Vol.XXV/2c Springer, Berlin 197o
[3] H.Haken, Z.Physik 181, 96 (1964)
[4] R. Graham and H.Haken, Z.Physik 213, 42o (1968); 237, 31 (197o)
[5] V.DeGiorgio and M.O.Scully, Phys.Rev.A2, 117o (197o)
[6] H.Haken in "Festkörperprobleme X" ed. O.E. Madelung Pergamon, Vieweg 197o
[7] H.Haken and H.Sauermann, Z.Physik 173, 261 (1963)
[8] H.Haken and H. Ohno, Opt.Comm. 16, 2o5 (1976)
H. Ohno and H.Haken, Phys.Lett. 59A, 261 (1976)
H.Haken and H. Ohno, Opt.Comm. 26, 117 (1978)
M. Renardy, Math.Meth.in the Appl.Sci. I, 194 (1979)
[9] E.N. Lorenz, J.Atmos.Sci. 2o, 13o (1963)
[1o] H.Haken, Phys.Lett. 53A, 77 (1975)
[11] For a review with further references see
R. Bonifacio and L. Lugiato, in "Pattern formation by dynamic systems and pattern recognition", ed. H.Haken, Springer 1979
[12] Compare Newell's article in "Pattern formation by dynamic systems and pattern recognition", ed. H.Haken, Springer 1979
[13] H.Haken Z.Physik B29, 61(1978); B3o, 423(1978)
[14] Bifurcation theory and applications in scientific disciplines, ed. O. Gurel and O.E.Rössler, Ann.N.Y.Academy of Sci. Vol.316 (1979)
[15] "Nonlinear nonequilibrium statistical mechanics", Proceedings of the 1978 Oji Seminar at Kyoto, Suppl.Progr. Theor.Phys. 64 (1978)
[16] "Structural stability in physics", ed.W.Güttinger and H.Eikemeier, Springer 1979
[17] G. Nicolis, Rep.Progr.Phys. 42, 225 (1979)

[illegible] follows that $\mu(0,0,u,0)$ is quasiperiodic.

4) A sufficient condition [illegible]

[illegible]

5) [illegible] advantage [illegible]

[illegible]

6) It will [illegible] should [illegible] needed for [illegible] theorems on linear operators. The group(s) [illegible] representations [illegible] unitary transformations [illegible] imagine [illegible] spectrum is a point spectrum. The [illegible] difficulties [illegible] additional [illegible], which I shall not present here.

(7) A phase space of solutions [illegible]

[illegible]

[illegible] Hasegawa, J. Phys. Soc. Jpn. [illegible] (1978) [illegible]
[10] Bifurcation theory and applications in scientific disciplines, ed. O. Gurel and O.E. Rössler, Annals N.Y. Academy of Sci. Vol. 316 (1979)
[illegible] Proceedings [illegible] 1978 [illegible] Seminar [illegible]
[illegible] Suppl. Prog. Theor. Phys. 64 (1978)
[13] Structural Stability in Physics, ed. W. Güttinger and H. Eikemeier, Springer 1979
[14] [illegible] (1979)

Part II

Equilibrium Phase Transitions

Critical Phenomena: Past, Present and "Future"

H.E. Stanley[1], A. Coniglio, W. Klein, H. Nakanishi, S. Redner, P.J. Reynolds, and G. Shlifer

Center for Polymer Studies and Department of Physics, Boston University
Boston, MA 02215, USA

1. Introduction

The opening talk of an interdisciplinary meeting should ideally start at "square one". In the present case this means I should assume no previous background in the field of equilibrium phase transitions. Although everyone in the audience has some background in this field, the background of no two people is identical. Hence I begin with a brief introduction to phase transitions. Accordingly, I shall organize this talk around three simple questions:

(i) "What happens?" That is to say, "What are the basic phenomena under consideration?"
(ii) "Why do we care?"
(iii) "What do we actually do?"

The answer to the third question will occupy the bulk of our presentation. We shall divide the answer into two parts. In the first part we will briefly review three of the many important themes which have served to provide the framework of much of our current understanding of critical phenomena:

(a) scaling,
(b) universality, and
(c) the concept of a fixed point Hamiltonian.

The second part of the talk concerns the "future". Of course, if we could predict the future we would not be doing physics research. Hence there is some guesswork and also some personal predilection reflected here. Due to limitations of time and format of presentation, we shall discuss only four future directions:

(a) Second-order phase transitions,
(b) First-order phase transitions and metastability,
(c) Tricritical and other multicritical-point phenomena, and
(d) Cooperative phenomena and phase transitions in extremely complex systems.

Our main point of optimism for the future is that in-depth studies of judiciously-chosen simple models, however idealistic they may seem at first sight, may nevertheless produce useful information. This is particularly true if our studies are broadened to include relatively simple generalizations of the model in question.

Before beginning, it is appropriate to state my own degree of indebtedness to a number of individuals. In addition to those of my colleagues who join me as co-authors of this chapter, I must also acknowledge fruitful interactions with R. Blumberg, M. Daoud, A. Gonzales, L. Lucena, S. Muto and P. Ruiz. The theoretical

[1] Talk delivered by HES.

models to be presented here were motivated strongly by discussions of experimental phenomena with C.A. Angell, R. Bansil, R.J. Birgeneau, T. Tanaka, and J. Teixeira.

2. Question (i): "What Happens?"

Suppose we have a simple bar magnet, as shown in Fig.0. We know it is a ferromagnet because it is capable of picking up thumbtacks, the number of which is some measure of the "order parameter". As we heat this system, the order parameter decreases and eventually, at a certain critical temperature T_c, it reaches zero: no more thumbtacks remain! In fact, the transition is remarkably sharp, since the order parameter approaches zero at T_c with infinite slope (Fig.1a). Such singular behavior is an example of a "critical-point phenomenon".

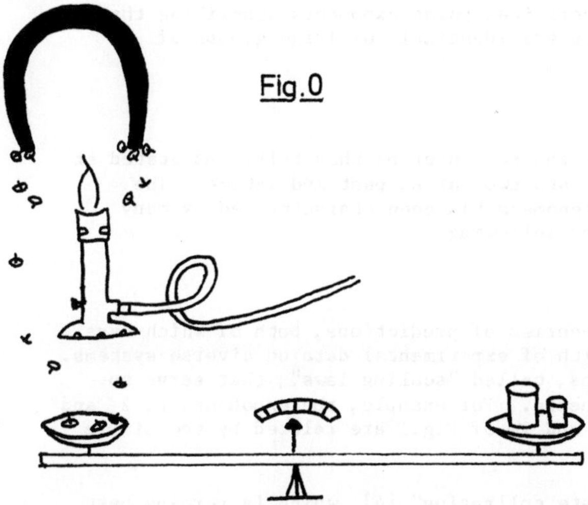

Fig.0

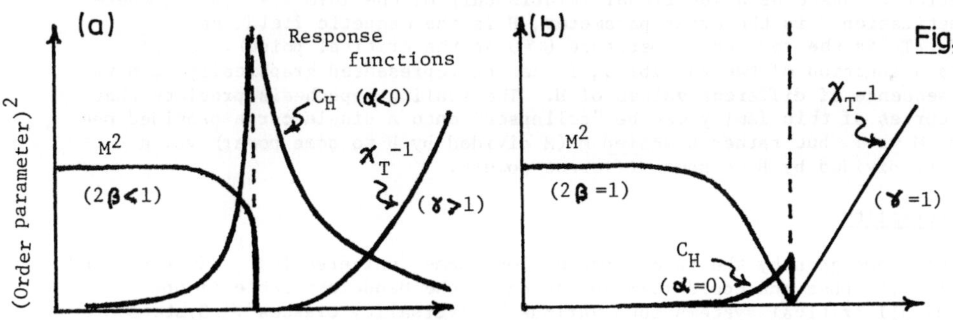

Fig.1

Critical phenomena are by no means limited to the order parameter. For example, Fig. 1a also shows two response functions, the constant-field specific heat C_H and the isothermal susceptibility χ_T. Both these response functions become infinite at the critical point!

3. "Question (ii): "Why Do We Care?"

One reason for interest in any field is that we simply do not fully understand the basic phenomena; i.e., for even the simplest three-dimensional system we cannot make exact predictions from any microscopic model at our disposal. Of the models which can be solved in closed form, most make the same predictions for behavior near the critical point as the classical mean-field theory or Husimi-Temperley model, in which one assumes that each magnetic moment interacts equally with every other magnetic moment in the entire system [1]. The predictions of the Husimi-Temperley model are sketched in Fig.1b. The square of the order parameter approaches zero linearly as $T - T_c$, rather than with infinite slope as in Fig.1a, the specific heat does not diverge, and the inverse susceptibility is simply a straight line. In fact, the mean-field theory cannot even locate the value of T_c to better than about 40% for some cases (e.g., Ising spins on a square lattice).

A second reason for our interest is the striking similarity in behavior near the critical point among systems that are otherwise quite different in nature. A celebrated example is the "lattice-gas" analogy between the behavior of a single-axis ferromagnet and a simple fluid, near their respective critical points [2]. Even the numerical values of the critical-point exponents describing the quantitative nature of the singularities are identical for large groups of apparently diverse physical systems.

4. Question (iii): "What Do We Do?"

The answer to this question will occupy the remainder of this talk. As stated at the outset, we shall divide the answer into two parts, past and future. The recent past of the field of critical phenomena has been characterized by many important themes, three of which are the following:

(a) Scaling

The scaling hypothesis [3] has two categories of predictions, both of which have been remarkably well verified by a wealth of experimental data on diverse systems. The first category is a set of relations, called "scaling laws", that serve to relate the various critical-point exponents. For example, the exponents α, 2β and γ describing the three functions C_H,M^2 and χ_T of Fig.1 are related by the simple scaling law $\alpha + 2\beta + \gamma = 2$.

The second category is a sort of "data collapsing" [4], which is perhaps best explained in terms of our simple example of a uniaxial ferromagnet. We may write the equation of state as a functional relationship of the form $M = M(H,\varepsilon)$, where the magnetization M is the order parameter, H is the magnetic field, and $\varepsilon \equiv (T-T_c)/T_c$ is the reduced temperature ($\varepsilon = 0$ at the critical point). Since $M(H,\varepsilon)$ is a function of two variables, it can be represented graphically as M vs. ε for a sequence of different values of H. The scaling hypothesis predicts that all the curves of this family can be "collapsed" onto a single curve provided one plots not M vs. ε but rather a scaled M (M divided by H to some power) vs. a scaled ε (ε divided by H to some different power).

(b) Universality

The second theme goes by the rather pretentious name "universality". It was found empirically [5] that one could form an analog of the Mendeleev table if one partitions all critical systems into various "universality classes". Systems belonging to the same universality class have the same critical-point exponents and the same "scaled" equation of state.

Many of the universality classes correspond to special cases of the Potts [6] and n-vector [7] hierarchies, which in turn are generalizations of the simple Lenz-Ising model of a uniaxial ferromagnet. This is indicated schematically in Fig.2, in which the Potts hierarchy is indicated as a North-South Metro line, while the n-vector hierarchy appears as an East-West line. The various stops along the respective Metro lines are labeled by the appropriate value of s (the number of discrete states of Potts model spins) and n (the dimensionality of the spin space of the n-vector spins). The two lines have an intersection at the Lenz-Ising model, where s=2 and n=1.

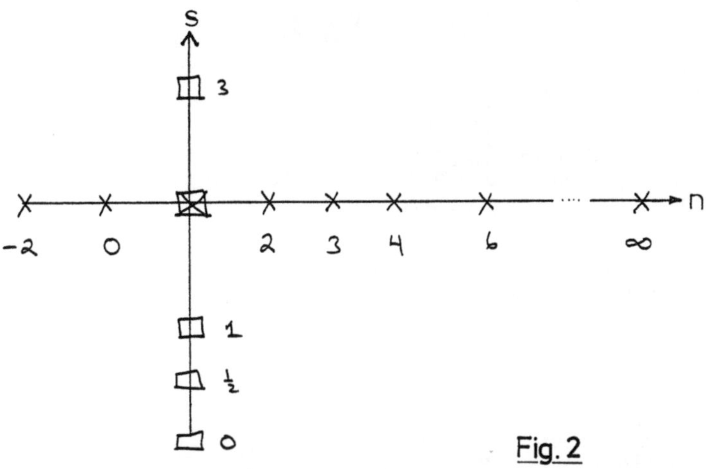

Fig. 2

The East-West Metro line, though newer, has probably been studied more extensively than the North-South line; hence we shall discuss the East-West line first. The cases n=2 and 3 have long been known to correspond, respectively, to ^{4}He near its λ-transition and to isotropic ferromagnets [1]. When the n-vector model was first proposed in 1968, its author noted [8] that the limit $n=\infty$ corresponds to the Berlin-Kac spherical model. Four years later, DE GENNES discovered that another apparently unphysical limit, n=0, corresponds to a system of dilute polymers [9]. The case n=-2 corresponds to the Gaussian model [10], while the cases n=4,6,8,... may correspond to certain antiferromagnetic orderings [11].

Along the North-South Metro line (the s-state hierarchy), the limit s=1 reduces to the random percolation problem [12], which may be relevant to the onset of gelation. Similarly, the limit s=0 corresponds to a type of tree-like percolation, which may be relevant to the behavior of branched polymers [13]. The case s=1/2 corresponds to a certain model of a spin glass [14]; s=3 has been demonstrated to be of relevance in interpreting experimental data on structural phase transitions and on absorbed monolayer systems [15].

(c) Fixed Points

The third theme stems from WILSON's essential idea that the critical point can be mapped onto a fixed point of a suitably chosen transformation on the system's Hamiltonian [16]. This resulting "renormalization group" description has:

(i) provided a conceptual foundation for understanding the themes of scaling and universality,
(ii) provided a calculational tool permitting one to obtain numerical estimates for T_c and for the various critical-point exponents, and
(iii) provided us with altogether new concepts not anticipated previously.

An example of item (iii) is the concept of the upper and lower marginal dimensionalities d_+ and d_- (cf. Fig.3, due originally to F. WEGNER [17]).

For $d \geq d_+$, the classical mean-field or Husimi-Temperley theory is quite adequate for obtaining critical point exponents. For $d < d_-$, fluctuations are so strong that the system cannot sustain long-range order for any $T > 0$. For $d_- < d < d_+$, we do not know exactly the properties of systems (in most cases) except when n approaches infinity. One can, however, develop expansions in terms of the parameters (d_+-d), $(d-d_-)$, and $1/n$ (one of which applies in each of the three different shaded regions of Fig.3).

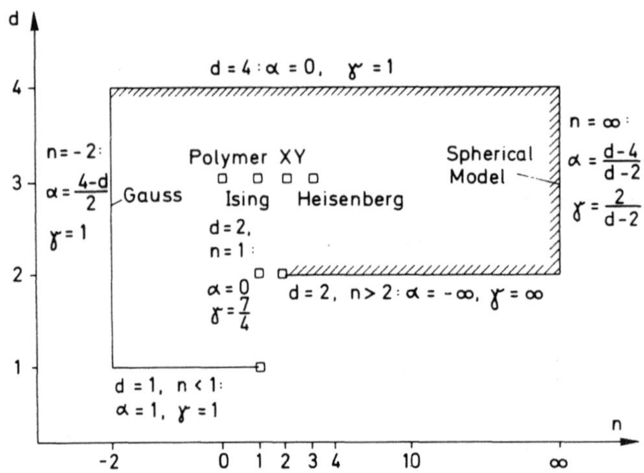

Fig. 3

5. Open Topic (i): "Ordinary" Second-Order Transitions

Concerning theme 1, scaling, there remains open the question of whether certain two-exponent ("hyperscaling") laws involving d, the system dimensionality, are in fact only approximately valid (with an accuracy of about 2%), as a variety of evidence suggests [18], or whether they are exact as the renormalization group seems to imply.

Concerning theme 2, universality, we should point out that the universality classes are by no means all identified, especially when it comes to time-dependent or dynamic critical phenomena [19]!

A particularly fascinating question concerns the sort of phase transition--if any--that takes place for systems described by the n-vector model with n=2, at the lower marginal dimensionality d = d = 2 [20]. There is a growing body of evidence, both theoretical and experimental, suggesting that this transition may be extremely rich and worthy of further study [21].

6. Open Topic (ii): First-Order Phase Transitions

Attention is beginning to be focused on the sort of singularity to be expected when one approaches a line of first-order phase transitions [22]. For the uniaxial ferromagnet, the phase diagram is shown schematically in Fig.4. There are reasons to believe that when the first-order transition is approached along the path shown, there exists--in addition to the jump discontinuity in the order parameter--an additional "essential singularity" in the free energy [23]. This has implications for the possibility of describing the metastable state by analytic continuation of the equilibrium thermodynamic functions.

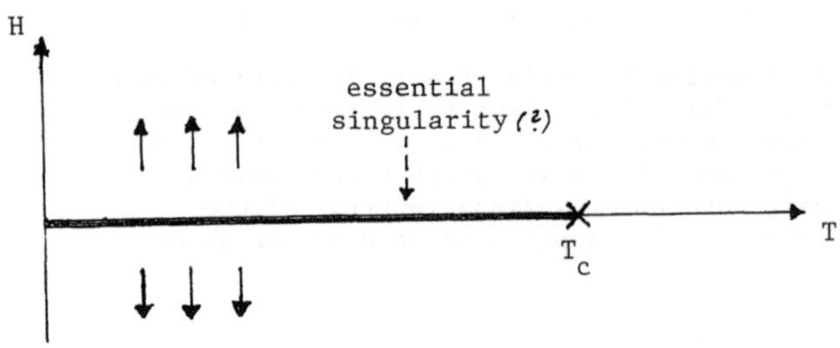

Fig. 4

7. Open Topic (iii): Tricritical and Other Multicritical Phenomena

The first of what is now an entire "zoo" of various species of multicritical points is a simple tricritical point [24]. For a one-component fluid, the simplest example is illustrated schematically in Fig.5. Part (a) shows an

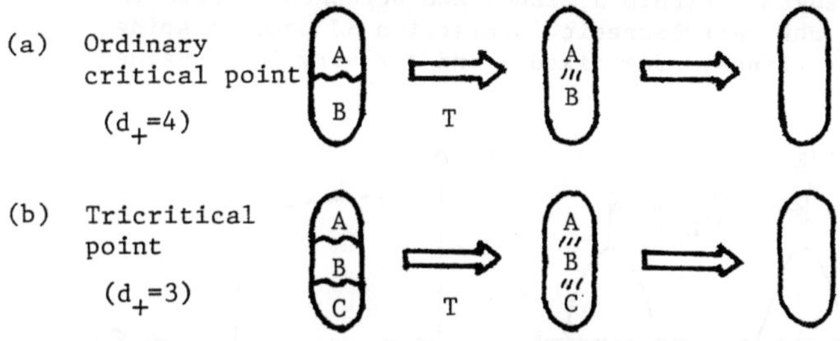

Fig.5

ordinary critical point at which two separate phases--A and B--become indistinguishable. Part (b) shows the analogous illustration for a tricritical point, where now three separate phases--A, B and C--simultaneously become critical [25]. In our example of a uniaxial or Ising magnet, a tricritical point can be obtained [26] by having antiferromagnetic nearest-neighbor interactions (Fig.6a) and introducing an additional ferromagnetic interaction between next-nearest-neighbor magnetic moments (Fig.6b). The resulting phase diagram displays a line of first-order phase transitions, which splits into three lines of second-order phase transitions at the tricritical point.

While d_+= 4 for most ordinary critical points, one finds d_+ = 3 for many tricritical points [27]. Experiments in three-dimensional systems demonstrate that classical theory indeed holds, modulo certain logarithmic corrections that are predicted by renormalization group theory [28].

Tricritical points have now been studied rather extensively. Do there exist still more exotic systems, displaying even more surprising and intricate multicritical point phenomena? If so, can exhaustive study of these phenomena be

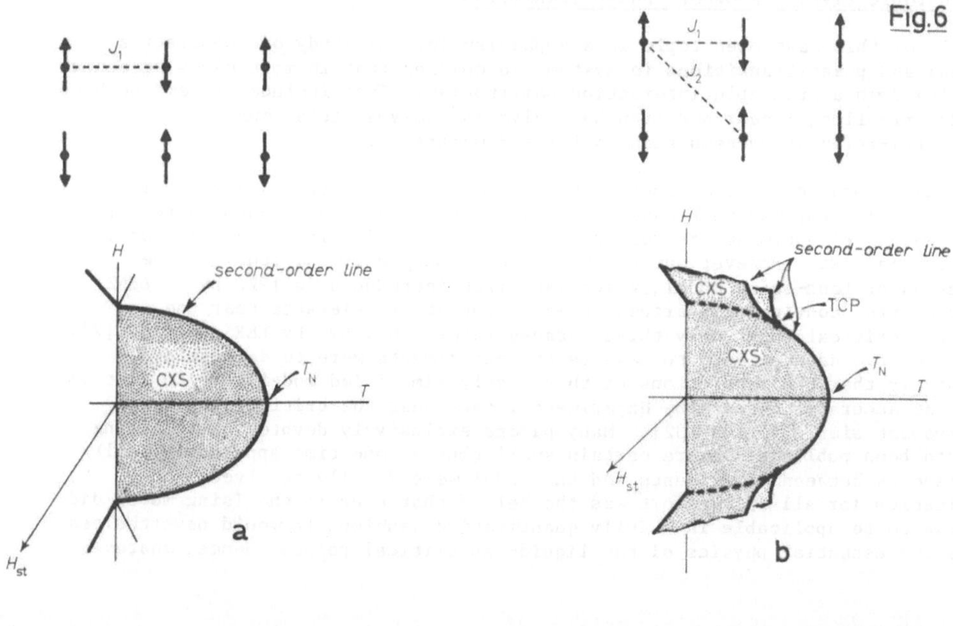

Fig.6

of utility in enlarging our understanding of critical points in general? We believe the answer to both of these questions is "yes".

Of potential future interest is the Lifshitz point, which is perhaps best explained by means of the simple example shown in Fig.7. Figure 7a schematically defines the R-S model [29], an Ising ferromagnet with nearest-neighbor ferromagnetic interactions of strength J within a plane, and strength RJ between planes; there is also a competing antiferromagnetic interaction SJ between spins belonging to next-nearest-neighbor planes. For fixed R and for very large values

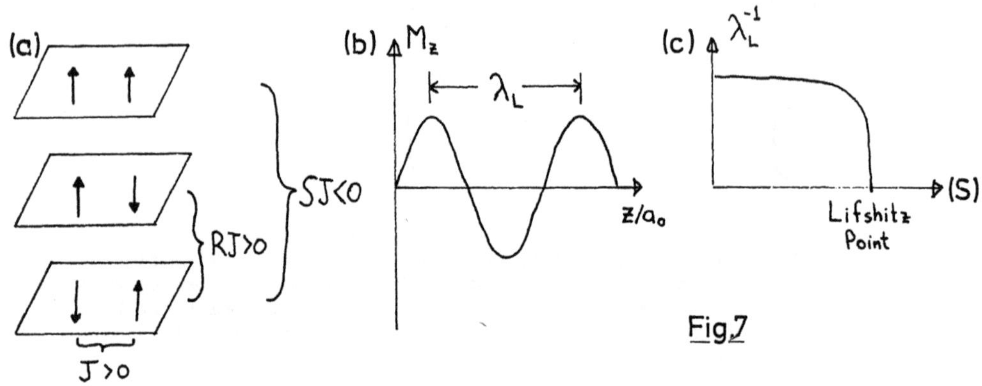

Fig.7

of $|S|$, the spins will assume a configuration in which the local order parameter on each plane oscillates from plane to plane with a characteristic period $\lambda_L(S)$ (cf. Fig.7b). As $|S|$ is decreased, $\lambda_L(S)$ increases, and in fact diverges with a characteristic exponent as $S \to S_c$ (cf. Fig.7c). Such a singularity is called a Lifshitz point [30]. The upper marginal dimensionality d_+ can assume a wide variety of values, depending on the type of Lifshitz point studied. Moreover, the universality classes for various Lifshitz points depend upon the precise nature of short-range competing interactions. Thus there is the potential for observing many new kinds of critical behavior.

8. Open Topic (iv): "X-Order" Phase Transitions

The title of this last open topic is a euphemism for the study of cooperative phenomena and phase transitions in systems so complex that in most cases we cannot even write down a tractable interaction Hamiltonian. They include systems such as colloids, micelles, linear and branched polymers, polymer gels, and biologically-relevant systems such as DNA and membranes.

One way to proceed in the study of such materials is to treat very simple model systems that may (or may not) have direct applicability. This approach requires a certain amount of optimism, as well as perhaps a love of simple physical models for their own sake. However our optimism can be buoyed by the story of the lattice-gas or Lenz-Ising model, which was first introduced in 1920 by W. LENZ [31], but first convincingly argued to be of potential relevance near the liquid-gas critical point only three decades later, in 1952, by LEE and YANG [2]. Yet another two decades were to pass before experiments were to demonstrate conclusively that the predictions of this highly simplified model were in fact of sufficient accuracy to describe experimental data near the critical points of one-component simple fluids [32]. Many papers exclusively devoted to the Ising model had been published before certain small (but at one time apparently real) discrepancies between experiments and the model were finally resolved. One justification for all this effort was the belief that even if the Ising model did not prove to be applicable in a fully quantitative fashion, it would nevertheless capture the essential physics of the liquid-gas critical point. Hence, whatever "lore" that could be gradually built up from exhaustive study of the "simplest case" would prove useful in determining an appropriate generalization. Fortunately, the Ising model itself appears to need no generalization to be of relevance to liquid-gas critical points. Moreover, straightforward generalizations--such as the s-state discrete hierarchy and the n-vector continuum hierarchy--are proving to be sufficient to describe an incredibly large fraction of observed critical-point phenomena!

FLUID NEAR "CRITICAL POINT"	GEL NEAR "GEL POINT"
Detailed V(r) unknown	Detailed potential unknown
Essential feature: Hard Core Repulsion Short Range Attraction ⇓ STUDY ISING MODEL	Essential feature: Connectivity ⇓ STUDY PERCOLATION

Fig. 8 (a) (b)

What is the analog of the Ising model for polymer gelation? Fig.8a suggests that the basis for the utility of the Ising model stems from the fact that even though the detailed form of the interparticle potential V(r) is not known, this model nevertheless captures the essential physical features of a hard-core repulsion and a short-range attraction betwen molecules. By analogy, we indicate in Fig.8b the hypothesis that even though the details of intermolecular interactions are not known near the gelation threshold, the disarmingly simple model of percolation [33] may be sufficient to capture the essential physical features of connectivity between polyfunctional monomers.

Let us consider a very simple example of percolation phenomena. Imagine we have an infinite piece of window screen which is constructed in an unusual fashion: a randomly chosen fraction p_b of the "links" are conducting, while the remaining fracton $(1-p_b)$ are insulating. Computer simulations of a finite (16 x 16) section of this screen are shown in Fig.9 for $p_b= 0.2$, 0.4, 0.6, and 0.8. Clearly for p_b small, as in part (a), the system consists of small clusters of conducting bonds. In (b), the conducting fraction p_b has doubled, yet the system still consists of only finite clusters, albeit larger than in (a). In (c), $p_b=0.6$ and the system is totally different: in addition to the finite clusters (only two in number in the example shown), there exists a single cluster that is infinite in spatial extent. (Of course, the window screen must be infinite if the cluster is to be infinite!) For $0.4 < p_b < 0.6$ there is some value of p_b, termed the percolation threshold p_b^c, at which the system changes suddenly. Below p_b^c, the screen cannot conduct electricity from its top to its bottom; above p_b^c it can. Thus its macroscopic properties change suddenly as a microscopic parameter p_b increases infinitesimally from $p_b^c - \delta$ to $p_b^c + \delta$ (Fig.10a).

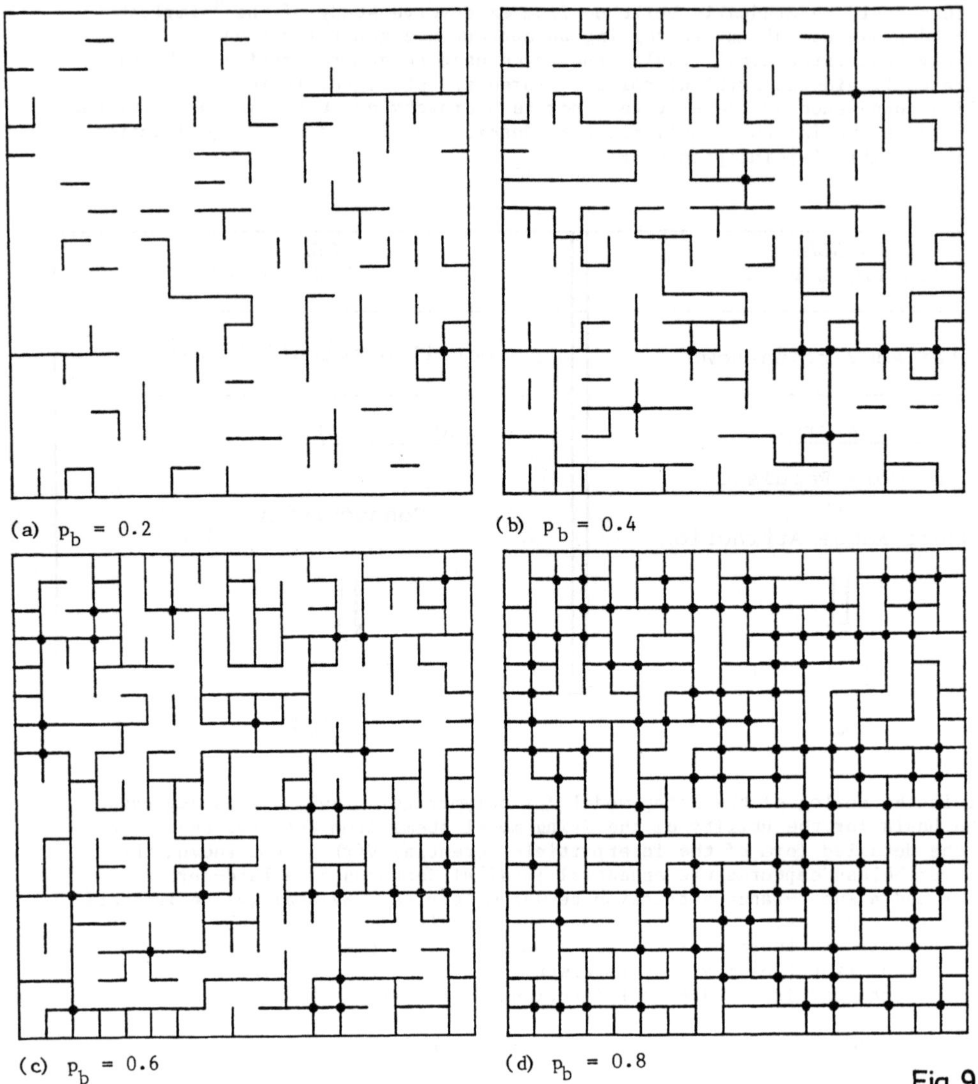
(a) p_b = 0.2
(b) p_b = 0.4
(c) p_b = 0.6
(d) p_b = 0.8

Fig.9

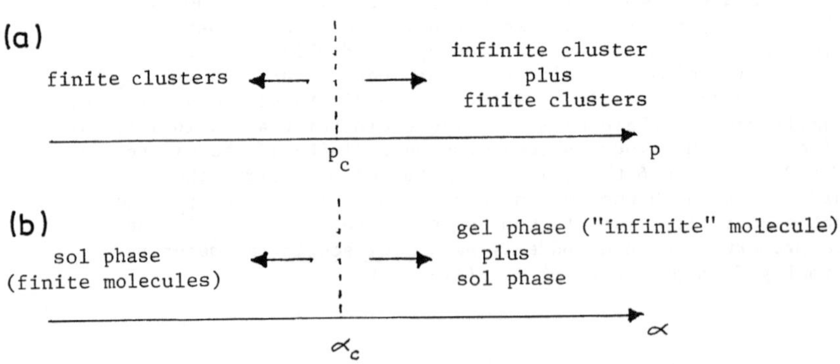
(a)
finite clusters
infinite cluster
plus
finite clusters
p_c
p
(b)
sol phase
(finite molecules)
gel phase ("infinite" molecule)
plus
sol phase
α_c
α

Fig.10

Similarly, below the gelation threshold, a system consists of only finite-size polymers. It cannot, e.g., propagate a shear stress; above the threshold it can. That is, the macroscopic properties change suddenly as a microscopic parameter α, the extent of reaction (or equivalently, the fraction of formed crosslinks) increases infinitesimally from $\alpha_c - \delta$ to $\alpha_c + \delta$ (Figs.10b and 11) [34].

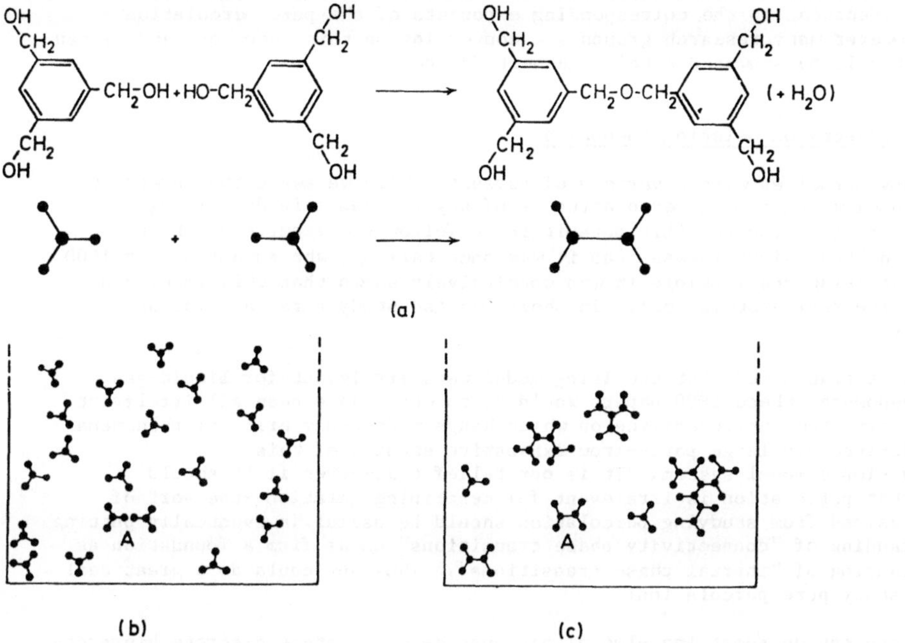

Fig.11

(a) THERMAL (b) PERCOLATION (c) GEL **Fig.12**

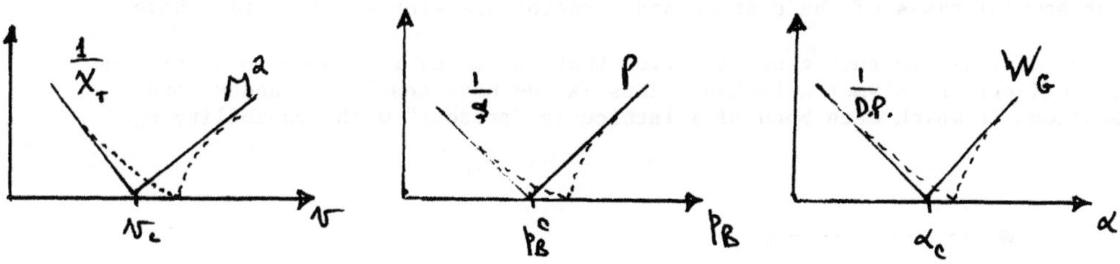

coupling constant	bond probability	"conversion"
$v_c = \tanh(J/kT_c)$	p_B^c = percolation pt.	α_c = gel point
droplet	bond cluster	polymer
M	P(p)	W_G = wt.fraction
χ_T	S(p)	D.P. = deg.of poly.
G(H,T)	$\langle N_{clusters} \rangle$	$\langle N_{molecules} \rangle$
Husimi-Temperley	Fisher-Essam	Flory
$d_+ = (2\beta + \gamma)/\nu = 4$	$d_+ = 6$	$d_+ = 6$

The analogy between random bond percolation and polyfunctional condensation was first pointed out almost two decades ago by the inventor of percolation, the British mathematician J.M. HAMMERSLEY [35]; it is summarized in Fig.12, based on work of DEGENNES and STAUFFER [36][2]. Thus far there is no analog of the HOCKEN-MOLDOVER experiments on fluid critical points [32] to demonstrate conclusively that the critical exponents found near the gelation threshold are numerically identical to the corresponding exponents of the pure percolation problem. However many research groups are converging on this question, and we can hope that it will be answered within the next decade.

9. Solvent Effects on Gelation Phenomena

We can occupy ourselves with a variety of pursuits while we await the answer to the question of whether pure percolation is of any relevance in describing the gelation threshold. One possible pursuit is to follow the example of those infected with "the Ising disease" (as it was once called), who authored over 1000 papers on the Ising model before it was conclusively shown that this model did indeed describe real systems well. In short, we can study pure percolation exhaustively.

Even had it transpired that the Ising model were irrelevant for liquid-gas critical phenomena, these 1000 papers would by no means have been all irrelevant. On the contrary, the entire edifice on which hangs modern-day critical phenomena theory is derived--in large part--from exhaustive studies of this appealingly simple model system. It is our belief that--even if it should transpire that percolation is irrelevant for describing gelation--the sort of experience gained from studying percolation should be useful in eventually putting our understanding of "connectivity phase transitions" on as firm a foundation as our understanding of "thermal phase transitions". Thus one could do a great deal worse than study pure percolation!

However one can do something else also: Just as the s-state discrete hierarchy and n-vector continuum hierarchy arose as natural generalizations of the pure Ising model, so also one can look for suitable generalizations of pure percolation. It is quite possible that some of these will prove as useful as various special cases of the s-state and n-vector hierarchies discussed above.

We now discuss one such generalization that may be of relevance in describing solvent effects on polymer gelation. Thus far we have described random-bond percolation, in which each bond of a lattice is "present" with probability p_b.

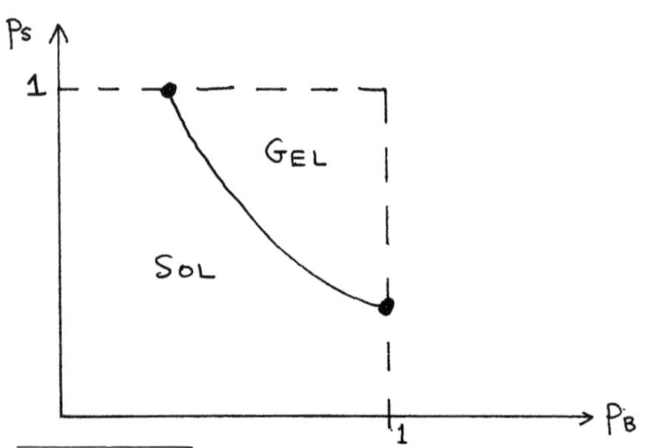

Fig.13

[2] Actually, the classic gel work of FLORY [37], four decades ago used percolation concepts and thus anticipated HAMMERSLEY's innovation by roughly two decades!

One can equally treat the complementary problem of random-site percolation, in which each site of a lattice is present with probability p_s [33]. Suppose we now consider a system in which both bonds and sites are present at random, with respective probabilities p_b and p_s [38]. The phase diagram for such a "bond-site percolation problem" is shown schematically in Fig.13; clearly the line p_s=1 corresponds to pure bond percolation, while the line p_b=1 corresponds to pure site percolation.

As noted above, a gel model in which every site is occupied by a polyfunctional monomer is isomorphic to pure bond percolation. Of course monomers are actually in a solvent, so there are two states for each site: occupied by a monomer or occupied by a solvent molecule. If we were to assume that the monomer sites are randomly distributed with probability p_s , then the polyfunctional condensation in a solvent could be modeled by "bond-site" percolation.

TANAKA and co-workers have recently made careful measurements of solvent effects in polymer gelation [39]. A typical T-ϕ phase diagram is shown in Fig.14; here ϕ is the volume fraction of polymer. We see that in addition to a line of connectivity phase transitions separating the high-temperature sol phase from the low-temperature gel phase, there exists also a phase separation curve characterized by a consolute temperature T_c below which the system splits up into two separate phases.

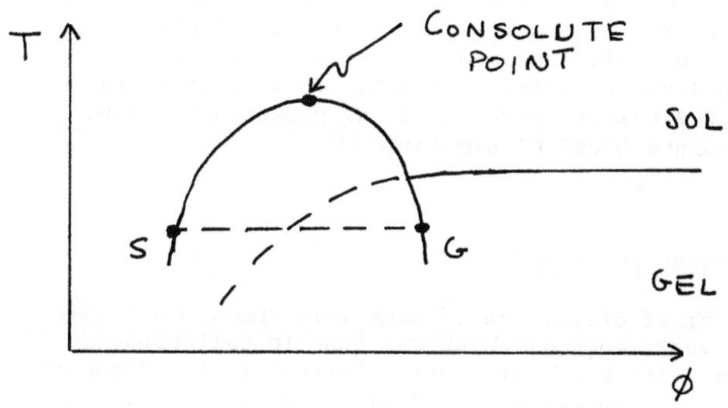

Fig.14

One cannot expect the "random bond-random site" percolation model presented above to be sufficient for the interpretation of TANAKA's data, since the presence of a consolute point means that monomer-monomer and solvent-solvent interactions are strong. Hence we generalize our "bond-site" model by incorporating a correlation between sites: if a given site is occupied by a monomer, then its neighbors are more likely to be occupied by a monomer. Also, we choose the bond probability $p_b=p_b(T)$ to be temperature-dependent, as one would expect on the basis of physical considerations.

The resulting model, "random bond-correlated site" percolation, like the original model of random percolation, has not been solved in closed form except in the approximation that all intramolecular interactions are neglected. This is the same approximation used by FLORY four decades ago in his solution of the original random bond percolation problem [37]; nowadays one often says that this approximation corresponds to an exact solution on a Cayley tree pseudolattice. Indeed, the Cayley tree solution is like a mean-field theory and gives the exact exponents for $d \geq d_+$ (d_+= 6 for percolation).

The agreement between TANAKA's data [39] and our calculations [40] is sufficiently encouraging that we are now motivated to begin the arduous task associated with incorporating intramolecular interactions as well.

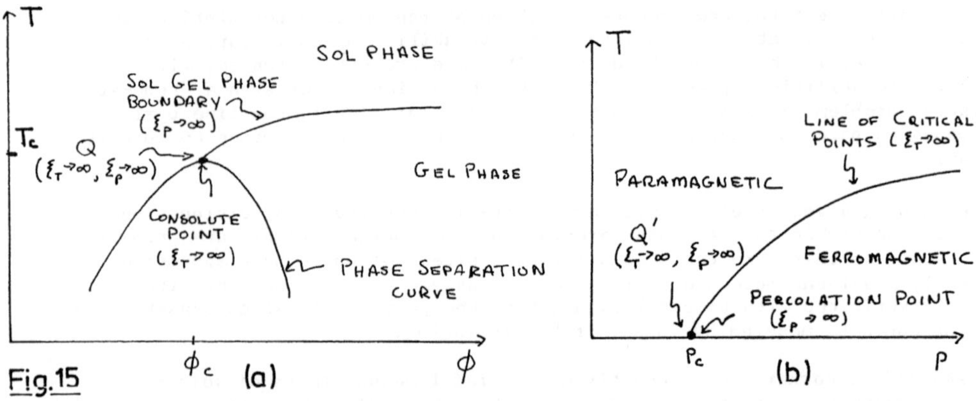

Fig.15

It is possible to adjust the experimental parameters in such a fashion that the gel-sol phase boundary intersects the consolute point. At this point, labeled Q in Fig.15a, we expect that connectivity fluctuations and concentration fluctuations are simultaneously critical; percolation is relevant to the former, while the Ising model is relevant to the latter. A particularly intriguing open question concerns the sort of phenomena that one might expect in the neighborhood of point Q. There is some evidence from renormalization group calculations [41] which indicates that the properties of this system at the point Q are more complicated than predicted by mean field theories. An analogously-defined point Q′ exists in the phase diagram of the dilute ferromagnet, as shown in Fig.15b. Are the phenomena occurring near points Q and Q′ similar?

10. Are Percolation Concepts Relevant to Water?

I shall conclude this talk with a brief discussion of some very recent work that is still in an early stage. This work, much of which was done in collaboration with J. TEIXEIRA, concerns the application of percolation (connectivity) concepts to elucidate the many unusual physical properties of liquid water [42]. Of course there have been many theories of liquid water, but most are nowadays considered essentially inadequate [43]. The present model may, in 10 years, belong to the same category. However the careful study of even an incomplete model is often useful, and for this reason we are now attempting to study exhaustively the properties of a simple "percolation model of water".

Perhaps the feature that most distinguishes water from a typical liquid is the fact that each water molecule forms up to four strong and highly directional hydrogen bonds with adjacent water molecules (Fig. 16) [44].

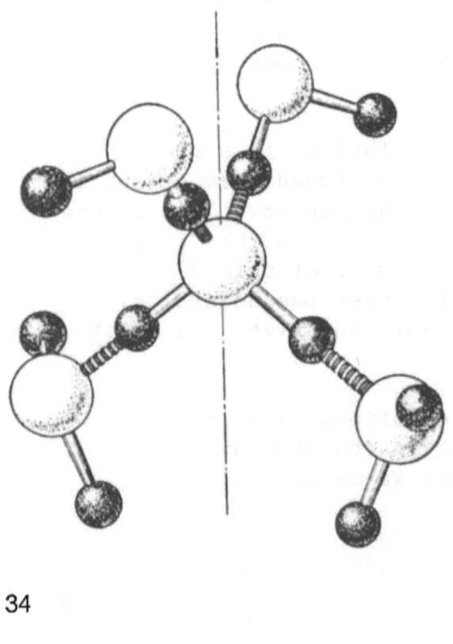

Fig. 16

In a system of N water molecules, a fraction p_b of the 2N possible hydrogen bonds may be considered to be intact. The intriguing fact is that p_b is remarkably large; although it is difficult to obtain p_b to an accuracy of better than 20-30% there seems to be a wide body of evidence supporting the possibility that p_b is at least 0.5, which is well above the percolation threshold for any three-dimensional system [44]. Thus water, at least at normal temperatures, consists of a single infinite network (or "gel molecule") as well as smaller, finite networks. The reason water does not "look like a gel" is that the mean hydrogen bond lifetime, about 10^{-12} seconds [45], is much shorter than the times characteristic of our sensory organs.

How is the presence of an infinite network related to the unusual properties of water? To answer this question, we re-examine the computer simulations of random-bond percolation shown in Fig.9. We have indicated by heavy dots those sites with all four bonds intact. Consider Fig.9d, for which $p_b=0.8$ (a value typical of water at low temperatures). At this value of p_b, the heavy dots form relatively large clusters; inside a cluster all water molecules are--by construction--maximally bonded. If we make the conjecture that local properties of patches differ from the global properties of the surrounding "gel", then we have a physical mechanism by which many (if not all) the unusual properties of liquid water can be interpreted [42]. It is interesting to note that although the <u>density</u> of four-coordinated water molecules is simply p_b^4, the <u>spatial distribution</u> of these "heavy dots" is far from random. For example, it is impossible to have a site surrounded by four neighbors, all of which are heavy dots, without the site itself being a heavy dot. Thus we have a correlated-site problem, where the correlation arises from a purely random process, that of random bond occupancy. The exponents for this novel correlated-site problem appear to be the same as those for the random-site problem [46].

The preceding picture is very much a zeroth order explanation. It will almost certainly be necessary to improve it with various generalizations, some of which might include considerations of continuum percolation (so that the lattice can be omitted), polychromatic percolation (so that three-bonded, two-bonded, and one-bonded species may be adequately taken into account), and Ising-correlated site percolation.

The important point is that to make progress on so difficult a problem as liquid water, one must first identify the essential physical features and the essential physical mechanism giving rise to these features. Since hydrogen bonds are the primary distinguishing feature of water, it seems reasonable to pursue the extent to which the connectivity between water molecules is relevant to providing insight into the extremely unusual properties of liquid water.

11. Summary

In this introductory talk, I have tried to give some feeling for the degree of development of the field of phase transitions and critical phenomena. I have also attempted to illustrate the fact that there remain many unsolved problems. To this end, I have discussed two examples where our greater understanding of connectivity (and phase transitions characterized by an abrupt change in the nature of the connectivity) could be useful in elucidating physical phenomena of fundamental importance. I am optimistic that in-depth study of relatively simple model systems can often provide the conceptual framework on which to build our understanding.

References

1. See, e.g., H.E. Stanley, <u>Introduction to Phase Transitions and Critical Phenomena</u>, Second Edition (Oxford University Press, London and New York, 1980), Chapter 6.

2. T.D. Lee and C.N. Yang "Statistical theory of equations of state and phase transitions. II. Lattice gas and Ising model" Phys. Rev. 87, 410-419 (1952).

3. The scaling approach to critical phenomena was proposed in 1965 by Widom and independently, by Domb and Hunter: it was extended to correlation functions in 1966 by Kadanoff (and by Patashinskii and Pokrovskii), and to dynamic phenomena in 1967 by Ferrell et al. and by Halperin and Hohenberg. The generalized homogeneous function approach to scaling is developed in A. Hankey and H.E. Stanley, "Systematic application of generalized homogeneous functions to static scaling, dynamic scaling, and universality", Phys. Rev. B6, 3515-3542 (1972).

4. Ref. 1, Chapter 9.

5. The concept of universality classes of critical behavior was first clearly put forth by Kadanoff, at the 1970 Enrico Fermi Summer School, based on earlier work by a large number of workers including Griffiths, Jasnow and Wortis, Watson, Fisher, Stanley and others.

6. The s-state hierarchy was proposed in R.B. Potts, "Some generalized order-disorder transformations", Proc. Camb. Phil. Soc. 48, 106-109 (1952).

7. The n-vector hierarchy was proposed in H.E. Stanley, "Dependence of critical properties on dimensionality of spins", Phys. Rev. Lett. 20, 589-592 (1968).

8. H.E. Stanley, "Spherical model as the limit of infinite spin dimensionality" Phys. Rev. 176, 718-722 (1968).

9. P.G. deGennes, "Exponents for the excluded volume problem as derived by the Wilson method", Phys. Lett. 38A, 338-340 (1972).

10. R. Balian and G. Toulouse, "Critical exponents for transitions with n=-2 components of the order parameter", Phys. Rev. Lett. 30, 544-547 (1973).

11. D. Mukamel and S. Krinsky, "Physical realizations of $n \geq 4$-component vector models. I. Derivation of the Landau-Ginzburg-Wilson Hamiltonians" Phys. Rev. B 13, 5065-5077 (1976).

12. P.W. Kasteleyn and C.M. Fortuin, "Phase transitions in lattice systems with random local properties", J. Phys. Soc. Japan Suppl. 26S, 11-14 (1969).

13. M.J. Stephen, "Percolation problems and the Potts model", Phys. Lett. 56A, 149-150 (1976).

14. A. Aharony and P. Pfeuty, "Dilute spin glasses at zero temperature and the 1/2-state Potts model", J. Phys. A 12, L125-L128 (1979).

15. A. Aharony, K.A. Müller, and W. Berlinger, "Trigonal-to-tetragonal transition in stressed SrTiO : A realization of the three-state Potts model", Phys. Rev. Lett 38, 33-36 (1977); S. Alexander, "Lattice gas transition of He on graphoil. A continuous transition with cubic terms.", Phys. Lett. 54A, 353-4 (1975).

16. K.G. Wilson, "Renormalization group and critical phenomena. I. Renormalization group and the Kadanoff scaling picture", Phys. Rev. B 4, 3174-83 (1972).

17. F. Wegner, "Phase transitions and critical behaviour", Festkörperprobleme 16, 1-14 (1976).

18. See, e.g., B.G. Nickel and B. Sharpe, "On hyperscaling in the Ising model in three dimensions", J. Phys. A 12, 1819-1834 (1979), and references therein.

19. See, e.g., the recent review P.C. Hohenberg and B.I. Halperin, "Theory of dynamic critical phenomena", Rev. Mod. Phys. 49, 435-479 (1977) and references therein.

20. The possibility of a phase transition to a novel low-temperature phase without infinite-range order for the case d=n=2, was first proposed in H.E. Stanley, "Critical properties of isotropically-interacting classical spins constrained to a plane", Phys. Rev. Lett. 20, 150-154 (1968).

21. In 1972, Kosterlitz and Thouless proposed that the order in the low-temperature phase is associated with the presence of bound vortex pairs, and the critical point is characterized by the unbinding of these vortices. Their proposal has very recently received widespread theoretical and experimental support. See, e.g., the review J.M. Kosterlitz and D.J. Thouless, "Two-dimensional physics", Prog. Low Temperature Physics 7B, 371-433 (1978).

22. K. Binder "Dynamics of first order phase transitions". In Fluctuations, Instabilities and Phase Transitions, Ed. T. Riste (Plenum Publ. Corp., New York 1975). pp. 53-86.

23. See, e.g., W. Klein, D.J. Wallace, and R.K.P. Zia, "Essential singularities at first-order phase transitions", Phys. Rev. Lett 37, 639-642 (1976).

24. Tricritical points were first studied in fluid systems by Soviet workers in the 1960's, based on an earlier (1926) proposal of Kohnstamm. Independently, in 1970, R.B. Griffiths introduced the concept; his work spurred tremendous interest in a wide range of multicritical-point phenomena. An accurate historical account, together with references to the original literature, is given in B. Widom "Tricritical points in three-and four-component fluid mixtures", J. Phys. Chem. 77, 2196-2200 (1973).

25. R.B. Griffiths and B. Widom, "Multicomponent-fluid tricritical points", Phys. Rev. A 8, 2173-2175 (1973).

26. See, e.g., M. Wortis, F. Harbus and H.E. Stanley, "Tricritical behavior of the Ising antiferromagnet with next-nearest-neighbor ferromagnetic interactions: Mean-field-like tricritical exponents?", Phys. Rev. B 11, 2689-2692 (1975) and references therein.

27. Ref. 1, Chapter 10, and references therein. See, especially, R. Bausch, "Ginzburg criterion for tricritical points", Z. Physik 254, 81-88 (1972).

28. See, e.g., the recent review E. Stryjewski and N. Giordano, "Metamagnetism", Adv. Phys. 26, 487-650 (1977) and references therein.

29. See, e.g., S. Redner and H.E. Stanley, "Helical order and its onset at the Lifshitz point", Phys. Rev. B 16, 4901-4906 (1977) and references therein. See also W. Selke and M.E. Fisher, "Monte Carlo study of the spatially modulated phase in an Ising model", Phys. Rev. B 20, 257-265 (1979).

30. R.M. Hornreich, M. Luban, and S. Shtrikman, "Critical behavior at the onset of k-space instability on the λ-line", Phys. Rev. Lett. 35, 1678-81 (1975).

31. W. Lenz, "Beitrag zum Verständnis der magnetischen Ersheinungen in festen Körpern", Phys. Z. 21, 613-615 (1920).

32. R. Hocken and M.R. Moldover, "Ising critical exponents in real fluids: an experiment", Phys. Rev. Lett. 37, 29-32 (1976).

33. Recent reviews on percolation, from rather different perspectives, include D. Stauffer, "Scaling theory of percolation clusters", Phys. Rep. 54, 1-74 (1979), J.W. Essam, "Percolation theory", Repts. Prog. Phys. (in press) and P. Pfeuty and E. Guyon (in preparation).

34. Figure 11 is adapted from work of J.M. Gordon.

35. See, e.g., the review article H.L. Frisch and J.M. Hammersley, "Percolation processes and related topics", J. Soc. Indust. Appl. Math. 11, 894-918 (1963).

36. P.G. deGennes, "Critical dimensionality for a special percolation problem" J. de Phys. (Paris) 36, 1049-1054 (1975). D. Stauffer, "Gelation in concentrated critically-branched polymer solutions: Percolation scaling theory of intramolecular bond cycles" J. Chem. Soc. Faraday Trans. II 72, 1354-1364 (1975).

37. P.J. Flory, "Molecular size distribution in three-dimensional polymers. I. Gelation", J. Am. Chem. Soc. 63, 3083-3090 (1941).

38. The properties of bond-site percolation are discussed in e.g., P. Agrawal, S. Redner, P.J. Reynolds, and H.E. Stanley, "Site-bond percolation: A low-density study of the uncorrelated limit", J. Phys. A 12, 2073-2085 (1979), and H. Nakanishi and P.J. Reynolds, "Site-bond percolation by position-space renormalization group", Phys. Lett. 71A, 252-254 (1979).

39. T. Tanaka, G. Swislow, and A. Ohmine, "Phase separation and gelation in gelatin gels", Phys. Rev. Lett. 42, 1556-1559 (1979).

40. A. Coniglio, H.E. Stanley, and W. Klein, "Site-bond correlated- percolation problem: A statistical mechanical model of polymer gelation", Phys. Rev. Lett. 42, 518-522 (1979); a more extensive manuscript is in preparation.

41. A. Coniglio and W. Klein, "Clusters and Ising critical droplets: A renormalization group approach", J. Phys. A (submitted).

42. H.E. Stanley, "A polychromatic correlated-site percolation problem with possible relevance to the unusual behaviour of supercooled H_2O and D_2O" J. Phys. A 12, L329-335 (1979); H.E. Stanley and J. Teixeira, "A qualitative interpretation of exerimental data on supercooled H_2O and D_2O: Tests of a percolation model", preprint.

43. D. Eisenberg and W. Kauzmann, The Structure and Properties of Water (Oxford University Press, New York, 1969), Chapter 5.

44. A. Geiger, F.H. Stillinger and A. Rahman, "Aspects of the percolation process for hydrogen-bond networks in water", J. Chem. Phys. 70, 4185-4193 (1979).

45. C.J. Montrose, J.A. Bucaro, J. Marshall-Croakley, and T.A. Litovitz, "Depolarized Rayleigh scattering and hydrogen bonding in liquid water", J. Chem. Phys. 60, 5025-5029 (1974).

46. R. Blumberg, G. Shlifer, and H.E. Stanley, "Tests of universality in a correlated-site percolation problem", preprint.

Critical Properties of Relativistic Bose Gases[1]

D.E. Miller[2], R. Beckmann, and F. Karsch

Fakultät für Physik, Universität Bielefeld
D-4800 Bielefeld 1, Fed. Rep. of Germany

1. Introduction

The basic phenomenon of condensation appears in a great many different but related forms throughout the sciences. After we briefly mention some of the physical properties of the special type of phase transition known as Bose-Einstein condensation (BEC) in direct connection with some of the main themes of this symposium in synergetics, we shall proceed to describe the particular collective phenomena relating to the relativistic generalization [1,2] of this problem in d spatial dimensions.

Although the concept of condensation in physics has for a long time been related to the gas-liquid phase transition below the critical point through the macroscopic clustering processes, it was first pointed out by A. EINSTEIN [3] in 1925 that such a situation existed for the ideal Bose gas. From the basic structure of a single quantum state, it was shown to be possible under the right thermodynamical conditions to have a finite fraction of the system drop into the ground state. A few years later this type of condensation was considered by F. LONDON [4] to relate to the then known properties of superfluidity, which even today has its place in the study of the collective phenomena using macroscopic models for superfluid helium and superconductivity. During the course of the years the general structure of BEC has shown itself in a great many fields of physics - in particular, wherever the symmetric nature of the wave function allows the particles at finite temperatures to go into the ground state. For instance, in nuclear matter where the massive pseudoscalar bosons (pions) dominate the strong interactions a type of pion condensation [5] is known to take place. It is often speculated that even at higher energies this type of condensation is important to the collective behavior of the strong-interacting particles. From these phenomena we are motivated to study the BEC in its relativistic form.

As we have already seen in other contributions to this symposium, the process of condensation is not in any way restricted only to the equilibrium phase transitions. Two very important examples have already been discussed here in relation to the interdisciplinary field of synergetics [6]. The laser [7] is a system far from equilibrium with known properties of phase transitions, which are effected by the quantum statistics [8]. The ordering [9] caused by the condensation of the photons can be readily established using a one dimensional model. Another example previously discussed here [10] arises from the nonequilibrium phase transitions in highly excited semiconductors where the electron-hole pairs accumulate into droplets through a condensation process [11]

---

1 Work supported in part by the Deutsche Forschungsgemeinschaft under contract Mi211/1

2 Presently at the Zentrum für Interdisziplinäre Forschung, Universität Bielefeld

similar to the usual BEC. These examples, to be sure, are only the beginning of a long list of possibly related condensation processes.

From this discussion we can see the great adaptibility of BEC in many-particle phenomena. In particular the ideal bose gas represents a simple model for the collective behavior due to the symmetry of the wavefunction. In this context we shall here discuss some of our work [1,2] which extends this model to include relativistic dynamics in arbitrary spatial dimensions d in such a way that the critical phenomena [12] can be calculated from limiting cases.

2. Relativistic Bose Gas in d Dimensions

The relativistic properties of a gas of spinless bosons with a mass m become important when the individual particles' velocities approach the speed of light. This situation clearly arises often at very high energies. We assume for simplicity a noninteracting relativistic energy spectrum $\varepsilon(p)$ of the form ($c = 1 = \hbar$)

$$\varepsilon(p) = \sqrt{m^2 + p^2} \quad , \tag{1}$$

where p represents the modulus of the momentum vector in a d dimensional space. In this section we want to investigate the collective behavior of a relativistic Bose gas confined to a d dimensional spatial volume V_d at a given inverse temperature β and relativistic fugacity A from a covariant thermodynamical formulation [13,14].

The relationship to the dynamics we shall first mention for two extreme cases, which may be derived directly out of (1) as the following limits:

1) The nonrelativistic limit $p \ll m$ or $m \to \infty$

$$\varepsilon(p) = p^2/2m \; ; \tag{2a}$$

2) The ultrarelativistic limit $p \gg m$ or $m \to 0$ with

$$\varepsilon(p) = p. \tag{2b}$$

We may combine these limiting cases into a more general form of an energy spectrum in d dimensions as follows:

$$\varepsilon_\sigma(p) = \sum_{i=1}^{d} c_i \, p_i^\sigma \quad , \tag{3}$$

where we see that (2a,b) are the isotropical forms of (3) with the values of σ = 1 and 2 respectively.

Now we shall briefly explain the treatment of the relativistic ideal Bose gas in d dimensions using the invariant phase space [13,14] in the grand canonical ensemble to derive the thermodynamical potential Ω of the form

$$-\beta\Omega = \frac{V_d}{(2\pi)^d} \int d^dp \, \ln(1-A\, e^{-\beta\varepsilon(p)}) - \ln(1-A\, e^{-\beta m}) . \tag{4}$$

The first term is simply the phase space integral, while the second term comes from the ground state contribution. When we use $\varepsilon(p)$ as given in (1), we find the equation

$$\Omega = - \frac{V_d (\sqrt{\pi})^{d-3} (2m)^{d'}}{2(2\pi)^{d-1} \beta^{d'}} \sum_{k=1}^{\infty} \frac{A^k}{k^{d'}} K_{d'}(km\beta) + \frac{1}{\beta} \ln(1-A\, e^{-\beta m}) \tag{5}$$

where $d' = (d+1)/2$ and $K_\ell(X)$ is the modified Bessel function of order ℓ.

For the average particle number density n a simple derivative on A with a multiplication by A gives to us

$$n = \frac{(\sqrt{\pi})^{d-3}(2m)^{d'}}{2(2\pi)^{d-1}\beta^{d'-1}} \sum_{k=1}^{\infty} \frac{A^k}{k^{d'-1}} K_{d'}(km\beta) + \frac{A\, e^{-m\beta}}{V_d(1-A\, e^{-m\beta})} \quad . \tag{6}$$

This expression for n contains the two physically distinct terms: the usual cluster contributions from the integral over finite momenta and the ground state or zero momentum term, which clearly comes from the logarithm added to (4).

A criterion for the relativistic BEC comes out of the definition [15] of the first relativistic cluster coefficient, which we have extended to d dimensions [1,2] in the following way:

$$L_d(m,\beta) = \frac{(\sqrt{\pi})^{d-3}2^{d'-1}(m\beta)^{d'}}{\beta^d\,(2\pi)^{d-1}} e^{m\beta} K_{d'}(m\beta) . \tag{7}$$

From this definition we find that the relativistic BEC takes place whenever

$$L_d^{-1}(m,\beta) \geq \sum_{k=1}^{\infty} \frac{e^{(k-1)m\beta}}{k^{d'-1}} \frac{K_{d'}(km\beta)}{K_{d'}(m\beta)} \tag{8}$$

in the limit of $V_d \to \infty$ and $A \to \exp\{m\beta\}$. Equality in (8) defines uniquely the critical density n_c and its corresponding critical temperature T_c. Our numerical computation [1] as shown in Fig. 1 shows the dimensional dependence of $L_d^{-1}(m,\beta)n_c$ for the various stated values of $m\beta$.

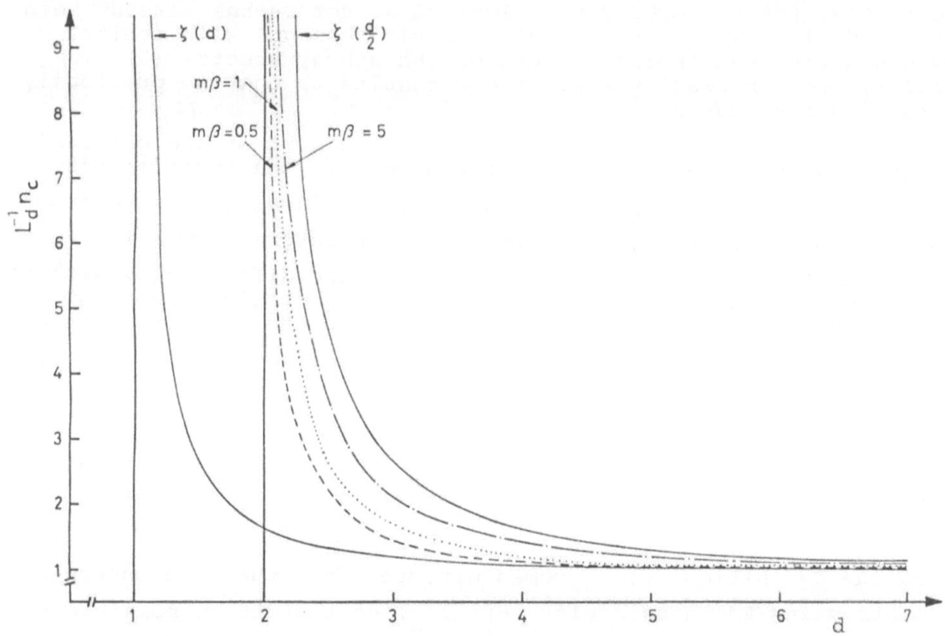

Fig. 1: The critical density n_c of the ideal relativistic Bose gas as a function of the spatial dimension d for various stated values of $m\beta$.

In fact, the limiting cases of the ultrarelativistic and nonrelativistic gases for $L_d^{-1} n_c$ are clearly seen on Fig. 1 to be indicated, respectively, by $\zeta(d)$ and $\zeta(\frac{d}{2})$. Furthermore, we see from these two limits why the ultrarelativistic gas is in its critical behavior distinct from all the relativistic massive systems, which follow basically in form the nonrelativistic limit. Thus we see that the massive Bose gas condenses for $d > 2$ while the massless system allows condensation for $d > 1$.

An important quantity for the critical properties of the relativistic gases is the specific heat at constant volume C_v. It was previously shown by P. LANDSBERG and J. DUNNING-DAVIES [16] that there exists a jump in C_v at T_c in the ultrarelativistic Bose gas in the usual three dimensional space. A corresponding jump for the nonrelativistic Bose gas in three dimensions is not found even though BEC does take place. In fact, for the massive relativistic Bose gases in general we have established [1,2] that this jump, which is analytically defined by

$$\Delta_d = |C_v(T_c+) - C_v(T_c-)|, \tag{9}$$

appears for the first time above four dimensions. However, these observations may be well summarized through the extreme cases by saying that for $\sigma = 1$, $\Delta_d \neq 0$ for $d > 2$, while for $\sigma = 2$, $\Delta_d \neq 0$ for $d > 4$. In the next part of this presentation we shall show how these facts can be incorporated in concrete statements about the critical exponents through the scaling relations.

3. Critical Phenomena

Here we shall look more explicitly into the properties of the critical exponents [12] for the BEC. The analytical structure has already been investigated [17-19] for the nonrelativistic Bose gases in arbitrary dimensions d with the isotropic form of the energy spectra (3). In particular, one can readily apply these results to the two previously mentioned limiting cases.

It was seen that in the nonrelativistic case most of the critical exponents with the exception of α associated with the specific heat ($C_v \sim (t)^{-\alpha}$) showed the expected behavior of a classical system above four dimensions, whereby t is the reduced temperature $(T-T_c)/T_c$. Furthermore, it was pointed out by C. HALL [19] that the difficulties with α can be remedied by introducing the two different critical exponents α_s and α_h related to two different contributions to the specific heat C_v. The exponent α_s is defined [17] by taking the m-th partial derivative, which is singular in the form

$$\frac{\partial^m C_v}{\partial T^m} \sim t^{-(\alpha_s+m)} \quad . \tag{10}$$

However, the definition for α_h comes directly from the free energy F after subtracting the nonsingular part F_{ns}, so that it is possible to find it simply by through the second derivative [18] as

$$\frac{\partial^2}{\partial T^2} \{F(T) - F_{ns}(T)\} \sim t^{-\alpha_h} \quad . \tag{11}$$

What was really found by C. HALL [19] was that these exponents α_s and α_h satisfy different scaling relations, which have been in general discussed by H.E. STANLEY [12]. The exponent α_h satifies the Rushbrooke equality in (12a) below as a scaling relation for all dimensions; while α_s does the same for the Josephsen equality in (12d). For the dimensions $d > 4$, however, α_s does not satisfy (12a) and α_h cannot be replaced in (12d). This means simply that $\alpha_s \neq \alpha_h$ for $d > 4$. The remedy for this situation comes through addition to all the other invalid scaling relations a factor of the form $\alpha_h - \alpha_s$. We list in Table 1 the values of the above discussed critical exponents as a function of d and σ. To this list we add [12] the following other exponents: β from the order parameter, γ from the compressibility, δ from the equation of state, η from the correlation function and ν from the correlation length.

Table 1: A summary of the critical exponents [12, 17, 18]

	α_h	α_s	β	γ	δ	η	ν
$d < \sigma$	–	–	–	–	–	–	–
$\sigma < d < 2\sigma$	$\frac{d-2\sigma}{d-\sigma}$	$\frac{d-2\sigma}{d-\sigma}$	$\frac{1}{2}$	$\frac{\sigma}{d-\sigma}$	$\frac{d+\sigma}{d-\sigma}$	$2-\sigma$	$\frac{1}{d-\sigma}$
$d > \sigma$	0	$\frac{2\sigma-d}{\sigma}$	$\frac{1}{2}$	1	3	$2-\sigma$	$\frac{1}{\sigma}$

The revised scaling relations [19] become

$$\alpha_h + 2\beta + \gamma = 2 \tag{12a}$$

$$\gamma = \beta(\delta-1) \tag{12b}$$

$$\delta = (2-\delta)\nu \tag{12c}$$

$$d\nu = 2 - \alpha_s \tag{12d}$$

$$d\nu - \gamma = 2\beta + (\alpha_h-\alpha_s) \tag{12e}$$

$$\nu(d-2+\eta) = 2\beta + (\alpha_h-\alpha_s) \tag{12f}$$

$$2 - \eta = d\gamma/(2\beta+\gamma+\alpha_h-\alpha_s) \tag{12g}$$

$$d - 2 + \eta = d(2\beta+\alpha_h-\alpha_s)/2 - \alpha_s \tag{12h}$$

We see in the last four equations that the combination $2\beta + \alpha_h - \alpha_s$ brings back the scaling relation when $d > 2\sigma$. In (12a) we have the inclusion of the scaling singularity of the free energy, while in (12d) the hyperscaling structure of the specific heat dominates. Only (12b) and (12c) are without any alteration always true.

In Table 1 we have included the effect of the energy spectrum (3) on the critical exponents [17, 18]. These results allow us to probe into one part of the dimensionality plane already discussed by STANLEY [12] relating to the $n = \infty$ line, where n is the dimension of the order parameter. The nonrelativistic limit corresponds directly to the upper corner of this line with $d = 4$. On the other hand the ultrarelativistic limit relates to the lower corner formed by the $n = \infty$ line with $d = 2$. Unfortunately, these limiting cases of our model only reach the corners of the line at infinity.

It is nevertheless possible to reach more of this line at infinity with a similar model through an alteration of the energy spectrum to the form

$$\varepsilon(p) = \sqrt{m^2 + c_\sigma p^\sigma} \tag{13}$$

This apparently slight modification changes quite drastically the symmetrical structure of $\varepsilon(p)$ in (1) as well as its physical meaning in relativistic gas theory.

4. Generalizations and Conclusion

There are a great number of motivations coming out of physical problems for further investigation of the relativistic BEC in high energy processes [20]. It has been shown how relativistic Fermi systems promote a condensation of the related interacting Bose particles [21]. As it was shown for these systems, one may derive a generalized energy spectrum in various special cases of the form

$$\varepsilon_g(p) = \sum_{n=-\infty}^{\infty} a_n \left(\sqrt{m^2+p^2}\right)^n, \tag{14}$$

which had been previously investigated as a mathematical model [22] for the collective behavior of hadronic matter. It is possible to see that such a generalization could greatly shift the dimensional relatenship between the critical exponents.

In a separate investigation the critical exponents for the nonrelativistic interacting Bose gas have been calculated by K.K. SINGH [23] using methods of the renormalization group [24]. Here the usual ε-expansion is used with $\varepsilon = 4 - d$.

There are, however, other known mechanisms which determine the critical dimensions of the BEC. It has been shown by D. ROBINSON [25] for the nonrelativistic BEC that the type of boundry conditions can change the structure of the BEC. For example, in $d = 2$ one can find BEC when attractive walls are present. This effect can be seen through other mechanisms. For the laser system it is known that [6,7] the photons possess a coupling to the atomic dipole moments, which themselves couple to the atomic inversion. In this system there has been found a condensation into a single mode [9] expressed in terms of a nonequilibrium phase transition using a one dimensional model. Finally, it is clear that the extensions of this type of investigation can bring many interesting aspects to the work in critical phenomena.

Acknowledgements

We would like to thank Prof. H. SATZ for continual support and interest as well as for many discussions on hadronic physics. We would also like to recognize discussions with Prof. H.E. STANLEY on critical phenomena. One of us (D.E.M.) would like to express his gratitude to Prof. H. HAKEN for many fruitful suggestions and conversations as well as his encouragement in taking an active part in this very successful symposium on synergetics.

References

1. R. Beckmann, F. Karsch and D.E. Miller, "Bose-Einstein Condensation of a Relativistic Gas in d-Dimensions," Phys.Rev.Lett., to be published
2. R. Beckmann, F. Karsch and D.E. Miller, "Analysis of the Critical Structure of Bose-Einstein Condensation for a Relativistic Gas in d-Dimensions," to appear; see also R. Beckmann, "Thermodynamische Eigenschaften hadronischer Materie für relativistische Gasmodelle in mehreren Dimensionen," Diplomarbeit, Universität Bielefeld, 1979
3. A. Einstein, Sitzber. kgl. preuss. Akad. Wiss., 261 (1924), 18 (1925); also in the book, Albert Einstein, Sein Einfluß auf Physik, Philosophie und Politik, P.C. Aichelburg and R.U. Sexl (Eds.) (Vieweg, Braunschweig, 1979)
4. F. London, Phys. Rev. 54, 947 (1938); J. Phys. Chem. 43, 49 (1939); for a general reference see F. London, Superfluids, Vol. I, II (Dover, New York, 1950, 1954)
5. A.B. Migdal, Rev. Mod. Phys. 50, 107 (1978); G.E. Brown and W. Weiser, Phys. Rep. 27c, 1 (1976)
6. H. Haken, Synergetics (Springer, Berlin, 1978), see especially Chapters 6 and 8 as well as the proceedings of this conference
7. H. Haken, Encyclopedia of Physics, Vol. XXV/2c, "Light and Matter 1c," (Springer, Berlin, 1970)
8. H. Haken, Proc. 13th IUPAP Conf. Stat. Phys., Haifa, Aug.,1977, Ann. Israel Phys. Soc. 2, 99 (1978)
9. R. Graham and H. Haken, Z. Phys. 213, 420 (1968)
10. H. Haug and S.W. Koch, "Nonequilibrium Phase Transition in Highly Excited Semiconductors," in these proceedings; Phys. Lett. 69A, 445 (1979)
11. E. Hanamura and H. Haug, Phys. Rep. 33C, 209 (1977)
12. H.E. Stanley, "Critical Phenomena, Past, Present and Future," these proceedings; general reference, H.E. Stanley, Phase Transitions and Critical Phenomena (Oxford, 1971)
13. B. Touschek, Nuovo Cimento 58B 295 (1968); B. Touschek and G. Rossi, Meccanica Statistica (Boringhieri, Torino, 1970), see Ch. 3
14. D.E. Miller and F. Karsch, "On the Theory of Relativistic Quantum Gases I. The Covariant Formulation for Equilibrium Systems," Bielefeld Preprint BI-IP 79/08, April, 1979
15. F. Karsch and D.E. Miller, "On the Theory of Relativistic Quantum Gases II. The Exact Equations of State for Ideal Systems," Bielefeld Preprint BI-IP 79/09, June, 1979
16. P.T. Landsberg and J. Dunning-Davies, Phys. Rev. 138A, 1049 (1965)
17. J.D. Gunton and M.J. Buckingham, Phys. Rev. 166, 152 (1968)
18. P. Lacour-Gayet and G. Toulouse. J. Phys. (Paris) 35, 425 (1974)
19. C.K. Hall, J. Stat. Phys. 13, 157 (1975)
20. H. Satz, Proc. 13th IUPAP Conf. Stat. Phys. Haifa, Aug., 1977, Ann. Israel Phys. Soc. 2, 255 (1978)
21. A.I. Akhieser et al. "On Condensation of Bose Fields in Relativistic Fermi Systems," Kiev Preprint ITP 78, 80E, Aug., 1978

22. D.E. Miller, Proc. 13th IUPAP Conf. Stat. Phys., Haifa, Aug., 1977, Ann. Israel Phys. Soc. 2, 940 (1978); J. Math. Phys. 20, 801 (1979)
23. K.K. Singh, Phys. Rev. B12, 2819 (1975); B13, 3192 (1976); B17, 324 (1978)
24. G. Toulouse and P. Pfeuty, Groupe de Renormalisation (Grenoble, 1975); D.J. Amit, Field Theory, the Renormalization Group and Critical Phenomena (McGrawHill, New York, 1978)
25. D.W. Robinson, Comm. Math. Phys. 50, 53 (1976)

Part III

Nonequilibrium Phase Transitions

Collective Effects in Rasers

P. Bösiger, E. Brun, and D. Meier

Institute of Physics, University of Zurich
CH-8001 Zurich, Switzerland

1. Introduction

The creation of ordered structures in nonlinear many body systems far from thermal equilibrium is a subject of increasing interest. In particular, the vast field of nonlinear optics has provided many examples of systems showing the effects of cooperative behaviour. The best known is the laser where the analogy with phase transitions at thermal equilibrium was first pointed out by GRAHAM and HAKEN [1] and DEGIORGIO and SCULLY [2]. A more recently discovered phenomena is optical bistability which occurs when light interacts with a nonlinear absorptive or dispersive medium inside an optical cavity [3]. In this case the systems involved exhibit a behaviour analogous to first or second order phase transitions depending on the nature of the relaxation mechanism [4].

In this paper we report on nuclear spin systems in solids which can be pumped to superradiant spin-flip states, thus emitting spontaneously radio frequency radiation at their natural resonance frequency. Such systems have many properties in common with a laser. We call them *RASER* (from radiofrequency amplification by stimulated emission of radiation). However, since the wave length of the radiation is many orders of magnitude larger than the superradiant device itself, the spin ordering can thus be treated as a purely temporal phenomenon.

After the realisation of the ruby raser [5,6] it became obvious that this system has properties making it suitable as a simple model system to study experimentally collective ordering processes. The synergetic features of its disorder-order transitions far from thermal equilibrium can be demonstrated with ease. In particular, the temporal evolution can be followed with high accuracy and reproducibility. This stems from the fact that the time of evolution from one significant macro-state to another lies within the electronically convenient time scale of microseconds to seconds. In addition, a theoretical model is available having parameters which can be measured or estimated to a high degree of reliability. Our present results concern:

(i) the dynamics of the evolution of a spin structure from a nonradiant to a coherently radiating state (spin superradiance);
(ii) the description of the macroscopic behaviour of nuclear spins by means of order parameters, and their abrupt or continuous changes when certain physical variables are changed;
(iii) the critical behaviour near the raser threshold;
(iv) the problem of bistability and the hysteresis effects of a raser, driven with a phase-locked, external resonant rf-field;
(v) the hopping mode patterns after a Q-switch procedure.

Thus, with rasers, we have systems under control which can contribute significantly to the illustration of synergetics. For a detailed analysis see [7].

2. Ruby-raser: experimental aspects

At liquid He temperature and in a strong magnetic field we have observed self-induced, coherent rf-oscillations of the ^{27}Al nuclear spins in $Al_2O_3:Cr^{3+}$ with an inverted spin population together with a resonant, laser-type feedback system to produce stimulated spin-flips. These oscillations can easily be detected by an oscilloscope directly connected to the coil of a LC-circuit which is tuned to one of the ^{27}Al NMR frequencies while powerful microwave radiation is supplied to the sample in the vicinity of a selected Cr^{3+} ESR line (Fig. 1).

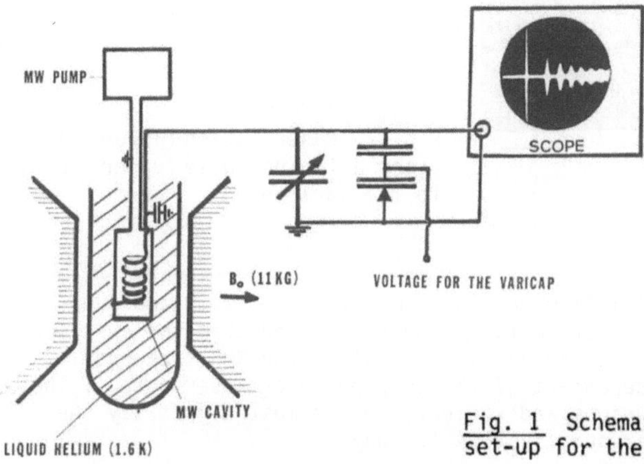

Fig. 1 Schematic diagram of the experimental set-up for the detection of raser signals.

The microwave source (~ 100 mW) acts as a pump causing the population-inversion of the nuclear Zeeman states by dynamic nuclear polarisation (DNP). The nuclear spin system is thus brought to a negative spin temperature θ_{Al} (~ mK) and an enhanced, negative, longitudinal nuclear magnetisation M_z is formed. In terms of synergetics, DNP produces an enhanced longitudinal Zeeman order. The coil of the LC-circuit provides the self-induced radiation field B_1^{ind} which causes phase-locked nuclear spin-flips. Due to this coherent slaving of the individual ^{27}Al spins, a rotating component M_v of the nuclear magnetisation is induced. In the sense of synergetics, we speak of a superradiant or transverse spin order. Hence, when superradiance occurs, longitudinal Zeeman order is continually transformed into transverse spin order.

Finally, an experimental detail should be mentioned: the tuning of the NMR coil is performed by an external capacitor which has a small Varicap in parallel. The electrically controlled capacitor can effectively be used for fast tuning, thus acting as a Q-switch. In this way raser transients can be induced which evolve into a cw-oscillation.

3. The free running single-mode ruby-raser

3.1 Single mode behaviour

If the LC-circuit with a Q = 60 and a filling factor $\eta \approx 0.6$ is tuned, e.g. to the (1/2, -1/2) line, a single mode behaviour can be observed. After a Q-switch procedure, transients are found which, typically, start with a strong delayed burst followed after a characteristic dead time by an amplitude-modulated oscillation as shown in Fig. 2.

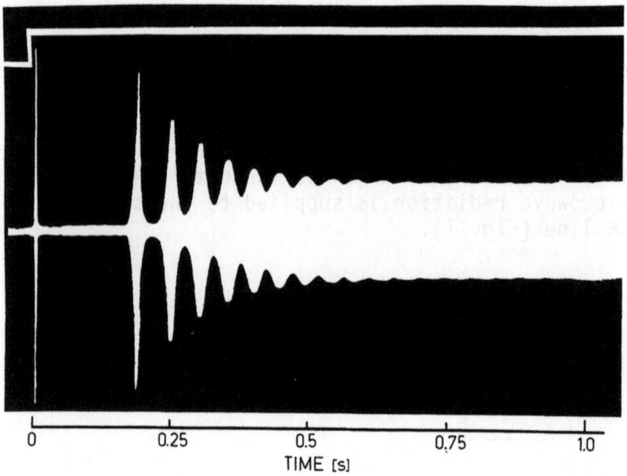

Fig. 2 Raser action (lower trace) after a tuning pulse (upper trace) to the Varicap.

With the microwave pump off, the oscillation decays slowly to zero; with the pump on, the cw-state is finally reached. Experimentally, we have found that the cw-raser-frequency is, within our accuracy of one part in 10^5, equal to the NMR frequency of the conventional central line and independent of the pump power or raser output. The dynamics of the transient is rather unique and reflects in the most direct way the coupling of the electronic and nuclear spin systems in ruby.

3.2 The reservoir model

Conventional studies of the spin dynamics in Al_2O_3:Cr^{3+} [6] have clearly demonstrated the important role of the electronic dipole-dipole interaction system $\{H_{SS}\}$ with regard to DNP and to spin-lattice relaxation. In particular, it has been shown that a thermal contact exists between the nuclear Zeeman system $\{H_{Al}\}$ and $\{H_{SS}\}$. Hence it is not at all surprising that this thermal mixing also shows up in the behaviour of the raser output. To explain this we introduce the thermodynamic model of coupled heat reservoirs (BUISHVILI [8], KOZHUSHNER [9], WENCKEBACH et al. [10]).

In essence, the dense ^{27}Al Zeeman reservoir $\{H_{Al}\}$ is broken up into five subsystems of fictitious spin-1/2 particles in correspondence with the five $\Delta m = \pm 1$ quadrupolar split NMR transitions. In a quasi-equilibrium state each subsystem $\{H^i{}_{Al}\}$ with the resonance frequency ω_i can be characterised by an inverse spin temperature and a heat capacity. The $\{H^i{}_{Al}\}$ are coupled to $\{H_{SS}\}$ each with a characteristic time constant $\tau^i{}_{Alss}$. The microwave pump heats $\{H_{SS}\}$; thermal mixing then leads to a negative spin temperature of $\{H^i{}_{Al}\}$. Hence, a negative longitudinal Zeeman order is formed. If we tune the LC-circuit to one of the NMR frequencies, say ω_k, then a transverse rotating magnetisation M_V^k may appear spontaneously, inducing an rf-voltage with the frequency ω_k. The system becomes superradiant. This single-mode operation becomes possible if the longitudinal magnetisation M_z^k of only that subsystem k to which the coil is tuned is pumped below the critical threshold value $M_z^{th} < 0$. Hence, only the spins belonging to $\{H_{Al}^k\}$ will perform a coherent precession, the remaining ones, belonging to $\{H_{Al}^j\}$, $j \neq k$, flip only incoherently. However, an appreciable amount of spin energy is stored in the $\{H_{Al}^j\}$ which, by thermal contact, can flow via $\{H_{SS}\}$ to $\{H_{Al}^k\}$. It is this stored energy and its flow which determines some of the most striking features of the raser dynamics, particularly the fast recovery after a superradiant burst (which drives M_z^k above the threshold value) and the subsequent amplitude-modulated relaxation oscillation.

3.3 The order parameter concept

Since we want to treat the appearance of spontaneous superradiance as a phase transition, we introduce the concept of the order parameter which has originally been coined by Landau to describe phase transitions at thermal equilibrium. Order parameters have the properties:

(i) to describe order of a many-body system on a macroscopic level (it is zero in the disordered state and acquires a maximum value in the completely ordered state);

(ii) to replace the direct interaction between the spins (which are due to dipole-dipole and hf-interactions), and it is on the one hand constructed from all spins and on the other hand it gives orders to the spins.

In this sense the spontaneous, transverse magnetisation M_v together with the properly rotating component B_1^{ind} of the induced rf-field are the order parameters of the raser. They can become critical by virtue of (ii). In contrast, M_z fulfils only the requirement (i). But we may also loosely call it an order parameter.

3.4 Bloch-type raser equations

The superradiant ^{27}Al spins of frequency ω_k form a fictitious spin-1/2 system. Thus we must transform the order parameters to fictitious variables. From experiment we know that ω_k is equal to the resonance frequency and that the LC-circuit has an unloaded ringing-time $T_c \approx 10^{-6}$ s, which is by far the shortest time-constant of the system. In consequence, we postulate that the transformed order parameters obey a set of Bloch-type differential equations and that B_1^{ind} responds promptly to M_v. These conditions are then expressed in terms of the true order parameters. For the central (1/2, -1/2) transition, in a frame of reference rotating with ω_k about B_0 and dropping the superscript k, we obtain the following dynamic equations:

$$B_1 = -\ 1/2\ \mu_o \eta Q M_v + F_B(t) \tag{1}$$

$$\dot{M}_v = 9\ \gamma M_z B_1 - M_v/T_2 \tag{2}$$

$$\dot{M}_z = -\ \gamma M_v B_1 - (M_z - M_e)/T_e. \tag{3}$$

Here, a possible dispersive contribution M_u has been discarded since $\dot{M}_u = -\ M_u/T_2$ with T_2 the spin dephasing time.

In equation (1) the term $F_B(t)$ represents the resonant fraction of the fluctuating field due to the thermal noise current of the coil. The term $(M_z - M_e)/T_e$ describes the pump and relaxation mechanisms. The effective magnetisation M_e to which M_z relaxes, as well as the effective relaxation time T_e depend in a complicated way on the spin dynamics of ruby. Within the limited accuracy of the reservoir model we have $T_e = \tau^k_{Alss}$, and we can relate M_e to the inverse spin temperature of the heat baths, their heat capacities and time constants, and further to the microwave pumping power [7].

If, at t = 0, a transient is induced by means of a Q-switch procedure, we can discriminate easily between the three time regimes: short, intermediate, and long in accord with $0 < t \ll \tau^k_{Alss}$, $t \sim \tau^k_{Alss} \sim 0.15$ s, and $t \gg \tau^k_{Alss}$, respectively.

3.5 Raser threshold and critical behaviour

As a limiting case we consider a situation where M_e has a constant value and where the fluctuations $F_B(t)$ can be discarded. B_1 can be eliminated in equations (2) and (3), and we find:

$$\dot{M}_v = -\,{}^9/_2\ \mu_o n Q\gamma M_z M_v - M_v/T_2 \tag{4}$$

$$\dot{M}_z = {}^1/_2\ \mu_o n Q\gamma M_v^2 - (M_z - M_e)/T_e . \tag{5}$$

If we define the threshold by $M_z^{th} = -\,{}^2/_9\ \mu_o n Q\gamma T_2 < 0$, we realise that $M_v^s = 0$ is stable for $M_e > M_z^{th}$. However, if the raser is pumped below M_z^{th}, then a transition to a new stable-state $M_v^s \neq 0$ takes place (Fig. 3).

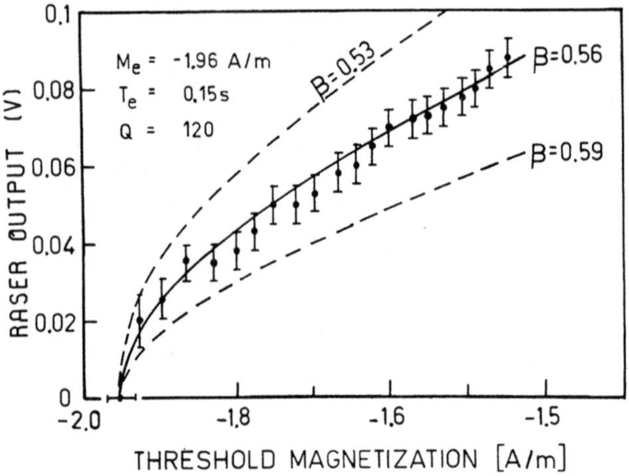

Fig. 3 Raser output at constant pumping power (M_e) versus threshold magnetisation M_z^{th} with three fits for critical Landau exponents β = 0.53, 0.56, 0.59.

If M_z is close to M_e, we can neglect $\dot{M}_z$ in equation (5). Thus M_z can be eliminated, and we end up in the adiabatic approximation with the single equation

$$\dot{M}_v = -\,\alpha M_v - \beta M_v^3 \tag{6}$$

with

$$\alpha = 1/T_2 + ({}^9/_2)\mu_o n Q\gamma M_e , \qquad \beta = ({}^9/_4)\mu_o^2 n^2 Q^2 \gamma^2 T_e ,$$

and the solution $M_v^s = \pm\,(-\,\alpha/\beta)^{1/2}$. The output voltage near threshold fulfils a power law with a critical Landau exponent $\beta = 1/2$. From preliminary measurements we have determined a value $\beta_{exp} = 0.56$ (Fig. 3). The discrepancy is most likely due to the fact that M_e is a weak function of the power output rather than a constant.

To demonstrate the close correspondence between this disorder-order transition far from thermal equilibrium with a Landau-type second-order phase transition at thermal equilibrium, we introduce by analogy with the free energy a raser potential, such that $\dot{M}_v = -\,d\phi/dM_v$:

$$\phi = (\alpha/2)M_v^2 + (\beta/4)M_v^4 . \tag{7}$$

In going from $\alpha > 0$ to $\alpha < 0$, the system passes through the critical point: for $\alpha = 0$, or $M_e = M_z^{th}$, we have a soft mode behaviour. The approach to the critical point from the non-radiant side ($\alpha > 0$) has a marked influence on the conventional NMR response, namely a narrowing and therefore critical growth of the NMR lines.

3.6 Transients

If a detuned raser is strongly pumped and subsequently tuned with a Q-switch, it finds itself in a state of unstable equilibrium. The fluctuations $F_B(t)$ are now be-

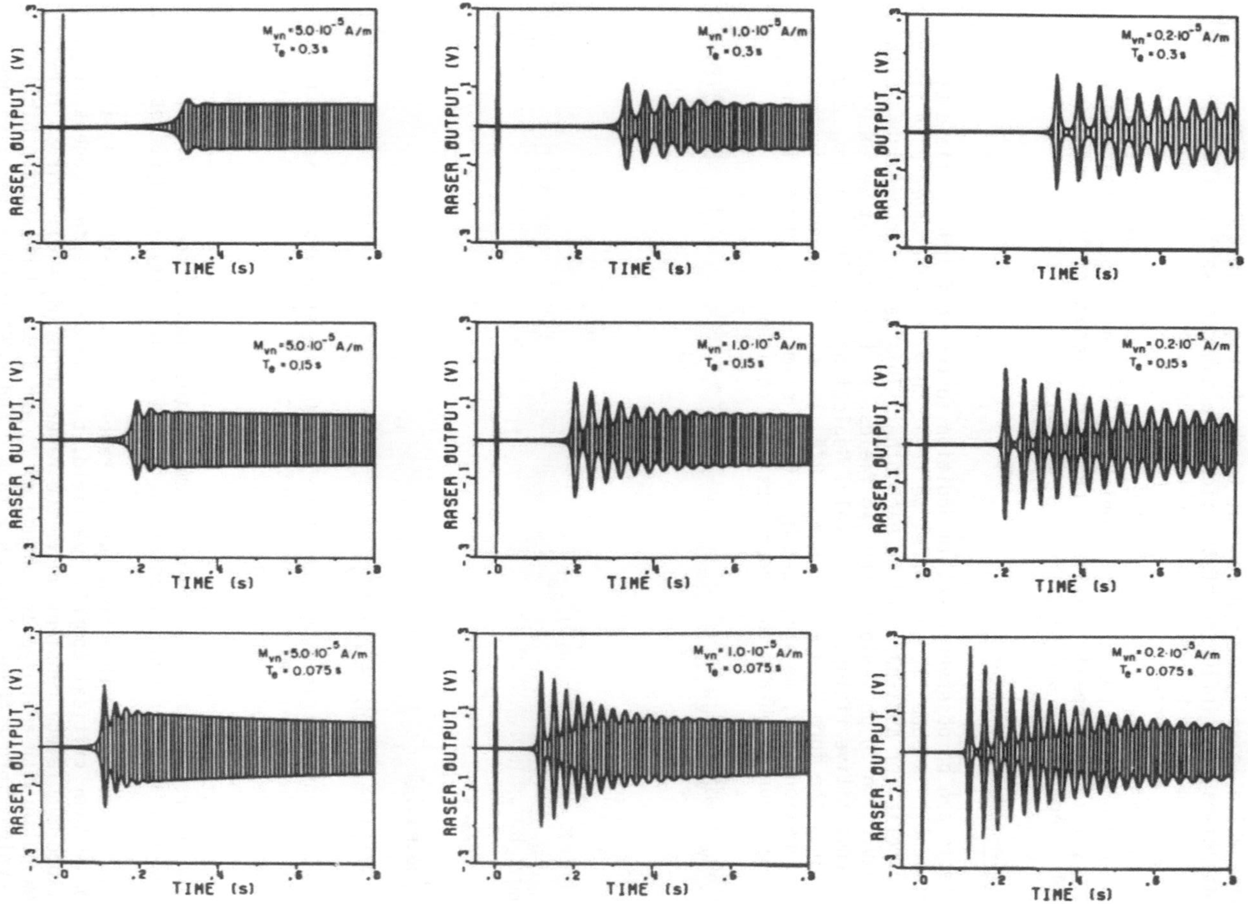

Fig. 4 Computed raser response after Q-switch procedure with different system parameters M_{vn} and Te.

coming essential for the ignition. Here the noise current in the NMR coil induces small fluctuations in the direction of the nuclear magnetisation. Random transverse components appear, having a finite mean square value $\overline{M_{vn}^2}$ which can be related to the Nyquist noise. In the short-time regime, the second term of the RHS of equation (5) can be dropped. The solution has approximately the shape of the well known inverse cosh-pulse which is delayed with respect to the tuning pulse at the Q-switch. For intermediate times, the influence of the microwave pumps as well as the spin-lattice relaxation can still be neglected. The transient signal is the same whether or not the microwaves are turned off after the Q-switch procedure. In this regime, however, the coupling of $\{H_{A1}^k\}$ and $\{H_{A1}^j\}$, $j \neq k$, to $\{H_{SS}\}$ must be taken into account. Then the computer produces a pulse sequence which evolves into a quasi-stationary spin oscillation, and which depends critically on the parameters of the system as shown in Fig. 4. There, $\{H_{A1}^k\}$ and $\{H_{A1}^j\}$ reach a common spin temperature.

4. Driven single-mode ruby-raser

Remarkable effects can be observed when, in addition to the self-induced field B_1^{ind}, an external field B_1^{ex} is superimposed. Here, we treat only the case where both fields remain either parallel or anti-parallel to each other. This can be accomplished with an electronic device which phase-locks the two fields. In this context we speak of cooperative or competitive fields, respectively. Neglecting the fluctuations, equation (1) then takes the form $B_1 = -\,1/2\,\mu_o \eta Q M_v + B_1^{ex}$. If we are interested in the adiabatic approximation only, we again set $\dot{M}_z = 0$ and eliminate B_1 and M_z in equations (2) and (3). Thus we obtain the order parameter equation

$$\dot{M}_v = -\rho - \alpha M_v - \delta M_v^2 - \varepsilon M_v^3 \qquad (8)$$

with

$$\rho = -\,9\,\gamma B_1^{ex} M_e,$$

$$\alpha = 1/T_2 + (9/2)\mu_o \eta Q \gamma M_e + 9\gamma^2 (B_1^{ex})^2 T_e,$$

$$\delta = -\,9\mu_o \eta Q \gamma T_e B_1^{ex},$$

$$\varepsilon = (9/4)\mu_o^2 \eta^2 Q^2 \gamma^2 T_e,$$

and the raser potential

$$\phi = \rho M_v + (\alpha/2)M_v^2 + (\delta/3)M_v^3 + (\varepsilon/4)M_v^4. \qquad (9)$$

Obviously, $\phi(M_v)$ is unsymmetric with respect to $M_v = 0$. The steady-state solutions M_v^s which correspond to the minima of ϕ are functions of B_1^{ex}. If two minima exist (one for a cooperative and one for a competitive B_1^{ex}), the one for $B_1^{ind} \uparrow\uparrow B_1^{ex}$ is lower (more stable) than the one with $B_1^{ind} \uparrow\downarrow B_1^{ex}$. With increasing absolute magnitude of B_1^{ex}, the first solution remains stable; the second one runs into an instability. When it occurs, the raser performs a jump to the stable mode, showing typical features of a first-order phase transition, namely bistability and hysteresis.

For example, if we start with a free running raser in a steady-state and, at $t = 0$, turn on an over critical competitive field, then we observe an exponentially growing amplitude-modulated transient until the raser stops ($M_v = 0$). This is followed by a new pulse with a 180° phase change of M_v and a damped relaxation oscillation which evolves into a cw-oscillation with B_1^{ex} and B_1^{ind} being cooperative now. The whole pattern can again be calculated from the generalised equations (1 - 3) as depicted in Fig. 5. If B_1^{ex} and B_1^{ind} are always kept in the competitive state the raser runs in a pulsation mode showing a periodic train of superradiant bursts (Fig. 6). One is tempted to call it a NMR pulsar!

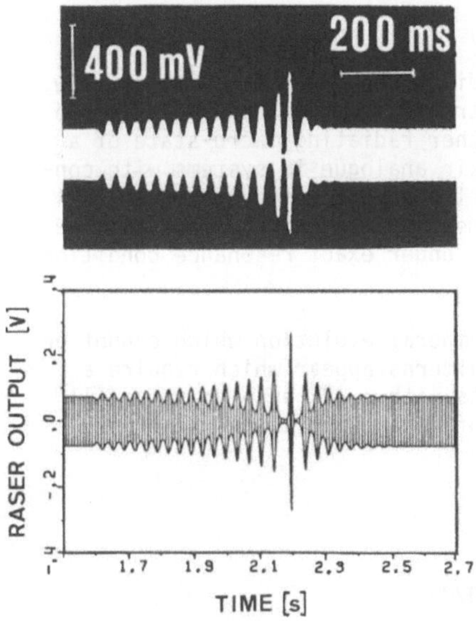

Fig. 5 Response of the free running raser after turning on an external competitive rf-field $B_I^{ex} > B_{1c}^{ex}$.

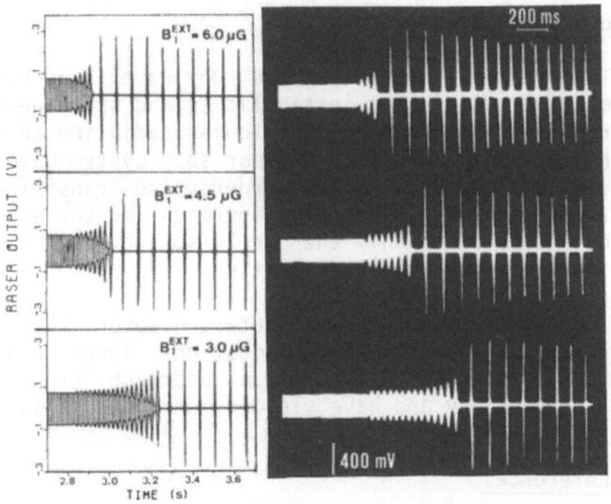

Fig. 6 Response of the free running raser after turning on an external, competitive field $B_I^{ex} > B_{1c}^{ex}$. By electronic means, the competitive field configuration is maintained in the subsequent pulsation mode.

5. Hopping modes

If the threshold of the raser is lowered by raising the Q of the tuned circuit, then more than one of the ^{27}Al subsystem $\{H_{Al}^{k}\}$ can become critical. The raser will emit radiowaves of different NMR frequencies. For simultaneous emission, beats can be seen with the beat frequencies equal to the NMR difference frequencies. If the emission proceeds sequentially, time-separated bursts are observed, each burst having its appropriate NMR frequency (Fig. 7).

Such hopping mode patterns can quantitatively be understood with sets of raser equations of the type (1 - 3), at least for the short-time regime. Giant raser pulses are obtained if different NMR lines overlap. In such a case we have produced raser pulses as high as 400 V peak-to-peak and 25 W peak power.

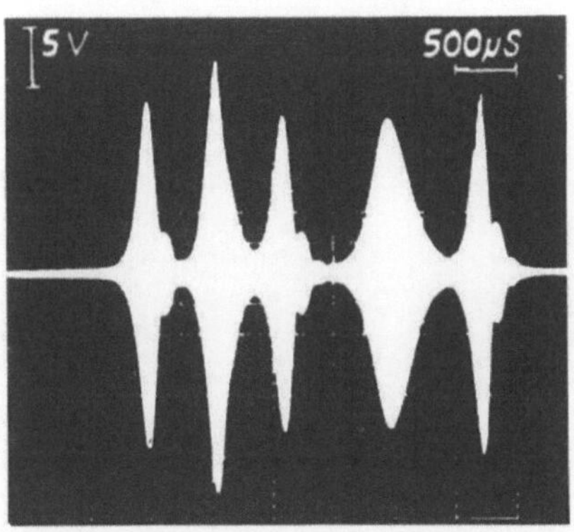

Fig. 7 Raser response in short-time regime with a high Q-coil showing resolved pulses with frequency-hopping.

6. Concluding remarks

The dynamical aspects of the ruby-raser have been discussed from the point of view of synergetics. The instabilities and spontaneous transition from non-radiating to radiating, or from one particular radiating to another radiating macro-state of a neglectively polarised nuclear spin system have their analogue in systems with conventional second or first order phase transitions. The good agreement between the theoretical Bloch equation approach and the experiment stems from the fact that we have considered only the purely absorptive response under exact resonance conditions together with a weak excitation.

Strongly pumped raser states, however, show a temporal evolution which cannot be treated with simple Bloch equations. There, beat patterns appear which require a more sophisticated theoretical approach. Experiments with a ^{19}F spin-½ raser [11] are under way and have yielded intriguing results of great theoretical interest.

References

1. R. Graham and H. Haken, Z. Phys. 237 (1970)
2. V. Degiorgio and M.O. Scully, Phys. Rev. A 2 (1970)
3. H.M. Gibbs, S.L. McCall, and T.N.C. Vankatesan, Phys. Rev. Lett. 36 (1975)
4. D.F. Walls, P.D. Drummond, S.S. Hassan and H.J. Carmichael, Prog. Theo. Phys. Supplement Nr. 64 (1978)
5. P. Bösiger, E. Brun, and D. Meier, Phys. Rev. Lett. 38, 602 (1977)
6. P. Bösiger, E. Brun, and D. Meier, Phys. Rev. A 18, 671 (1978)
7. P. Bösiger, E. Brun, and D. Meier, Phys. Rev. A, September (1979)
8. L.L. Buishvili, Sov. Phys. JETP 22, 1277 (1966)
9. M.A. Kozhushner, Sov. Phys. JETP 29, 136 (1969)
10. W.Th. Wenckebach, T.J.B. Swanenburg, and N.J. Poulis, Phys. Rep. (Phys. Lett. C) 14, 131 (1974)
11. B. Derighetti, H. Marxer, and E. Brun, unpublished

Nonequilibrium Phase Transition in Highly Excited Semiconductors

H. Haug and S.W. Koch

Institut für Theoretische Physik der Universität Frankfurt
Robert-Mayer-Straße 8
D-6000 Frankfurt, Fed. Rep. of Germany

I. Introduction

In the last ten years the properties of highly excited semiconductors have been studied thoroughly [1,2,3,4]. If the frequency of the exciting laser light is sufficiently high, electrons from the valence band are lifted into the conduction band. Thus, in the process of optical excitation one generates electron-hole pairs (e-h pairs) i.e. electrons in the conduction band and holes or missing electrons in the valence band. In the low density regime electrons and holes form bound pair states, called excitons, due to their attractive Coulomb interaction. In an intermediate density regime one observes besides free carriers and excitons also higher bound states such as trions (e-e-h or e-h-h), biexcitons and even multiexcitons. In the high density regime bound states are no longer stable, because the attractive interaction between an electron and a hole is screened by the other electronic excitations. In this limit a plasma is formed.

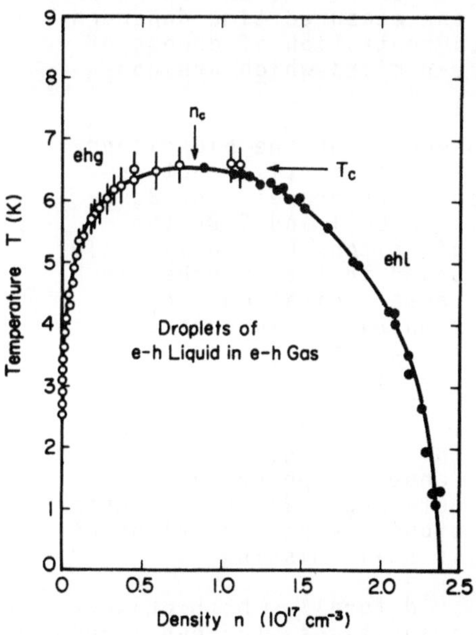

Fig.1:
Phase diagram of Ge [5]

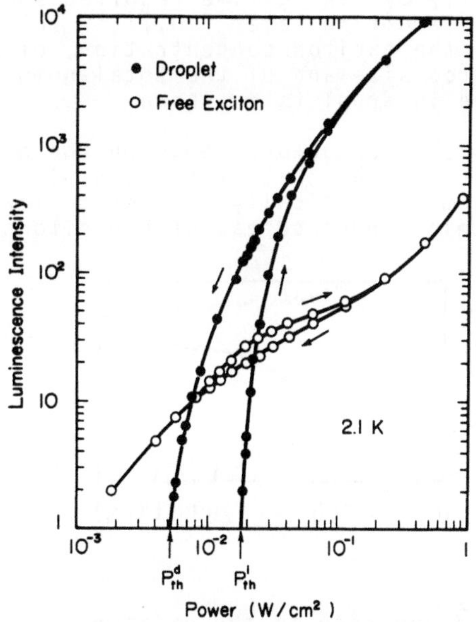

Fig.2:
Hysteresis in the droplet luminescence of Ge [9]

The lifetime of the elctronic excitations varies from 10^{-3} to 10^{-9} sec for the various types of semiconductors. Especially well investigated are the classical semiconductors Ge and Si, in which the lifetime is of the order of several microseconds. In these substancies, and more recently in many other semiconductors too, one observed that at low temperatures the transition between the low density phase, which consists mainly of excitons, and the conducting high-density plasma phase takes place in a discontineous way, which resembles strongly a first order gas-liquid phase transition. It was e.g. possible to measure experimentally the phase diagram for Ge (see Fig.1) with a critical temperature of about 6.5 K and a critical density of 8 x 10^{16} cm^{-3} [5]. The high density phase is a conducting two-component Fermi-fluid. Microscopic calculations showed indeed that for Ge at T=0 the mean energy of an e-h pair in the liquid phase is lower than that of an exciton [6,7]. In the coexistence region one observed droplets of the electron-hole liquid in a gas of excitons. In the experiments the size of the droplets has been determined by means of the Rayleigh scattering of infrared light [8]. If one applies a none-uniform stress, one can generate via deformation potentials energy minima, in which a large drop can be formed. Photographic images of drops with radii up to several hundred μm have been obtained and gave a spectacular demonstration of the phase transition in Ge [2]. Interestingly, one detected a hysteresis in the droplet luminescence (see Fig.2) [9]. If one increases the intensity of the exciting laser light one observes the typical droplet luminescence above a certain threshold intensity p(i). If one reduces the intensity of the laser beam again one is able to follow the droplet luminescence down to a threshold intensity $p(d) < p(i)$.

The dynamics of the droplet nucleation have been measured rather directly by taking time resolved spectra following the switching on of the exciting laser [10,11]. In Fig.3 the measured time dependence of the exciton concentration, of the concentration of drops, of the drop size and of the total number of e-h pairs which are condensed in droplets is shown.

In semiconductors in which the binding energy of the biexcitons is larger than that of the e-h plasma, one expects still another quantum-statistically degenerate low temperature phase: the Bose-Einstein condensed gas of biexcitons [3]. In CuCl and CuBr the biexcitons are so strongly bound that even under intense excitation the low temperature luminescence spectra are dominated by the recombination lines of the biexciton. Recently, experimental evidence for the existence of such a condensed phase has been obtained [12], but an unambiguous experimental proof is still missing.

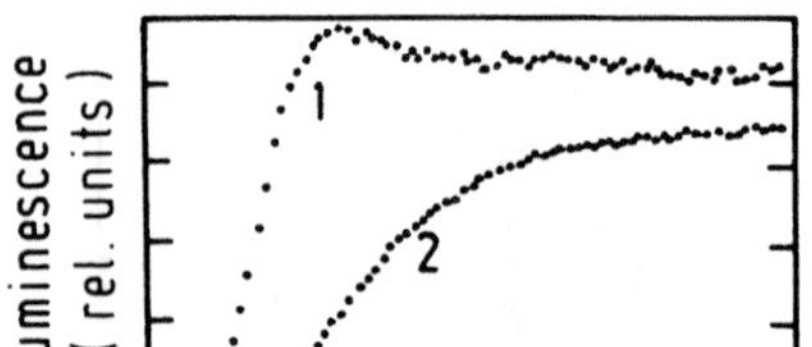

Fig.3:
Time dependence of the exciton luminescence (1) and of the droplet luminescence (2) of Ge [10]

Naturally, both phase transitions, the e-h liquid condensation and the expected Bose-Einstein condensation of biexcitons are nonequilibrium phase transitions, which have to be described by indeterministic, dynamic equations. In the following we will show, how the e-h droplet condensation, which is an example for a first-order phase transition can be treated by the methods, which have been developed in Synergetics [13].

2. Hydrodynamical Model

Because in a gas-liquid phase transition the relevant order parameter is the change in density, the hydrodynamical equations are the natural way to describe such a phase transition dynamically [14,15]. In order to simplify the description we assume that the nucleation process is isothermal and that charge neutrality holds in the whole system of electronic excitations (i.e. we neglect thermal nonaccommodation effects and plasma oscillations).

In the continuity equation for the density of e-h pairs ρ we have to take into account the finite lifetime $\tau=\tau(\rho)$ of the free carriers and the generation rate $\overline{\rho}/\tau$:

$$\partial\rho/\partial t + \mathrm{div}(\rho\vec{v}) = \frac{(\overline{\rho} - \rho)}{\tau} + F_0(\vec{r},t) \quad . \tag{1}$$

F_0 are the Langevin fluctuations which describe the shot noise of the dissipation processes of decay and pumping

$$< F_0(\vec{r},t)F_0(\vec{r}',t')> = m \frac{(\overline{\rho} + \rho)}{\tau} \delta(\vec{r} - \vec{r}')\delta(t - t') \quad . \tag{2}$$

Next we have to formulate a Navier-Stokes equation, which still holds in the presence of sharp gradients in the density which occur at the gas-liquid interface. This problem can be solved by deriving the reversible parts of the hydrodynamic equation from a Hamilton principle [17] and by adding a surface energy term [18] to the corresponding Lagrangian density. The resulting Navier-Stokes equation is

$$\rho(\partial/\partial t + \vec{v}\cdot\vec{\nabla})\vec{v} + \vec{\nabla}p - A\rho\vec{\nabla}\Delta\rho = -\rho\beta\vec{v} + \vec{F} \quad . \tag{3}$$

The constant A is related to the surface energy. On the r.h.s. of (3) we included viscous damping of the free carriers due to phonon scattering. The second moment of the corresponding Langevin force is

$$< F_i(\vec{r},t)F_j(\vec{r}',t')> = 2kT\beta\rho\,\delta(\vec{r} - \vec{r}')\delta(t - t')\delta_{ij} \ . \tag{4}$$

Next we assume that local quasiequilibrium exists within the system of electronic excitations, so that the quantum statistical methods can be used to calculate the equation of state $p=p(\rho,T)$. These microscopic calculations indeed give isotherms which resemble the pressure versus density curves of a classical Van der Waals-gas [19]. In order to get a qualitative understanding of the e-h droplet nucleation, we will use simply a modified Van der Waals-equation of state (Fig.4) [see Ref. 15 and 16]. To find the unstable modes, we make a linear analysis of the mean equations (1) and (2). The resulting complex eigenfrequencies are plotted in Fig.5 against the wavevector k in the system. Below T_c one mode becomes unstable for densities which correspond practically to the spinodal curve of the Van der Waals diagram, i.e. we find with the linear analysis only the points of the mechanical instability of the system. However, we know that the first-order gas-liquid phase transition already takes place at lower densities. The actual transition point can not be found by a linear analysis because it requires macroscopic (not infinitesimally small) fluctuations.

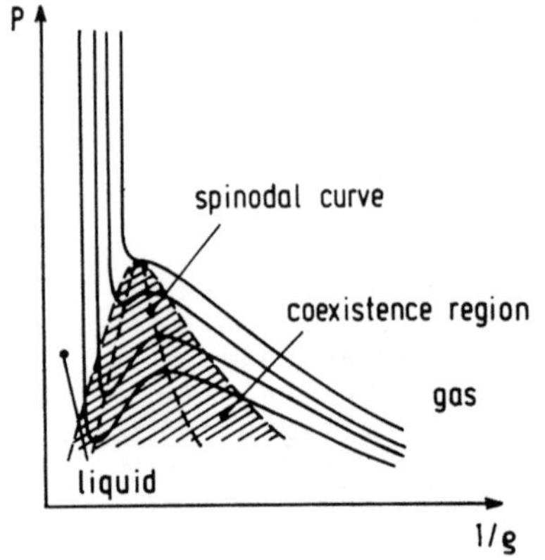

Fig.4:
Van der Waals phase diagram

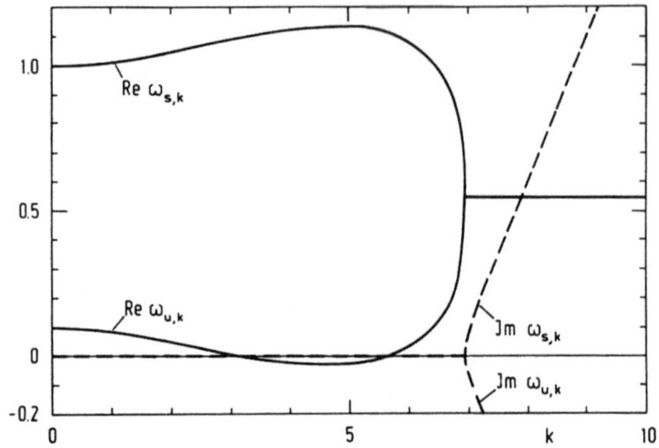

Fig.5:
Damping constant (Re ω) and energy (Im ω) of the stable and unstable eigenmodes [16]

Close to T_c however, the transition is only weakly of first order. In this range one can expand the nonlinear solution of (1) and (3) into the linear eigenmodes. The linear eigenmodes for a spherical droplet are proportional to spherical Bessel functions $j_0(kr)$, where k is the wave number of the solution. Even if one makes the crude single-mode approximation (using only one k value) one gets a qualitative description of the density and velocity field in and around a droplet, as a comparison of a numerical stationary solution of (1) and (2) with the single-mode approximation shows (Fig.6) [16]. Naturally, the strong density gradients at the droplet surface cannot be described correctly by a single-mode approximation. Fig.6 shows that a current from the gas phase into the droplet must exist, which is necessary to compensate the decay losses in the droplet. Contrary to the classical droplet nucleation, one gets due to the finite lifetime finite, stable droplet sizes. Using the concept of the generalized Ginzburg-Landau equations for nonequilibrium phase transitions [13] we derived an equation of motion for the expansion amplitude γ_k which can be written in the form [16]

$$\dot{\gamma}_k = - \partial V/\partial \gamma_k + F_k \quad , \tag{5}$$

where V is the generalized Ginzburg-Landau potential and F_k is again a statistical force. Fig.7 and 8 show the calculated potential for

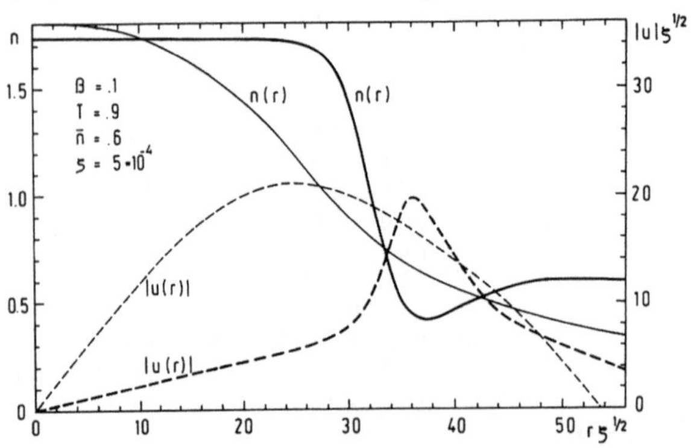

Fig.6:
Density and velocity profile of the droplet
thick curves: numerical solutions,
thin curves: single-mode approximation [16]

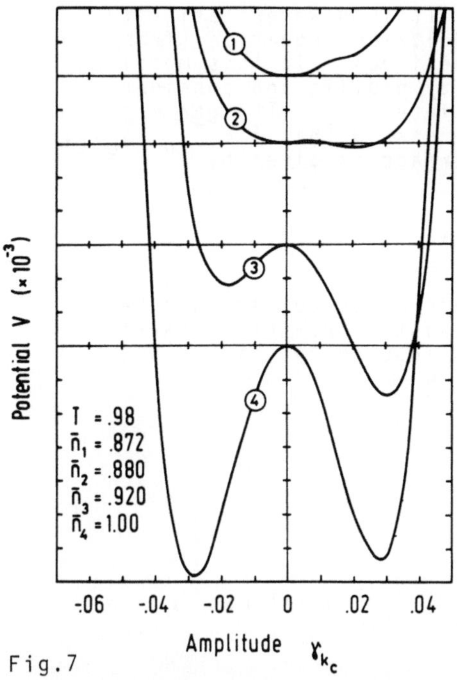

Fig.7

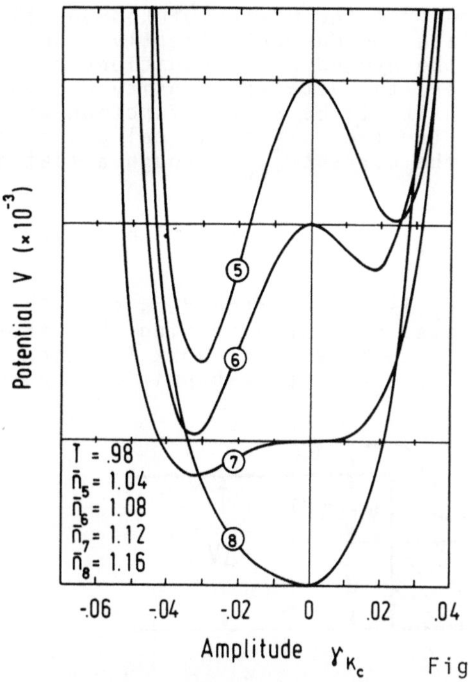

Fig.8

Fig.7 and 8:
Ginzburg-Landau potentials for various pumping intensities. (All quantities are normalized with respect to the critical values n_c and T_c)

increasing excitation intensities [2o]. Below the phase transition only $\gamma_k=0$ is stable. This corresponds to a homogeneous gas phase. With increasing excitation a second minimum develops at $\gamma_k > 0$. In this situation we have droplets in the gas. At still higher excitations the homogeneous solution becomes unstable. The two minima correspond to a mixture of bubbles and droplets. Just before the homogeneous liquid phase is reached one has a situation with gas bubbles in the fluid. Thus, the simple single-mode approximation gives a good qualitative description of the whole phase transition region.

3. Fokker-Planck treatment

In order to get quantitative results for the dynamics of the droplet nucleation, we simplify the description by deriving from the hydrodynamical model rate equations. This equations govern the change of electronic excitations. Integrating (1) over the droplet volume we get the following mean equation for the number n of e-h pairs in a droplet

$$\partial n/\partial t + \int d\vec{\sigma} \cdot \vec{j} = (\bar{n} - n)/\tau_d \quad . \qquad (6)$$

The second term gives the current through the droplet surface $\sigma = 4\pi r^2$. The radius of the droplet and n are related $4\pi r^3 \rho_0/3 = n$ so that $\sigma \propto n^{2/3}$. The current $\vec{j}$ has to be calculated from the Navier-Stokes equation. It is however easier to calculate the current directly from the Boltzmann equation (from which the hydrodynamic equations can be derived). In this model we disregard the detailed density and velocity profile and consider the density to be a step function at the interface. The gas phase can be treated as a classical gas of excitons so that $\vec{j} = \vec{j}_{in} - \vec{j}_{out}$, where the current $\vec{j}_{in}$ through a flat interface is given by

$$j_{in,1} = \frac{1}{V} \sum_{p_1} (p_1/m) f(\vec{p}) = n_x \left(\frac{kT}{2\pi m} \right)^{\frac{1}{2}} , \tag{7}$$

where n_x is the exciton density. The evaporation current can be calculated similarly, noting that there exists a potential barrier at the interface. The work function ψ is reduced by the surface energy $sn^{-1/3}$ per e-h pair.

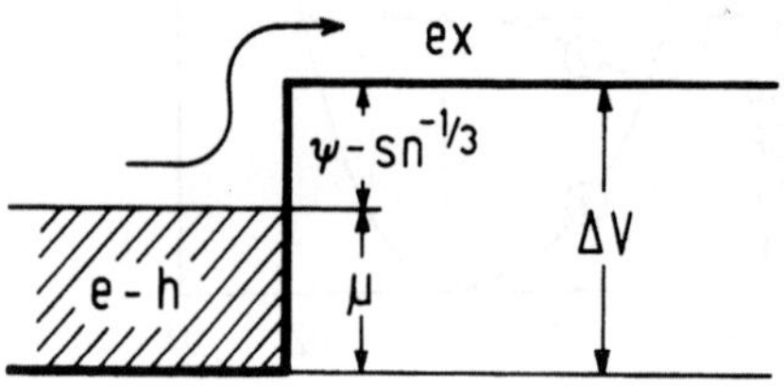

Fig. 9 Potential barrier at the drop surface

The e-h pairs can only evaporate from the droplet if their momentum $p \geq p_{min} = (2\, m\, \Delta V)^{\frac{1}{2}}$.

Therefore,

$$j_{out,1} = \frac{1}{V} \sum_{p_{min,1}}^{\infty} (p_1/m) \frac{1}{\exp(\beta (p^2/2m - \mu)) - 1} \tag{8}$$

which gives the Richardson-Dushman formula

$$j_{out,1} = \frac{4\pi m}{(2\pi\hbar)^3} (kT)^2 \exp(-\beta(\psi - sn^{-1/3})) .$$

The collection and evaporation currents give rise to gain and loss terms in the droplet rate equation. Adding the corresponding shot noise terms, eq.(6) becomes

$$\partial n/\partial t = g_n - \ell_n + F_n \tag{9}$$

with

$$g_n = \sigma\, j_{in,r} \quad , \quad \ell_n = \sigma\, j_{out,r} + n/\tau_d$$

and

$$\langle F_n(t) F_n(t') \rangle = (g_n + \ell_n)\delta(t - t') \quad ,$$

where we have neglected the unimportant pump term into the droplet. The Fokker-Planck equation which belongs to the statistical rate equation (9) is

$$\frac{\partial f_n}{\partial t} = - \frac{\partial}{\partial n} J_n \quad , \tag{1o}$$

where the probability current J_n is given by

$$J_n = (g_n - \ell_n)f_n - \frac{1}{2}\frac{\partial}{\partial n}\{(g_n + \ell_n)f_n\} \quad . \tag{11}$$

From the hydrodynamic continuity equation (1) we get still another equation, which gives the total energy conservation. Integrating (1) over the total crystal volume we find

$$\partial n_x/\partial t + \partial n_d/\partial t = - n_x/\tau_x - n_d/\tau_d + G(t) \quad , \tag{12}$$

where n_d is the number of all e-h pairs which are condensed in droplets and G(t) is the laser generation rate. τ_x and τ_d are the exciton and e-h droplet lifetime, respectively. It is important to note that a critical droplet size n_c exists for which the collection rate compensates the evaporation losses. We call all clusters larger than n_c 'droplets', and the small clusters we count as 'excitons':

$$x_0 = \int_{n_c}^{\infty} dn\, f_n \qquad \text{density of droplets,}$$

$$n_d = \int_{n_c}^{\infty} dn\, f_n\, n \qquad \text{density of e-h pairs in all droplets,}$$

$$n_x = \int_1^{n_c} dn\, f_n\, n \qquad \text{density of excitons.} \tag{13}$$

Now, the Fokker-Planck equation (1o) and the conservation equation (12) form a closed set of equations. The time dependent distribution function $f_n(t)$ has to be calculated for a given generation rate which we assume to be of the form $G(t) = G_1\{1-\exp(-t/t_r)\}$.

The Fokker-Planck equation, which is usually derived from the master equation, has been treated in various approximations [1o, 15,22]. We obtained an approximate solution by constructing from (1o) and (12) a closed set of equations for the moments [23]

$$x_\alpha = \int_{n_c}^{\infty} dn\, f_n\, n^\alpha \tag{14}$$

and the exciton density. The numerical solution for this equations is shown in Fig.1o and 11 for the case of slow and fast nucleation. Note, that the nucleation rate changes very suddenly when a certain supersaturation in the exciton gas is reached. The results are in agreement with the experimental findings shown in Fig.3 . In the region, where the onset of nucleation occurs, large fluctuations in the droplet size show up. From the knowledge of the time-dependent moments we determined numerically the distribution function, which at each time gives the known moments [23]. The ansatz for the distribution function we got from an eigenvalue analysis for the Fokker-Planck equation. The resulting droplet distribution is given in Fig.12 for the case of fast nucleation. These curves show the narrowing of the droplet distribution as it approaches the stationary situation.

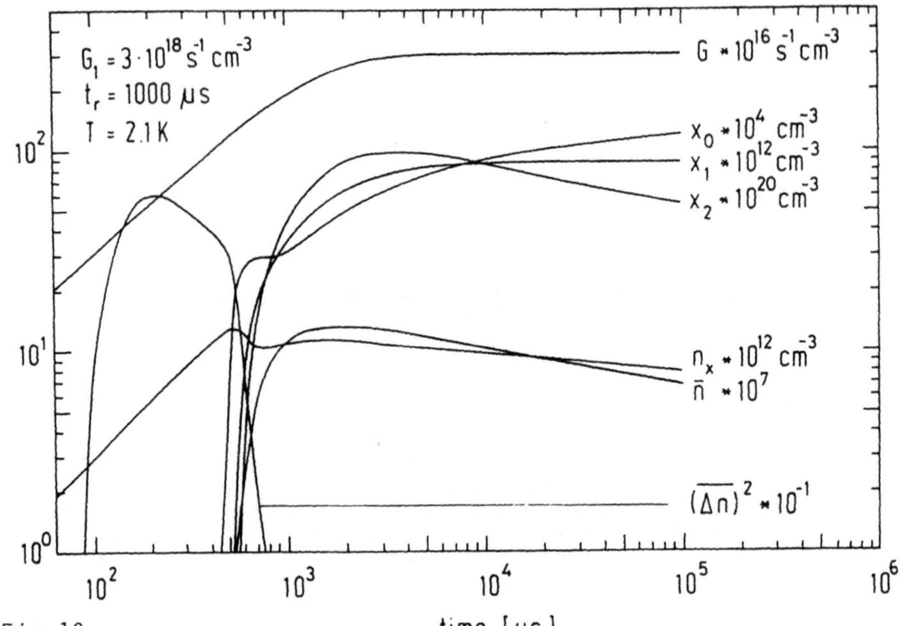

Fig. 10

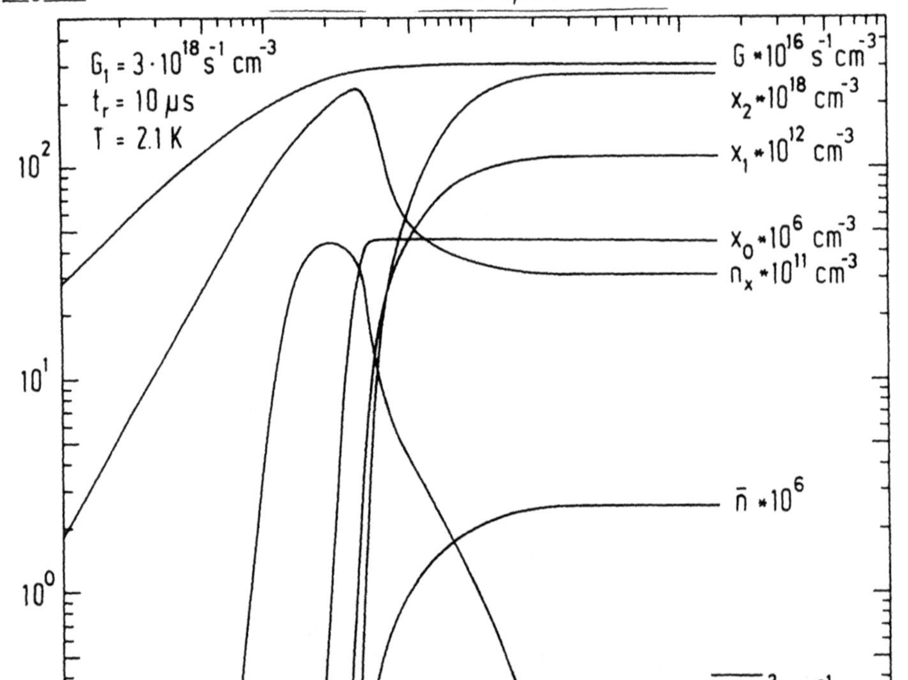

Fig. 11

Figure captions see opposite page

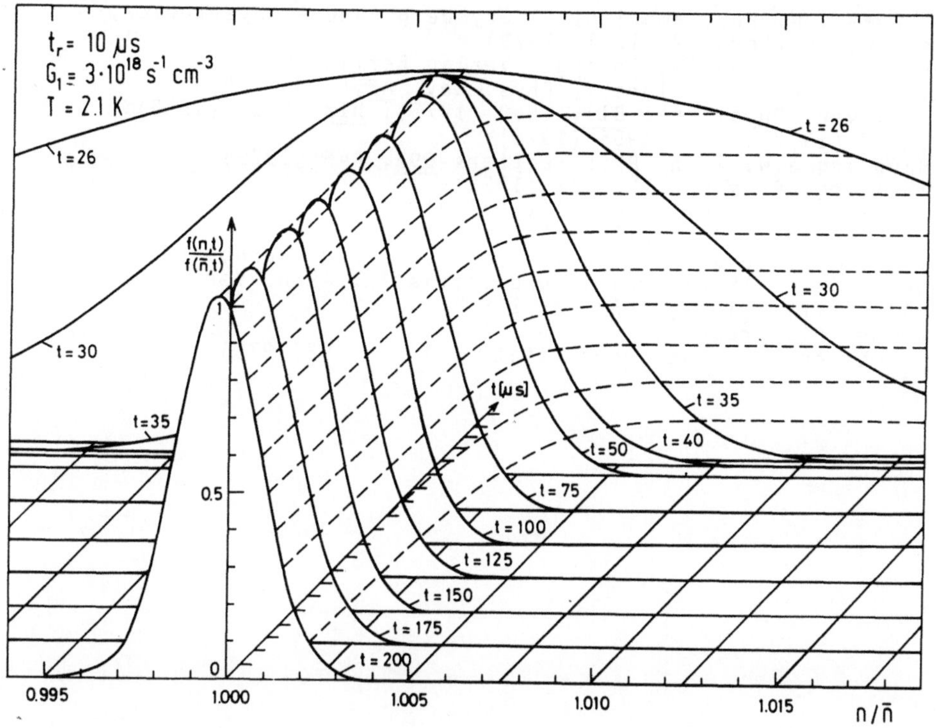

ig.12: Time development of the droplet distribution function [23].

References

1. Excitons at High Densities, eds. H. Haken and S. Nikitine, Springer Tracts of Modern Physics 73 (1975)
2. C.D. Jeffries, Science 189, 955 (1975)
3. E. Hanamura and H. Haug, Phys. Reports 33C, 2o9 (1977)
4. T.M. Rice, Solid State Physics 32, 1 (1977)
5. G.A. Thomas, T.M. Rice, and J.C. Hensel, Phys. Rev. Letters 39, 219 (1974)
6. M. Combescot and P. Nozières, J. Phys. C5, 2369 (1972)
7. W.F. Brinkman and T.M. Rice, Phys. Rev. 87, 15o8 (1973)
8. Ya.E. Pokrovskii and K.I. Svistunova, JETP Letters 13, 212 (1971)
9. R.M. Westervelt, T.K. Lo, J.L. Staehli, and C.D. Jeffries, Phys. Rev. Letters 32, 1o51 (1974)
1o. J.L. Staehli, phys. stat. sol. (b) 75, 451 (1976)
11. V.S. Bagaev, N.V. Zamkovets, L.V. Keldysh, N.N. Sibel'din, and V.A. Tsvetkov, Sov. Phys. JETP 43, 783 (1976)

◀ Fig.1o and Fig.11:
Calculated time dependence of the moments x_0, x_1, x_2 of the droplet distribution function for Ge. n_x, $\bar{n}$ and $\overline{(\Delta n)^2}$ are the exciton concentration, the mean number of e-h pairs per drop and the mean square of the droplet distribution, G(t) is the laser generation rate [23].

12. L.L. Chase, N. Peyghambrian, G. Grynberg, and A. Mysyrowicz, Phys. Rev. Letters 42, 1231 (1979)
13. Synergetics, H. Haken, Springer Verlag Berlin 1977
14. J.S. Langer and L.A. Turski, Phys. Rev. A8, 323o (1973)
15. R.N. Silver, Phys. Rev. B11, 1569 (1975); B12, 5689 (1975); B16, 797 (1977); B17, 3955 (1978)
16. H. Haug and S.W. Koch, Phys. Letters 69A, 445 (1979)
17. A. Thellung, Physica XIX, 217 (1953)
18. B.U. Felderhof, Physica 48, 541 (197o)
19. G. Mahler, Phys. Rev. B11, 4o5o (1975)
2o. S.W. Koch, Doktorarbeit, Univ. Frankfurt 1979, unpublished
21. A. Sommerfeld, Thermodynamik und Statistik, Akad. Verlagsgesellschaft, Leipzig 1965, p. 238
22. M. Combescot, preprint
23. S.W. Koch and H. Haug, phys. stat. sol. (b) 95(1979)

Nonequilibrium Transitions Induced by External White and Coloured Noise

W. Horsthemke

Service de Chimie Physique II, C.P. 231, Université Libre de Bruxelles
B-1050 Brussels, Belgium

1. Introduction

It is a very general observation that the environment of nonequilibrium systems, be they natural systems or laboratory systems, is at best only constant on the average. The environment is very complex and thus unavoidably its temporal behaviour appears to be random. In other words, the system perceives the environment as a noise source. A good illustration of this is the population of, say, some animal species. Its growth is determined by factors such as the birth rate, mortality, abundance of predators etc. which are strongly influenced by climatic conditions and therefore show a more or less pronounced variability. More generally, the fluctuations of the environment, which are called external noise, can be understood as the expression of a turbulent or chaotic state of the surroundings.

This very common situation, that nonequilibrium systems are coupled to a rapidly fluctuating environment, motivates the following question: Do nonlinear systems always adjust their macroscopic behaviour to the average properties of the environment or can one find situations in which the system responds in a certain, more active way to the randomness of the environment?

The external noise will of course have a certain influence on the behaviour of the system in the neighbourhood of instabilities, just as the internal fluctuations. Of much greater importance is however the fact, that nonequilibrium systems are, by their very nature, closely dependent on their environment. Thus one can imagine that the interplay of the *nonequilibrium*, of the *nonlinearity* of the system and of the *environmental randomness* might lead to some drastic modifications of the macroscopic behaviour of the system, even outside the neighbourhood of an instability of the phenomenological equations. So far this viewpoint had rather been neglected in the literature. The following theoretical analysis will establish that external noise can indeed deeply alter the macroscopic behaviour of a system, namely induce new transition phenomena which are quite unexpected from the usual phenomenological description. We will see that under the right conditions selforganisation in nonlinear systems can be the response to environmental randomness. Some experimental evidence for such transitions is presented in the last section of this paper.

2. Transitions Induced by External White Noise

2.1. The General Formalism

In the following I will restrict myself to spatially homogeneous systems. This implies that either the diffusion is very rapid compared to the reaction kinetics or that the system is well-stirred, the usual setup in the lab. Furthermore I will only consider systems that can be described by one relevant variable, i.e. a kinetic equation of the type:

$$\dot{x}=h(x) + \lambda g(x) = f(x,\lambda) \tag{1}$$

λ represents a parameter, e.g. a growth rate, that is essentially determined by the state of the environment. I will not inquire into the exact origin of the environmental fluctuations but use as a starting point only the experimental fact that the environment is noisy and can be described by a random process. Thus the fluctuations of the environment appear on the level of the phenomenological equation via the fact that λ becomes a stationary stochastic process λ_t with mean value $E\lambda_t=\lambda$ and variance $E(\lambda_t-E\lambda_t)^2= E\delta\lambda_t^2 =\sigma^2$. The assumption of stationarity for the external noise is made, in order to separate the effects of the environmental fluctuations clearly from those effects due to a systematic evolution of the surroundings. We obtain thus the following stochastic differential equation (SDE):

$$\dot{x}_t = h(x_t) + \lambda_t g(x_t) \tag{2}$$

Let us first consider the situation that the correlation time τ_{cor} of the external noise λ_t, defined e.g. by the relation:

$$\tau_{cor} = \sigma^{-2} \int_0^\infty E(\delta\lambda_t\, \delta\lambda_{t+\tau})d\tau \tag{3}$$

is very short on the typical macroscopic time scale of (1):

$$\tau_{cor} \ll \tau_{macro} \tag{4}$$

This allows us to pass to the idealisation of the so-called white noise, i.e. we set

$$E(\delta\lambda_t\delta\lambda_{t+\tau}) \sim \delta(\tau), \quad \text{i.e.} \quad \tau_{cor} = 0 \tag{5}$$

This idealisation is very appealing from a mathematical point of view, since the process x_t is markovian, if and only if the external noise is white. To be specific we set $\lambda_t= \lambda + \sigma\,\xi_t$, where ξ_t is a generalised Gaussian process with $E\xi_t=0$ and $E(\xi_t\xi_{t+\tau}) = \delta(\tau)$, and obtain:

$$\dot{x}_t = h(x_t)+\lambda g(x_t)+\sigma g(x_t)\xi_t = f(x_t,\lambda)+\sigma g(x_t)\xi_t \tag{6}$$

Multiplying formally with dt finally yields the standard form

$$dx_t = f(x_t,\lambda)dt + \sigma g(x_t)dW_t \tag{7}$$

W_t is the Wiener process, i.e. the usual Brownian motion, and its derivative, in the sense of generalised functions, is the Gaussian white noise ξ_t. In integral form (7) reads:

$$x_t = x_0 + \int_0^t f(x_s,\lambda)ds + \sigma \int_0^t g(x_s)dW_s \tag{8}$$

Since the sample paths of the Brownian motion are extremely irregular, the last integral cannot be understood as an ordinary Riemann integral [1]. In the literature two definitions are widely used, that due to Ito:

$$\int dW_s\, G(W_s) = \text{l.i.m.} \sum G(W_{s_{i-1}})\,(W_{s_i} - W_{s_{i-1}}) \tag{9}$$

and that by Stratonovic:

$$\int dW_s\, G(W_s) = \text{l.i.m.} \sum G(\tfrac{1}{2}(W_{s_i} + W_{s_{i-1}}))\,(W_{s_i} - W_{s_{i-1}}) \tag{10}$$

There are two guidelines to decide which one is more appropiate in a concrete application: If (7) is the continuous time limit of a discrete time problem, then Ito is the right choice. On the other hand, the theorem of WONG and ZAKAI [2] states, that if (7) is the white noise limit of a real noise problem, then the Stratonovic version is more appropiate. The results I am going to present are qualitatively the same in both versions and therefore I will discuss this point no further.

For our purpose it is convenient to switch to the equation governing the temporal evolution of the probability density $p(x,t)$ of the process x_t. This is the so-called Fokker-Planck equation (FPE) and it reads in the Ito interpretation of (7):

$$\partial_t\, p(x,t) = -\partial_x\, f(x)p(x,t) + \tfrac{1}{2}\sigma^2\, \partial_{xx}\, g^2(x)p(x,t) \tag{11}$$

and in the Stratonovic version

$$\partial_t\, p(x,t) = -\partial_x\, \{f(x) + \tfrac{1}{2}\sigma^2 g'(x)g(x)\}p(x,t) + \tfrac{1}{2}\sigma^2\partial_{xx}\, g^2(x)p(x,t) \tag{12}$$

with $g'(x)=\partial_x g(x)$. The coefficient of the second derivative, the diffusion, coincides in the two versions. Only the coefficients of the first derivative, the drift, differ if the diffusion is process-dependent. The drift describes the infinitesimal systematic evolution of x_t and the diffusion the variance around it. In the following we will only be interested in the stationary behaviour of the system. The stationary solution of the FPE is given by

$$p_s(x) = \mathcal{N}\, g^{-\nu}(x) \quad \exp\left\{ \frac{2}{\sigma^2} \int^x \frac{f(x')}{g^2(x')}\, dx' \right\} \quad , \quad \begin{matrix} \nu=1 \text{ Strat.} \\ \nu=2 \text{ Ito} \end{matrix} \tag{13}$$

if the stationary probability density $p_s(x)$ is normalisable. In this case the process x_t is ergodic, i.e. we can interpret $p_s(x)$ as a measure for that part of the time, an arbitrary realisation of x_t spends in an infinitesimal vicinity of x. It is for this reason that the extrema of p_s are usually identified with the macroscopic stationary states: The maxima, where the process spends relatively much time, as the stable stationary states and the minima, which the process leaves rather quickly, as unstable states. The extrema of $p_s(x)$ are easily calculated from the relation

$$h(x) + \lambda g(x) - \tfrac{\nu}{2}\sigma^2 g(x)g'(x) = 0 \tag{14}$$

Note that there is only a quantitative difference between the Ito and Stratonovic version, namely a factor 2 in the variance of the noise. Eq. (14) is the basic equation for an analysis of the influence of external white noise on the macroscopic behaviour of nonequilibrium systems. We have to distinguish two cases: i) $g(x) \equiv$ const, ii) $g(x) \not\equiv$ const. The first case is the so-called additive noise. The influence of the environmental fluctuations does not depend on the state of the system. In this case the extrema of p_s coincide with the deterministic steady states: additive external white noise does not modify the stationary macroscopic behaviour of one-variable systems. The second case is the so-called multiplicative noise. The effect of the environmental fluctuations does depend on the state of the system. We expect the following behaviour: If σ^2 is sufficiently small, the roots of (14) will not differ in any essential way from the deterministic steady states. If however the intensity σ^2 of the noise increases and if $g(x)$ is nonlinear in a suitable way, the extrema of p_s can be essentially different from the deterministic steady states. The *external noise* can thus deeply modify the macroscopic behaviour of the system, namely *induce new transition phenomena.*

Using the above formalism we have studied the influence of external white noise in some physico-chemical and biological systems, e.g. the VERHULST equation in population dynamics [3] , a problem in tumor immunology [4,5] , and an illuminated chemical system, the NITZAN-ROSS model [6] . In all these systems it was found that white noise induces indeed new nonequilibrium transitions. This has the consequence that the behaviour of those systems is much richer in a fluctuating environment than in an unvarying one. In the following I will discuss in detail the effects of external noise on a simple genetic model. Its advantage is that on the one hand it displays in a nice way the principal features of noise induced transitions. On the other hand it is sufficiently simple, so that most calculations can be done by hand and that, as it will turn out, even the case of nonwhite noise can be analysed in a satisfactory way.

2.2. A Simple Genetic Model

We consider a haploid population and are interested in a particular locus on the chromosome for which there are two possible alleles in the population, say A and a, with frequencies x resp. 1-x. We suppose that the size of the overall population does not change in time, and that there are only two mechanisms that change the frequency of A and a, namely mutation between the two alleles and natural selection. Then the kinetic equation for x reads [7] :

$$\dot{x} = 0.5 - x + \beta x(1-x) \tag{15}$$

where β is the selection coefficient and where for simplicity I have chosen the case that the the mutation rates are equal. Let us briefly look at the deterministic steady states. They are given by the relation

$$\beta = (x_s - 0.5)/\{x_s(1-x_s)\} \tag{16}$$

which is plotted in Fig. 1, curve labeled 0. It is easily verified that every steady state is globally asymptotically stable. In other words, in a constant environment this system does not exhibit any instability and looks rather uninteresting.

Let us however see what the situation is like if the environment fluctuates, i.e. the selection coefficient becomes a random quantity [8]. First we consider situations, where the correlation time of the external noise is very short and the white noise idealisation thus justified. Using the formalism presented above, we obtain the following SDE:

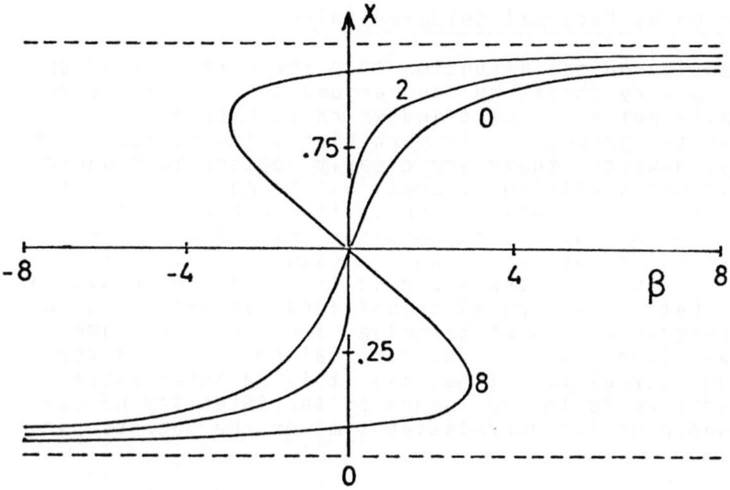

Fig. 1 Plot of the roots of (19) for three values of $\sigma^2/2$ (0,2,8) versus the (mean) selection coefficient β

$$\text{(S)} \quad dx_t = \{0.5 - x_t + \beta x_t(1-x_t)\}dt + \sigma x_t(1-x_t)dW_t \tag{17}$$

The stationary probability density is given by

$$p_s(x) = \frac{N}{x(1-x)} \exp\{\frac{2}{\sigma^2}(- \frac{1}{2x(1-x)} - \beta \ln(\frac{1-x}{x}))\} \tag{18}$$

and its extrema are the roots of

$$0.5 - x_m + \beta x_m(1-x_m) - \tfrac{1}{2}\sigma^2 x_m(1-x_m)(1-2x_m) = 0 \tag{19}$$

For the sake of simplicity let us first discuss the case β=0, i.e. no allele is favored on the average and the steady state of (15) is x=0.5. From (19) we obtain

$$x_{m1} = 0.5 \quad \text{and} \quad x_{m\pm} = \{1 \pm \sqrt{1-(4/\sigma^2)}\}/2 \tag{20}$$

We have the following situation: For $\sigma^2<4$ x_{m1} is a maximum. At $\sigma_c^2=4$ x_{m1} is a triple root and for $\sigma^2>4$ x_{m1} becomes a minimum and two maxima appear at $x_{m\pm}$ which tend to 0 resp. 1, i.e. the asymptotes of $x_s(\beta)$, cf. (16), as $\sigma^2 \to \infty$. Approaching the critical point from above $\sigma^2 \downarrow \sigma_c^2$ the distance between x_{m+} and x_{m-} tends to zero like $\sqrt{(\sigma^2-\sigma_c^2)}$. For $\beta \neq 0$ the transition is a hard one as can be seen from Fig. 1. For say β > 0 the peak corresponding to the steady state of (15) moves towards 1, the asymptote of $x_s(\beta)$ for $\beta\to\infty$, with growing σ^2 and if σ^2 depasses $\sigma_{th}^2(|\beta|)>4=\sigma_c^2$ a second peak appears at a finite distance from the original one, near the other boundary of the phase space and corresponds to the asymptote of $x_s(\beta)$ for $\beta\to -\infty$. These facts show that, as to the extrema of $p_s(x)$, in the (β,σ^2)-half plane we have a cusp catastrophe with critical point at (0,4).

3. Transitions Induced by External Coloured Noise

So far we have only considered the situation that the correlation time of the external noise is very short. We have argued that in such a case the idealisation of white noise is justified which is extremely useful since only in this case the process x_t is markovian allowing for a complete analytical study. However, there are clearly applications where the correlation time is not sufficiently small on the macroscopic time scale and hence the white noise idealisation is inadequate. There are two further reasons which motivate us to consider the influence of real noise on nonlinear systems: i) after all white noise is an idealisation which does not occur in nature and therefore it is desirable to have an explicit confirmation that noise induced transitions, as described in section 2, are not artefacts due to white noise, i.e. that the same transition phenomena are indeed also found for real noise with a very short, but nonvanishing, correlation time; ii) It is an interesting problem to study in addition to the influence of the intensity of external noise the influence of the correlation time on the macroscopic behaviour of the system.

Since under the influence of real noise the temporal evolution of the system is nonmarkovian, concrete results for nonlinear systems are almost impossible to obtain as discussed in [8]. Some approximation procedures, to determine e.g. the stationary probability density $p_s(x)$, have been proposed for the case that the external noise λ_t is nonwhite, but markovian. Such a noise is called coloured noise. More specifically the ORNSTEIN-UHLENBECK noise, i.e. a Gaussian Markov process with $E\lambda_t=0$, $E\lambda_s\lambda_t=(\sigma^2/2\alpha)\exp\{-\alpha|t-s|\}$ and which obeys the SDE

$$d\lambda_t = -\alpha\lambda_t dt + \sigma dW_t \quad , \tag{21}$$

has been considered. The approximation schemes cover either the immediate neighbourhood of white noise [9], namely $0\neq\tau_{cor}<<1<\tau_{macro}$ and qualitatively the same transition phenomena are observed as in the white noise analysis, or the opposite limit $\tau_{cor}>>\tau_{macro}$ [8]. Both approximation procedures have of course some drawbacks, the most serious is that their range of validity is not exactly known, and in view of the unexpected phenomena external noise might induce, exact analytical results for coloured noise are clearly desirable.

It turns out, that for a general nonlinear one-variable system exact analytical results can be obtained even for coloured noise, if one pays a certain price, namely a drastic restriction of the state space of the noise [10]. The dichotomous Markov process or random telegraph signal I_t is a process whose state space is $\{-\Delta,\Delta\}$ and whose temporal evolution is described by the following master equation for its probability

$$\begin{pmatrix} P_+(t) \\ P_-(t) \end{pmatrix} = -\frac{\gamma}{2}\begin{pmatrix} 1 & -1 \\ -1 & 1 \end{pmatrix}\begin{pmatrix} P_+(t) \\ P_-(t) \end{pmatrix} \tag{22}$$

where $P_\pm(t) = P(I_t=\pm\Delta)$. If $P_\pm(-\infty)=1/2$, I_t is a stationary process with

$$EI_t = 0 \ , \quad EI_tI_{t+\tau} = \Delta^2\exp(-\gamma\tau) \ , \quad \tau_{cor} = 1/\gamma \tag{23}$$

i.e. as far as mean value and correlation function are concerned it coincides with the ORNSTEIN-UHLENBECK process. Furthermore, as the latter, it converges in the limit $\Delta\to\infty$, $\gamma\to\infty$ such that $\Delta^2/\gamma=\text{const}=\sigma^2/2$ to the Gaussian white noise: $I_t \to \sigma\xi_t$. In [10] we have shown that the probability density $p(x,t)$ of the system subjected to the dichotomous Markov noise, that is

$$\dot{x}_t = h(x_t) + \lambda g(x_t) + I_t g(x_t) = f(x_t,\lambda) + I_t g(x_t) , \quad (24)$$

obeys the following equation:

$$\partial_t\, p(x,t) = -\partial_x\, f(x,\lambda)p(x,t) + \Delta^2\, \partial_x\, g(x) \int_{-\infty}^{t} dt'\, \exp\{-(\gamma+\partial_x f)(t-t')\}\, \partial_x g(x) p(x,t') \quad (25)$$

Its nonmarkovian character is clearly displayed by the memory kernel. Eq. (25) reduces in the white noise limit to the FPE (12), as it should according to WONG and ZAKAI's theorem. The stationary solution of (25) is given by

$$p_s(x) = \mathcal{N} \frac{g(x)}{\Delta^2 g^2(x) - f^2(x)} \exp\{\gamma \int^x \frac{f(x')}{\Delta^2 g^2(x') - f^2(x')} dx' \} \quad (26)$$

if the deterministic system is stable, i.e. $x(t)$ does not blow up. Roughly speaking the zeros of $\Delta^2 g^2(x)-f^2(x)=(\Delta g-f)(\Delta g+f)$, which correspond to the deterministic steady states for $\lambda\pm\Delta$, constitute the boundaries of the interval U outside which $p_s(x) = 0$. U is called the support of $p_s(x)$. The extrema of $p_s(x)$ are given by the equation:

$$(h(x_m)+\lambda g(x_m)) - (\Delta^2/\gamma) g(x_m) g'(x_m) + (2/\gamma)(h(x_m)+\lambda g(x_m))\cdot(h'(x_m)+\lambda g'(x_m)) - (1/\gamma)(h(x_m)+\lambda g(x_m))^2 \frac{g'(x_m)}{g(x_m)} = 0 \quad (27)$$

for $x_m \varepsilon$ U. This equation has a very interesting structure: The first term equal to zero yields the deterministic steady state. In the white noise limit only the first two terms survive, yielding indeed (14), in the Stratonovic interpretation. The last two terms are corrections due to the fact that γ is finite. This shows that indeed the effects induced by white noise are also observed under the influence of real noise with a sufficiently short correlation time. However, there can be additional modifications of the macroscopic behaviour of the system if the correlation time increases, i.e. γ decreases, as reflected by the last two terms of (27).

Before discussing the influence of coloured noise on the genetic model, I would like to consider briefly a slightly simpler system, the VERHULST equation:

$$\dot{x} = \lambda x - x^2 \quad (28)$$

This system has in a nonfluctuating environment a transition point at $\lambda=0$, where the trivial steady state $x_s=0$, stable for $\lambda<0$, becomes unstable and a new branch of steady states $x_s=\lambda$, which is stable for $\lambda>0$, emerges. The case of external white noise on the growth parameter, $\lambda_t=\lambda+\sigma\xi_t$, is discussed in [3]. It is shown that in addition to the deterministic transition point $\lambda=0$, a second transition point occurs at $\lambda = \sigma^2/2$, which is entirely due to the external noise: For $\lambda<0$ $p_s(x)=\delta(x)$; for $0<\lambda<\sigma^2/2$ a genuine probability density exists, i.e. $p_s(x) \neq 0$ for all positive x, but x=0 is still the most probable value. Only for

$\lambda > \sigma^2/2$ has $p_s(x)$ a maximum near the nontrivial deterministic steady state $x_s = \lambda$. For the case of the dichotomous Markov noise, we consider only the case $\lambda > 0$. Then the above formulae can be used to construct a "phase diagram" of the VERHULST system in the (Δ,γ)-plane and which is shown in Fig. 2

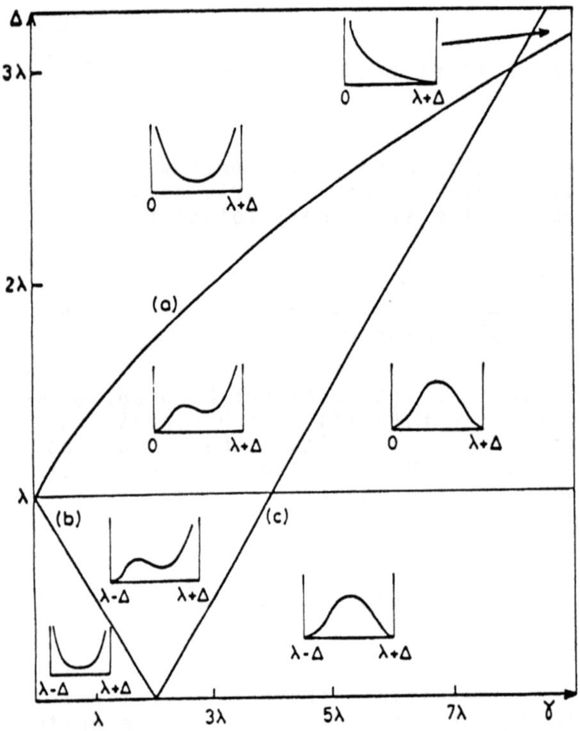

Fig. 2 "Phase diagram" for the steady state behaviour of the VERHULST system as a function of the intensity Δ and the inverse of the correlation time γ of the external dichotomous Markov noise for fixed positive λ. The shape of $p_s(x)$ for the different regions is sketched. Curve (a) is given by $\lambda(1+\lambda/\gamma)=\sigma^2/2$

This "phase diagram" shows that either by increasing the amplitude Δ of the external noise or by changing its correlation time $1/\gamma$, the system undergoes a series of transitions, which are completely unexpected from the usual phenomenological description. The neighbourhood of white noise corresponds of course to the upper right corner of the diagram and there we have on the curve (a) $\lambda \simeq \sigma^2/2$. Clearly, the results of the white noise analysis are explicitly confirmed.

Let us now come back to the genetic model and consider the influence of a dichotomous Markov noise in the selection coefficient β:

$$\dot{x}_t = 0.5 - x_t + \beta x_t(1 - x_t) + I_t x_t(1 - x_t) \tag{29}$$

Most formulae are a lot more complicated than in the VERHULST case and cannot be evaluated by hand. Let us however look at the equation for the extrema of the stationary probability density for the special case $\beta=0$ as in the white noise case. We obtain from (27):

$$\begin{aligned}(0.5-x_m)x_m(1-x_m) - (\Delta^2/\gamma)x_m^2(1-x_m)^2(1-2x_m)\\ -(2/\gamma)x_m(1-x_m)(0.5-x_m) - (1/\gamma)(0.5-x_m)^2\,(1-2x_m) = 0\end{aligned} \tag{30}$$

Obviously $x_m=0.5$ is a root of (30) and it is trivially in U, being the deterministic steady state. It is easy to check that it is a triple root, i.e. $\beta=0, x_m=0.5$ is a critical point (induced by the noise), if the condition

$$\Delta^2 = 2\gamma - 4 , \qquad \gamma>2 \tag{31}$$

is fulfilled. Remember that in the white noise limit $\Delta^2/\gamma = \sigma^2/2$, so that the above condition means

$$\sigma^2 = 4 - 8/\gamma = 4 - 8\tau_{cor} , \qquad \tau_{cor} < 0.5 \tag{32}$$

In [9], using an asymptotic representation valid for ORNSTEIN-UHLENBECK noise with $0\neq\tau_{cor} \ll 1$, the following result was obtained:

$$\sigma^2 = 4/(1+\tau_{cor}) = 4 - 4\tau_{cor} + O(\tau^2_{cor}) \tag{33}$$

Though no upper limit for τ_{cor} can be determined by this method, (33) shows that our result (32) does not depend qualitatively on the nature of the coloured noise.

(32) is a very interesting result for two reasons: i) the intensity σ^2, necessary to induce a critical point at $\beta=0, x=0.5$, decreases as the correlation time of the coloured noise increases; ii) there is an upper limit for the correlation time of the noise, beyond which the phenomena induced by white noise disappear. In our case, the upper limit is $\tau_{cor}=0.5$ which is of the order of the macroscopic relaxation time, which for $\beta=0$ is $\tau_{macro}=1$. This gives for this system a precise meaning to how much shorter the correlation time has to be for the white noise idealisation to yield qualitatively the right behaviour. It is rather large in this specific case. From the analysis of this and other systems we have arrived at the rule of thumb, that τ_{cor} should be about $0.1\tau_{macro}$ or less for the white noise results to hold.

4. Experimental Evidence and Conclusions

To my knowledge the first experimental evidence for noise induced transitions has been obtained by KABASHIMA et al. [11,12] on an electrical system, namely a parametric oscillator that undergoes a second order transition from nonoscillatory to oscillatory behaviour. The evolution of its amplitude can be described by

$$\dot{x} = lx - kx^3 \tag{34}$$

i.e. it displays qualitatively the same transition phenomena as the VERHULST equation. The electrical circuit was coupled to a wide band noise generator making l a fluctuating quantity. KABASHIMA's results confirm completely the theoretical predictions of our white noise analysis. More recently KABASHIMA [13] realized an electrical circuit to simulate the genetic model. Here again the experimental results, obtained with a wide band noise, agreed very well with the white noise analysis using the Stratonovic interpretation.

I would now like to present the main results of an experimental study of a chemical system, I did together with DE KEPPER at the Centre de Recherche Paul Pascal, Bordeaux [14]. We studied the BRIGGS-RAUSCHER reaction in a continuous stirred tank reactor. This reaction has been extensively studied since it constitutes like the BELOUSOV-ZHABOTINSKII reaction an example for dissipative structures. The following major reactants are fed into the reactor: $CH_2(COOH)_2$, KIO_3, H_2O_2, $HClO_4$ and $MnSO_4$. This reaction is extremely photosensitive, due to the intermediate I_2 which absorbs around 460 nm and is split into two radicals. Though the reaction mechanism is known only in a very limited way, there is good evidence that the following step

$$CH_2(COOH)_2 + I_2 \rightarrow I^- + CHI(COOH)_2 + H^+ \qquad (35)$$

plays a major role in the BRIGGS-RAUSCHER reaction. The important point is that it obeys a Michealian type of kinetics

$$[I_2] \sim \frac{[CH_2(COOH)_2]\cdot[I_2]}{k + [I_2]} \qquad (36)$$

and that the incident light acts on it by photolysing I_2. In [4,5] we have shown that a Michealian kinetic can display noise induced transitions. That is why we studied the effects of a noisy light source on this reaction.

The main experimental result is given in Fig. 3. There the optical density of the reacting mixture at 460 nm is plotted versus the (mean) intensity of the incident light. The solid lines correspond to a nonfluctuating intensity, the broken lines to a noisy light intensity. It was obtained by sending the incident light beam through a small box containing little polystyrene balls (radius ∿ 1mm) agitated in a turbulent air stream. This device generated a noise with a Gaussian probability distribution, relative variance 13%, and with an exponentially decreasing correlation function, $\tau_{cor} \simeq 0.05$ sec. Fig. 3 depicts a region of bistability between an oscillating state A of low concentration in I_2 (the vertical bars represent the amplitude) and a stationary state B of high concentration in I_2 for a constant light intensity. The constraints correspond to a situation near the critical point, the hysteresis being rather small. Using the fluctuating light source leads to a broadening of the region of bistability and

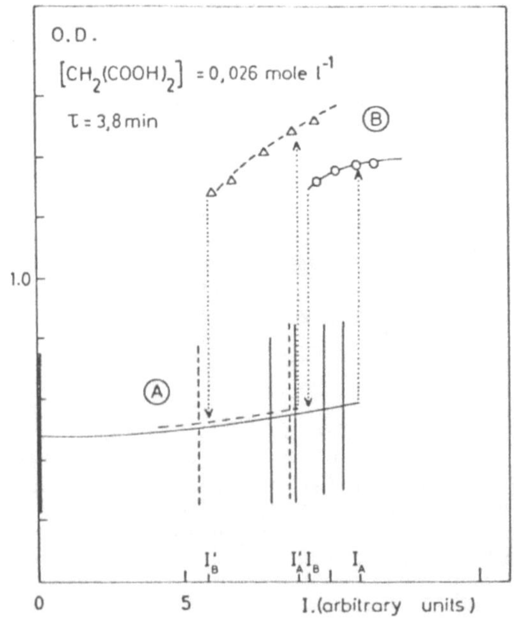

Fig. 3 Plot of the optical density (at 460 nm) of the BRIGGS-RAUSCHER system versus the (mean) incident light intensity. The concentrations of the reactants, fed into the reactor, the temperature and the renewal time τ of the reactor (volume 30 cm³) are indicated

to such a shift to lower (mean) intensities that it is disjunct with that of the nonfluctuating case. This implies that for intensities between I'_A and I_B the system undergoes a transition if, keeping the mean intensity constant, the noise is switched on or off.

The theoretical analysis of the influence of white and coloured noise as well as the experimental results reported above demonstrate that environmental fluctuations can deeply modify the macroscopic behaviour of a nonequilibrium system. When the fluctuations grow in amplitude or when the correlation time changes, various transition phenomena can occur that are quite unexpected from a deterministic point of view. They correspond to situations in which the system no longer adjusts its macroscopic behaviour to the average properties of the environment but responds in a definite way, that was examplified by the systems discussed in this paper, to the external noise. The interplay between the external noise and the nonlinearities of the system can for instance as in the genetic model lead to the "stabilisation" of macroscopic steady states which have no analog in a constant environment. This implies that external noise, even if it has such a highly incoherent character as the white noise does, should not in all cases be considered as a nuisance "unfortunately" perturbing an "ideal situation", but on the contrary could play an important role by allowing for phenomena that would be impossible in a constant environment. In my opinion it is especially important to explore its relevance in biological systems. As a general rule the natural processes of development and selforganisation that take place in living systems are subjected to extremely complex environmental conditions and are controlled by a multitude of environmental factors. Because of this complexity, fluctuations of the environmental state are unavoidable and in many instances one may look into the effect of these fluctuations on the functioning and organisation of biological systems as a whole. The results presented above suggest that biological systems do not only feel and respond to the average properties of the external world but that some macroscopic manifestations of biological order might be "forced" by environmental randomness.

Acknowledgement: This work was supported by the Instituts Internationaux de Physique et de Chimie, fondés par E. Solvay, and by the Belgian Government, Actions de Recherche Concertées, convention n° 76/81II3.

References

1. L. Arnold, Stochastische Differentialgleichungen, Oldenbourg, München - Wien 1973
2. E. Wong and M. Zakai, Ann. Math. Stat. 36, 1560 (1965)
3. W. Horsthemke and M. Malek-Mansour, Z. Physik B 24, 307 (1976)
4. W. Horsthemke and R. Lefever, Phys. Lett. 64A, 19 (1977)
5. R. Lefever and W. Horsthemke, Bull. Math. Biol. 41, 469 (1979)
6. R. Lefever and W. Horsthemke, Proc. Natl. Acad. Sci. USA 76, 2490 (1979)
7. M. Kimura and T. Ohta, Theoretical Aspects of Population Genetics, Princeton University Press, Princeton N.J. 1971
8. L. Arnold, W. Horsthemke and R. Lefever, Z. Physik B 29, 367 (1978)
9. J. M. Sancho and M. San Miguel, preprint 1979
10. K. Kitahara, W. Horsthemke and R. Lefever, Phys. Lett. 70A, 377 (1979)
11. S. Kabashima, Ann. Israel Phys. Soc. 2, 710 (1978)
12. T. Kawakubo, S. Kabashima and Y. Tsuchiya, Progr. Theor. Phys. Suppl. 64, 150 (1978)
13. S. Kabashima, preprint 1979
14. P. De Kepper and W. Horsthemke, C. R. Acad. Sc. Paris Ser. C 287, 251 (1978)

to such a shift to lower (mean) intensities that it is disjunct with that of the nonfluctuating case. This implies that for intensities between d' and d, the system undergoes a transition if, keeping the mean intensity constant, the noise is switched on or off.

The theoretical analysis of the influence of white and colored noise as well as the experimental results reported above demonstrate that environmental fluctuations can highly modify the macroscopic behaviour of a nonequilibrium system. When the fluctuations grow in amplitude or when the correlation time changes, various transition phenomena can occur that are quite unexpected from a deterministic point of view. They correspond to situations in which the system no longer adjusts its macroscopic behaviour to the average properties of the environment but responds in a definite way, that was exemplified by the systems discussed in this paper, to the external noise. The interplay between the external noise and the nonlinearities of the system can, for instance as in the genetic model, lead to the "stabilisation" of macroscopic steady states which have no analog in a constant environment. This implies that external noise, even if it has such a highly incoherent character as the white noise does, should not in all cases be considered as a nuisance "unfortunately" perturbing an "ideal situation", but on the contrary could play an important role by allowing for phenomena that would be impossible in a constant environment. In my opinion it is especially important to explore its relevance in biological systems. As a general rule the natural processes of development and self-organisation take place in [illegible] and are subjected [illegible]

[illegible]

[illegible] as a whole [illegible] and [illegible] to the average properties [illegible] as an [illegible] manifestations of biological order might be "forced" on environmental randomness.

Acknowledgement. This work was supported by the Institut International de Physique et de Chimie, [illegible] Solvay [illegible]

[illegible]
2. [illegible]
3. [illegible]
4. [illegible] and [illegible], Phys. Lett. 61A, 19 (1977)
5. R. Lefever and W. Horsthemke, Bull. Math. Biol. 41, 469 (1979)
6. R. Lefever and W. Horsthemke, Proc. Natl. Acad. Sci. USA 76, 2490 (1979)
7. M. Kimura and T. Ohta, Theoretical Aspects of Population Genetics, Princeton University Press, Princeton N.J., 1971
8. L. Arnold, W. Horsthemke and R. Lefever, Z. Physik B 29, 367 (1978)
9. J. M. Sancho and M. San Miguel, preprint 1979
10. K. Kitahara, W. Horsthemke and R. Lefever, Phys. Lett. 70A, 377 (1979)
11. S. Kabashima, Ann. Israel Phys. Soc. 2, 710 (1978)
12. T. Kawakubo, S. Kabashima and Y. Tsuchiya, Progr. Theor. Phys. Suppl. 64, 150 (1978)
13. S. Kabashima, preprint 1979
14. P. De Kepper and W. Horsthemke, C. R. Acad. Sc. Paris Ser. C 287, 251 (1978)

Part IV

Spatio-temporal Organization of Chemical Processes

Chemical Waves in the Oscillatory Zhabotinskii System. A Transition from Temporal to Spatio-temporal Organization

M.L. Smoes

Department of Chemistry, University of Michigan
Ann Arbor, MI 48109, USA

Introduction

The Zhabotinskii system [1], [2], is an excellent example of chemical synergetics [3]. When four or five chemical compounds are mixed in the appropriate concentration ranges and at the appropriate temperature, the Zhabotinskii system spontaneously organizes itself into temporal or spatio-temporal dissipative structures of macroscopic dimensions [4]. In this chemical reaction, at least twenty intermediates are formed. The chemical mechanism involved is so complex that almost all theoretical work is performed on models rather than on the best rate equations available today [5]. A first type of model involves "macrokinetic" steps rather than elementary ones. This type includes the model of ZHABOTINSKII and his collaborators [1] which attempts to reproduce both waveforms and periods of oscillations. It includes also the many versions of the Oregonator [6] which are designed to reproduce the waveforms of a few intermediates. Other models are of the heuristic-topological type, according to the PACAULT [7] classification. Two of them are the well-known PRIGOGINE-LEFEVER model [8] (or Brusselator) and the analytic BAUTIN system [9], [10] (or DREITLEIN-SMOES model). It is unnecessary to emphasize here the role played by the PRIGOGINE-LEFEVER model as a research tool in the theory of dissipative structures. The BAUTIN system is less well known in spite of several attractive features: this system is solvable in closed form; it exhibits a limit cycle and bistability; there is a saddle-node transition between steady-state and finite amplitude oscillations [2]. We will mention these theoretical results in the last part of this paper, but begin now with some experimental results, first on the temporal organization, next on the spatio-temporal organization, and finally on the transition between these two types of dissipative structures. Two new results will be emphasized. The first concerns the peculiar behavior of the Zhabotinskii waves at low concentrations of malonic acid ([MA]) [11]. The second concerns the properties of the centers of chemical waves [12]. The discussion of chemical waves will follow the approach of the one-wave theory [4,13,14], without artificial distinction of trigger and phase waves. All Zhabotinskii waves obey the same mechanism given by a unique set of partial differential equations which includes diffusion and probably thermal conduction.

A. Some Experimental Results

1. Temporal Organization (See reference [2])

Numerous variations of the Zhabotinskii system exist. They differ slightly in terms of the chemical compounds involved and of course also in their mechanism. In this work, we define the Zhabotinskii system as a mixture of sodium bromate ($NaBrO_3$), malonic acid ($COOH$-CH_2-$COOH$ or MA), sulfuric acid (H_2SO_4), and ferroin which is the tris-1,10 phenanthroline iron complex ion. Initially, this system also contains a trace of sodium bromide due to an impurity of 0.05% by weight in the sodium bromate. The same ferroin-catalyzed systems are used to observe oscillations and waves. Synchronous oscillations (i.e., of the same period and phase) are observed in the lumped parameter system, in which constant and adequate stirring is provided. When the stirring is interrupted, a distributed parameter system is generated in which waves may occur and which will be described in the next section. Since the chemical waves are observed through a change in color of the catalyst, we have monitored the homogeneous oscillations using the same color change, for direct comparison. The transmittance at 546 nm was therefore followed as a function of time. It increases when the concentration of ferriin increases. (Ferriin refers to the oxidized, blue form of the catalyst, while ferroin refers to the reduced, red form. The word ferroin is also used to designate both forms of the catalyst.) We obtained three main results: an expression for the period of oscillations as a function of the initial concentrations, an increase in the oxidized fraction of the period at low concentrations of malonic acid ($[MA]_0$) and the existence of an oxidized steady-state at still lower $[MA]_0$.

We have found the following relationship for the homogeneous period

$$T\ (s) = 0.22\ [BrO_3^-]_0^{-1.6}\ [H_2SO_4]_0^{-2.7}\ [MA]_0^{-0.27}\ sM^{4.57} \qquad (1)$$

when $[\mathrm{Ferroin}]_0 = 2.27 \times 10^{-3}$ M and Temp. = 24.9°C. $[A]_0$ is the initial concentration of A, and M is a unit of concentration, 1 mole per liter. This relationship holds for a wide range of initial concentrations (Table 1 in [2]). The period is not always stable and reproducible. When the period is long, at the approach of the zero frequency bifurcation, the period is clearly unstable. This is also true at high catalyst concentrations.

The initial concentrations also determine the waveforms or shapes of the oscillations. At high concentrations of sulfuric acid, the oscillations have a very small amplitude and are almost sinusoidal (See Fig. 6 in [2]). When $[H_2SO_4]_0$ decreases, we see an increase in the amplitude and in the relaxational character of the oscillations. The oxidized fraction of the period decreases, i.e., the time interval when the catalyst is mostly oxidized does not increase as much as the time interval when the catalyst is mostly reduced. At a sufficiently low concentration of sulfuric acid, a bifurcation takes place in which the high amplitude, low frequency oscillations cease abruptly. The steady-state is "reduced" inasmuch as the catalyst is mostly in its reduced (red) form. When $[BrO_3^-]_0$ is varied instead of $[H_2SO_4]_0$, we observe a similar behavior with a similar bifurcation to a "reduced" steady-state at low $[BrO_3^-]_0$ (Fig. 7 and 8a in [2]). The dependence on $[MA]_0$ is almost exactly reversed (Figs. 9 and 8b in [2]). When $[MA]_0$ is decreased, the width of the oxidized peak increases faster than the period. This produces an increase in the oxidized fraction of the period which culminates in an oxidized steady-state. It seems that this oxidized steady-state has not been recognized before. Its importance in terms of the chemical waves has certainly escaped notice until now. There is a parallelism between the oxidized fraction of the period in the lumped system and the oxidized fraction of the wavelength in the distributed one. Also, since malonic acid is consumed during the Zhabotinskii reaction, its concentration decreases progressively without external intervention and the system bifurcates spontaneously. The waves observed near and at this low [MA] bifurcation are described in the next section. Note that the bifurcation at low [MA] is also a transition from high amplitude, low frequency oscillations to a steady-state. This type of transition is sometimes interpreted as evidence of the presence of an unstable limit cycle. However, there are other possible interpretations. For instance, the saddle-node transition seen in the asymmetric Bautin system also corresponds to a transition from finite amplitude, low frequency oscillation to a steady-state. Saddle-node transitions will be discussed again in the last part of this paper. (See also the Appendices of [2]).

2. Spatio-temporal Organization

Chemical waves are observed in one, two, and three dimensions. We will limit ourselves to a discussion of the concentric waves in two-dimensional systems. Generalization to one and three dimensions does not present any difficulty. In the Zhabotinskii system, the (nonspiral) waves can exhibit strikingly different appearances [4]. Some wavefronts propagate with constant speed for some distance before a sudden and remarkable acceleration. Confronted with these accelerations, Winfree [15] has insisted on the "need" to distinguish two kinds of waves. FIELD [16] and NOYES [17] have adopted the same attitude and furtherdeveloped a two-wave theory for the Zhabotinskii waves. In contrast, we have always thought that a one-wave theory [18] was appropriate, and that all Zhabotinskii waves obey a single mechanism, or in other words a single set of partial differential equations. Since the Zhabotinskii waves may indeed look very different, the two wave theory is widely accepted, at least in the United States. However, it has not led to any increase in our understanding of the waves and seems to have even impaired further research. Some recent works [19], [20] and especially a paper by REUSSER and FIELD [16] illustrate the sterility of this type of theory. In contrast, the one-wave theory is simple and elegant [18]. More importantly, it facilitates further research on the waves as will be evidenced by our new results.

We will now describe the relevant properties of the Zhabotinskii waves in some detail, using entirely our own, direct observations, unless otherwise stated. The description will be free from the two-wave concept of trigger and phase waves. Chemical waves are observed only when one terminates the forced synchronization of the oscillations by interrupting the stirring. This corresponds then to a distributed system. In our case, the fluxes are null at the boundaries. Also, the initial concentrations are such that the distributed system shows bulk oscillations [21] with a bulk period very close to that predicted by (1) for the lumped system (see Fig. 1). It should be noted that the bulk frequency is an order parameter [22] for the oscillatory Zhabotinskii system. The distributed systems are prepared in the following way: the oscillatory solution is sandwiched between two flat, transparent plastic plates [23], less than 1 mm apart. The inverted top of a Petri dish is used as bottom plate, with the bottom of the Petri dish as top plate. For quantitative measurements, the Petri dish is placed on top of a flat, horizontal, metallic slab kept at fixed temperature (usually 25°C). This temperature control is needed because the period of oscillations depends on temperature [24] as well as concentrations. Since the system is oscillatory and since stirring is not provided, spontaneous desynchronization occurs. Hence, before putting the top plate in position, the system is carefully stirred until the color is completely homogeneous. This initial synchronization of the phases may require prolonged stirring of the solution, especially in the case of long bulk period. Best results are obtained when the stirring is done at the time of bulk oxidation, when the solution becomes blue in a fast oxidation. As soon as the solution is homogeneous, the top is positioned and the system is left undisturbed thereafter. The top plate must be in place before any new desynchronization occurs. Consequently, this operation must

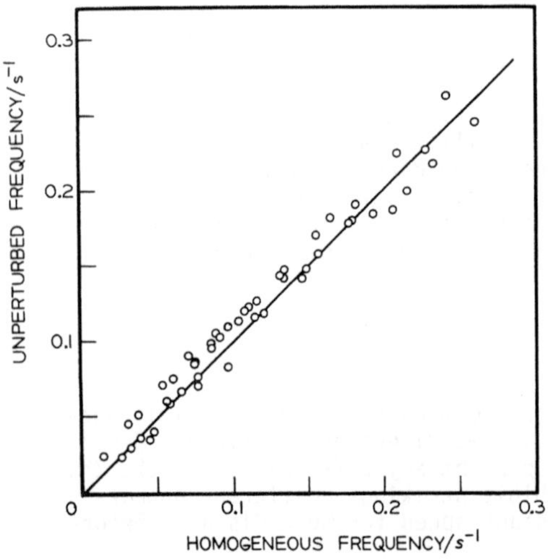

Fig.1 Unperturbed (or bulk) frequency as a function of the homogeneous frequency. Homogeneous frequency is calculated from (1), assuming no dependence on $[Ferroin]_0$. Bulk frequency is calculated from the observed time interval between two bulk oxidations in the distributed system. The solid line represents the equality of frequencies, which are angular frequencies ($2\pi/T$)

be performed very quickly if the system oscillates with a small period. If the initial homogenization and the positioning of the plate have been performed successfully, any subsequent desynchronization of the system takes place through the formation of punctual centers which propagate concentric waves. In case of failure, desynchronization appears as patches and filaments rather than punctual centers. Spiral waves will generally ensue in such cases. If they are not of interest, the synchronization procedure may be repeated at the time of the next bulk oxidation.

When punctual centers are obtained, different situations are observed according to the initial concentration of malonic acid ($[MA]_0$) in the system. The case of a high $[MA]_0$ will be discussed first. As previously stated, if the bulk period is long enough, the homogeneous oscillations have a large amplitude and a pronounced relaxational character, with a small oxidized fraction of the period. The homogeneous oscillations resemble a succession of sharp oxidation peaks (blue) separated by reduced plateaux (red). The same characteristics are evident for the bulk oscillations; one observes indeed that the distributed system remains red for most of the bulk period (Fig. 2). In this red background, blue centers may appear at unpredictable times and locations. The volume element where the center appears oscillates out of phase with the bulk. Within this volume element, the period is usually shorter than the period of bulk oscillation [25]. The formation of oxidized centers in the reduced background corresponds to the formation of gradients in concentrations and probably in temperature. The concentration gradients and phase gradients are clearly evident in this oscillatory system. The temperature gradients are deduced from an observation by KOROS and his collaborators [26]. They found that in the lumped system, the rate of heat production is not constant within one cycle; it is higher during the fast oxidation. In the distributed system, by analogy, we expect the rate of heat production to be higher at the blue center and along the blue wavefronts as they propagate from the center. Because of these gradients in concentrations and temperature, the formation of centers is followed by processes of diffusion and heat conduction. Both processes contribute to the propagation of the oxidizing blue waves. At each appearance of the blue center, a new wave propagates outwardly until it is annihilated. Annihilation occurs when a wave reaches the boundary of the dish or when two waves from different centers interfere.

Between formation and annihilation of a wavefront, its speed of propagation may or may not be constant and the speed may vary between wavefronts from different centers or even between successive wavefronts from the same center. Near the time of bulk oxidation, one may observe a sudden and quite impressive acceleration in some outermost wavefronts. From these wavefronts, the oxidation spreads very rapidly, like a spill, but without fluid motion. In Fig. 2, we see the acceleration of a wavefront at the approach of the first and second bulk oxidations. (The experimental conditions and bulk period for this system are listed in Table 1). In the two-waves theory, the accelerating wave is called a pseudo-wave or WFN (Winfree-Field-Noyes) phase wave. Before its fast acceleration, the wavefront propagating at constant speed is called a trigger wave in the same theory. Consider the narrow wavefront propagating from the west side of the dish (plate 1). This wavefront comes from three centers along the boundary. Each center has propagated a single wavefront. Only the envelope of the wavefronts remains, due to the annihilation between interfering wavefronts. The envelope is open on the left because of annihilation of the waves at the boundary. The thin wavefront propagates with constant speed but starts accelerating in the second plate, when the system approaches its bulk oxidation time. The third plate, which is taken only 8.8 seconds later, shows the magnitude of this acceleration (compare the accelerating wavefront to the other wavefronts in the system.) The quickly accelerating wave continues and sweeps all the available area. It stops at the boundary of the dish, and at the outermost wavefronts propagating from other centers, where the usual annihilation occurs (plate 4). After a time interval similar to the bulk period, a second bulk oxidation is due and we observe the second case of quick acceleration of the previously thin wavefronts. These accelerations are no longer observed when the Petri dish area has become filled with concentric wave patterns. At this point, the whole system is sharply out of phase.

Zhabotinskii Waves at High Concentrations of Malonic Acid

Fig. 2

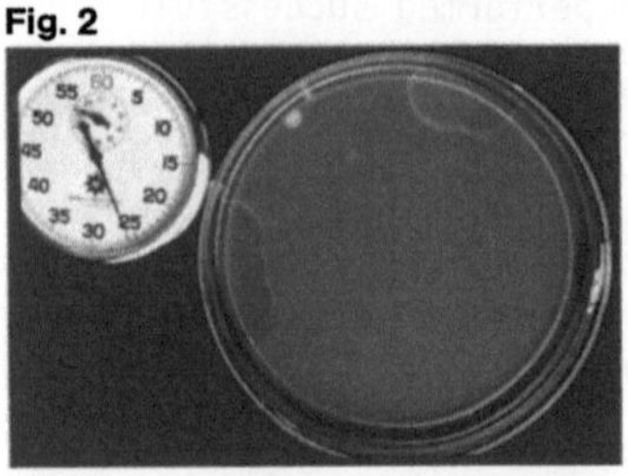

Plate 1

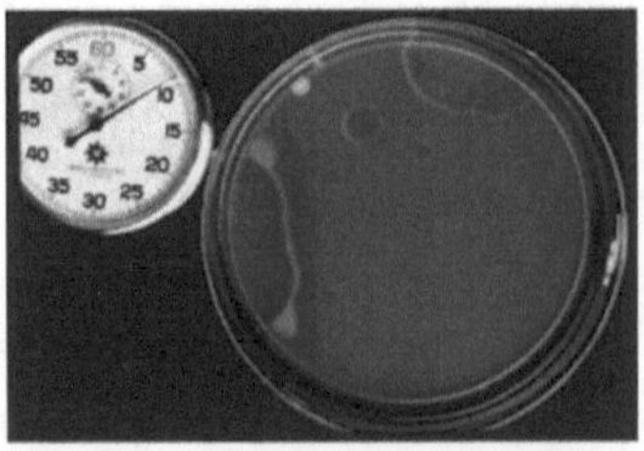

Plate 2

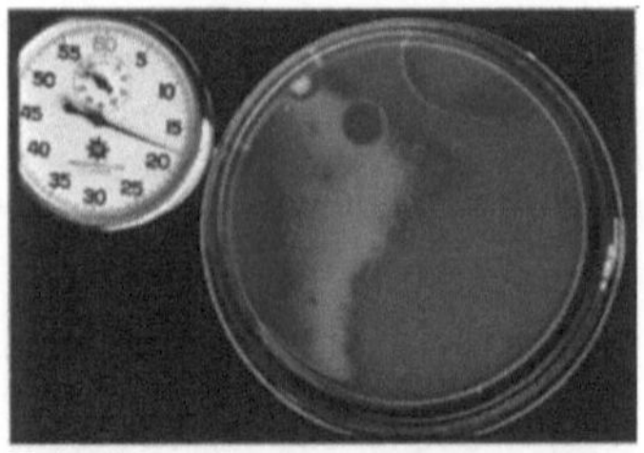

Plate 3

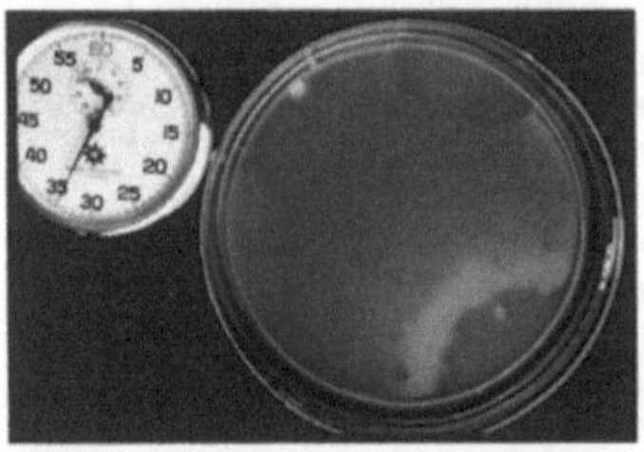

Plate 4

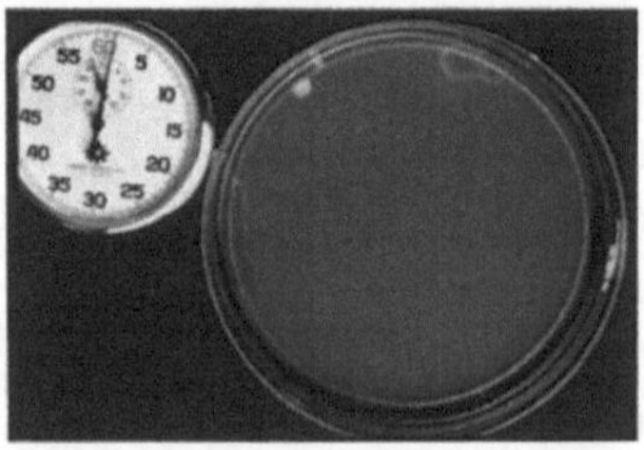

Plate 5

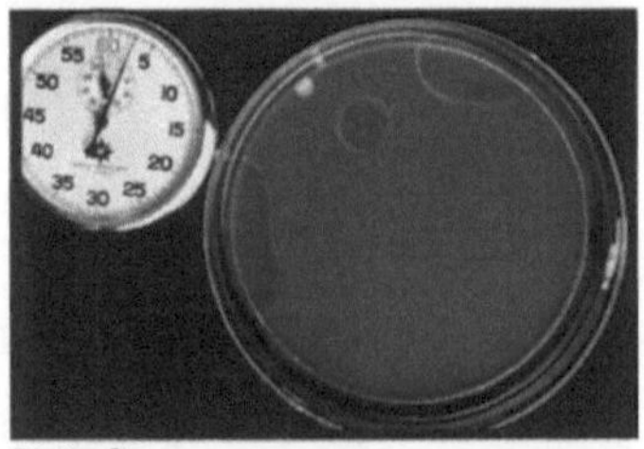

Plate 6

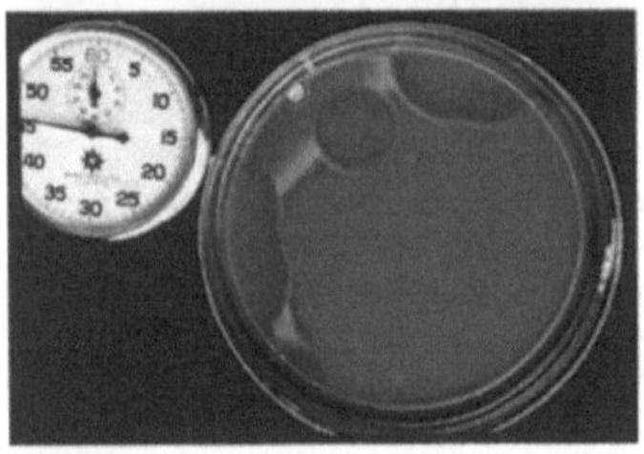

Plate 7

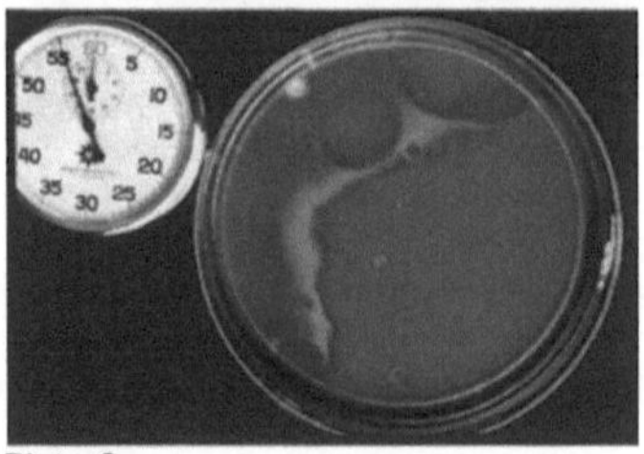

Plate 8

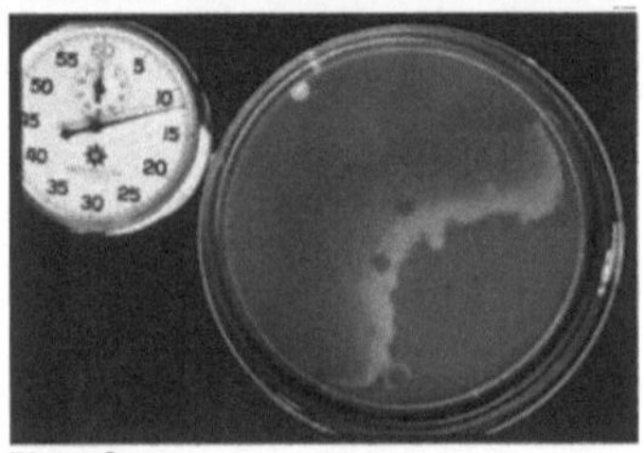

Plate 9

Fig. 3

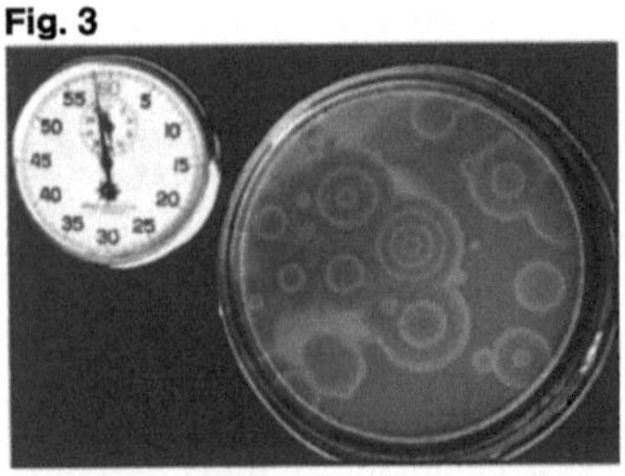

Plate 1

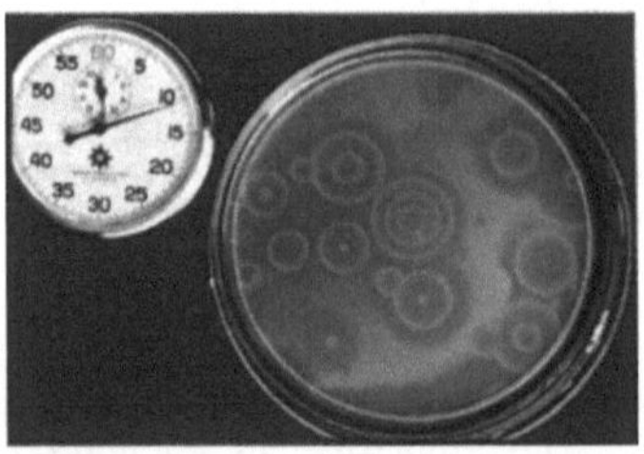

Plate 2

Plate 3

Fig. 11

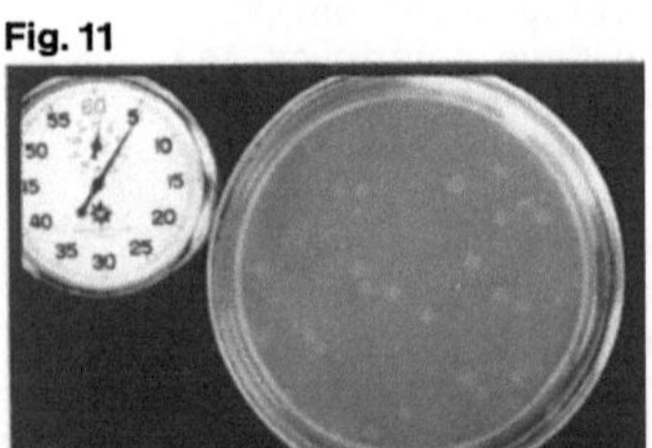

Plate 1

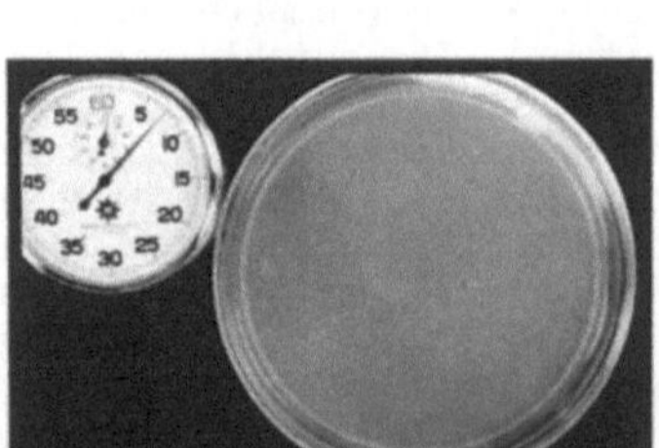

Plate 2

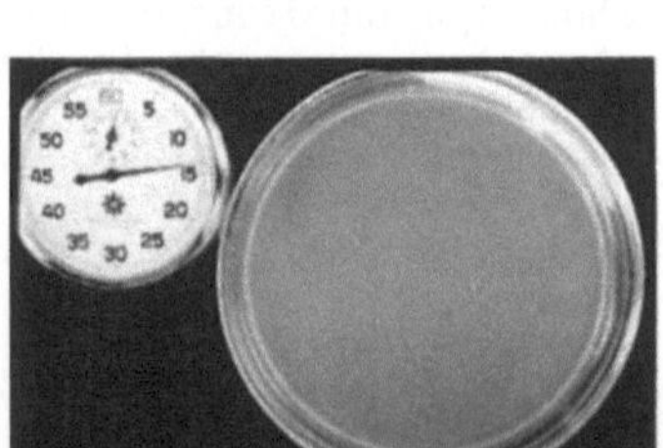

Plate 3

Zhabotinskii Waves at Low Concentrations of Malonic Acid

Fig. 4

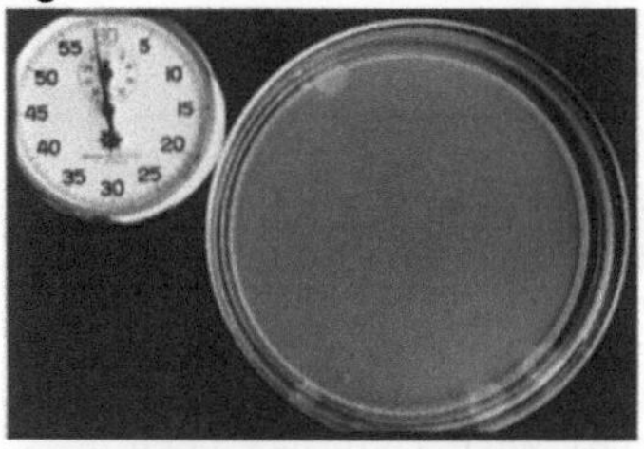

Plate 1

Plate 2

Plate 3

Plate 4

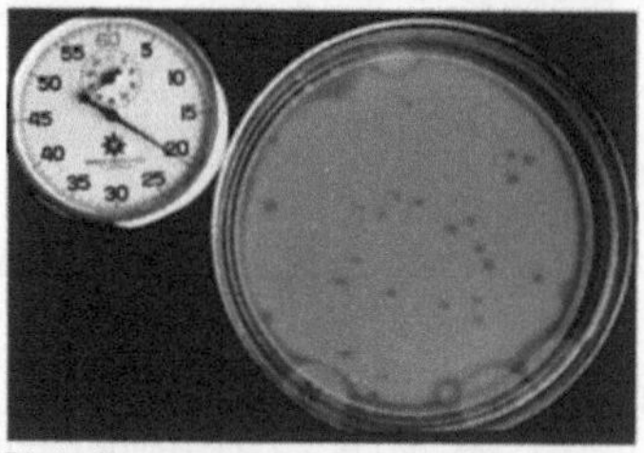

Plate 5

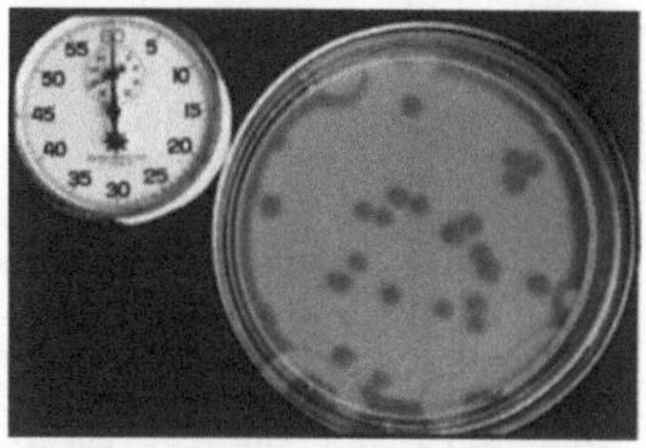

Plate 6

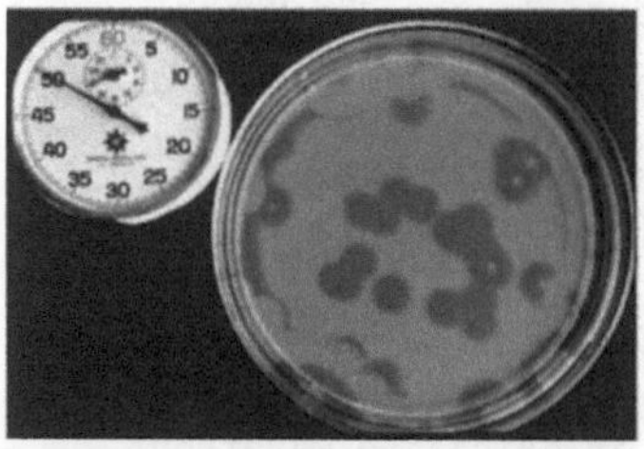

Plate 7

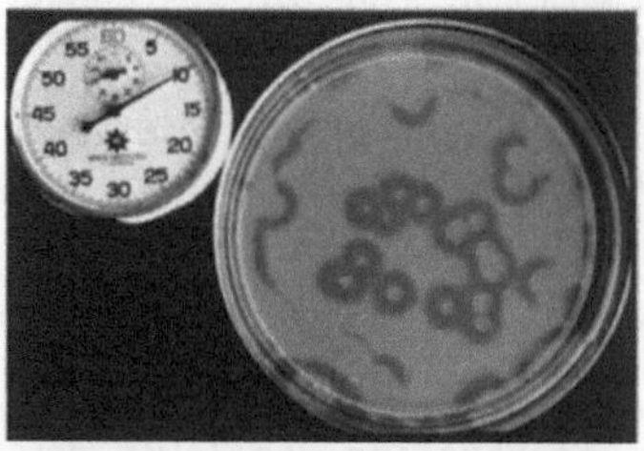

Plate 8

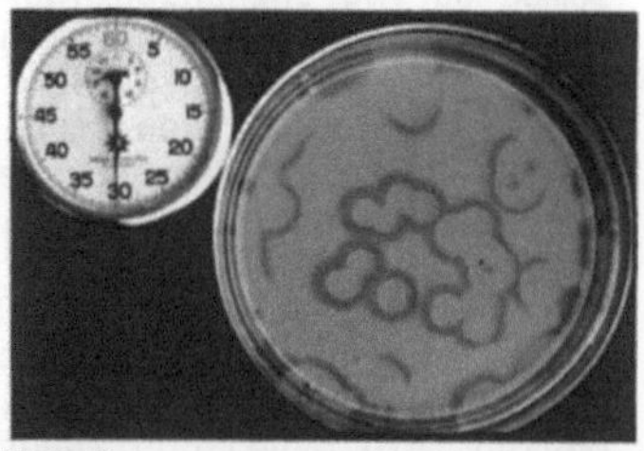

Plate 9

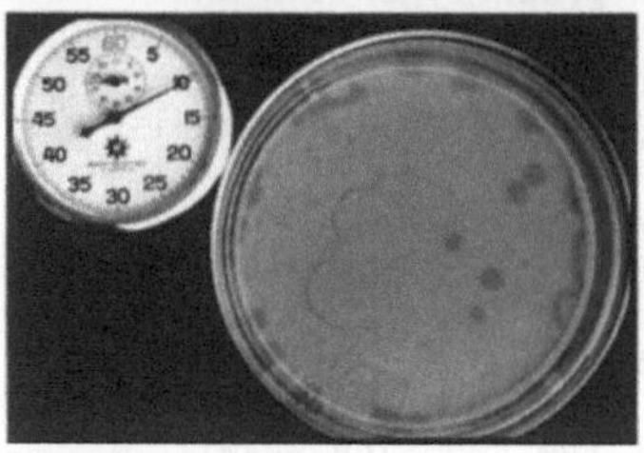

Plate 10

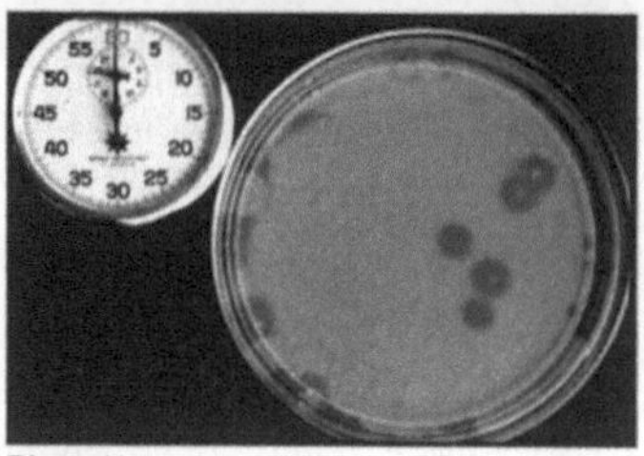

Plate 11

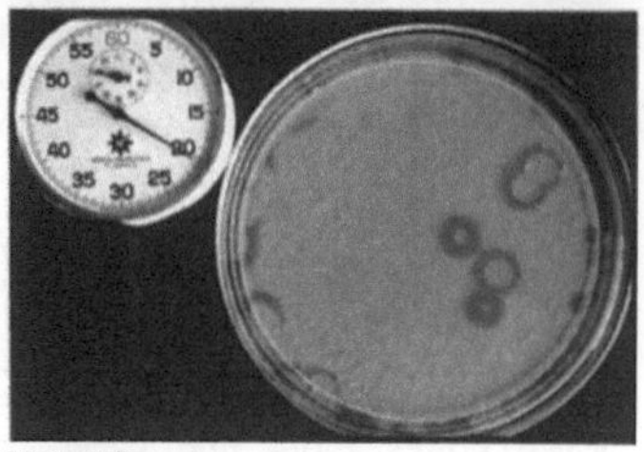

Plate 12

Fig. 7

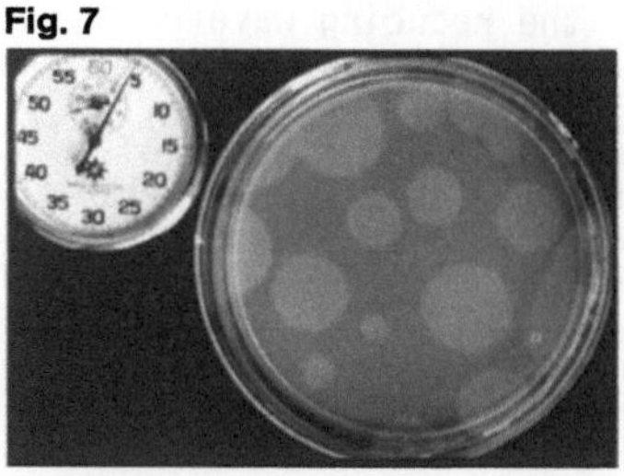

Fig. 8

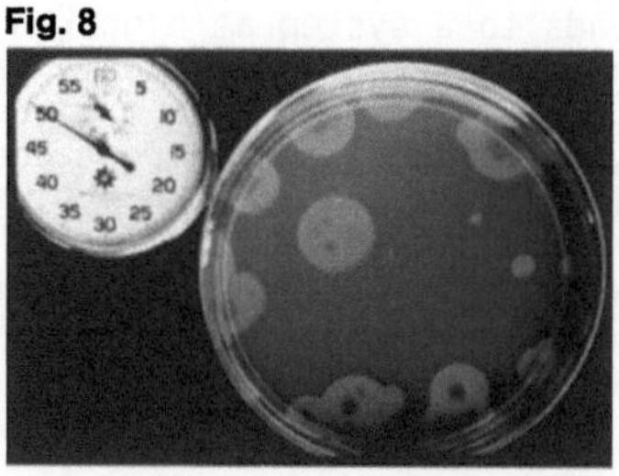

Fig. 9

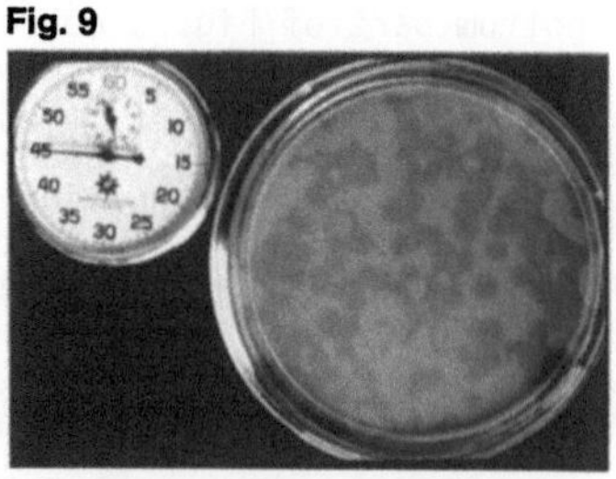

Figure 3 (plates 1 to 3) represents another set of initial concentrations, also at high $[MA]_0$ but with shorter bulk period, 93.2 s instead of 296 s. The experimental conditions are given in Table 1. This system will illustrate another type of acceleration of the wavefronts. Besides the large, obvious acceleration of the outermost wavefronts, there is also a subtle, hard to detect, acceleration of the inner wavefronts. Consider the central center in Fig.3. In plates 1 and 2, the system is going through its second bulk oxidation. The central center has already propagated five wavefronts (one of them annihilated at the first bulk oxidation). Here, the speeds of the inner wavefronts show a small acceleration. Near and during the second bulk oxidation (plates 1 and 2), the speeds of the wavefronts are, from the outermost to the innermost, 0.114, 0.092, 0.081 and 0.081 mm s^{-1}. At the next bulk oxidation, 93 seconds later (plate 3), the fastest external wavefront has been annihilated and the others now propagate with speeds of 0.092 (no change), 0.093 (15% increase) and 0.087 (7% increase) mm s^{-1}. Because of the small acceleration, these waves are undefined in the two-wave theory. However, since the acceleration is hard to detect, they would easily be confused with "trigger waves." There are no such identity problems with the one-wave theory.

Finally, we will describe the peculiar properties of the Zhabotinskii waves near the low $[MA]_0$ bifurcation. Figure 4 (plates 1 to 12) contains only 0.024 M malonic acid initially. This may be compared to the higher concentration of 0.080 M in the two previous systems (Fig. 2,3). Other initial concentrations are listed in Table 1. To understand the system of Fig. 4, we have to remember that malonic acid is consumed during the Zhabotinskii reaction. One of its end products is CO_2 gas whose bubbles are clearly seen in the last plates. The system of Fig. 4 starts with a $[MA]_0$ of 0.024 M which is slightly above the bifurcation value. We observe the last oscillatory oxidation in plate 2. At the next oxidation, in plate 4, the oxidized steady-state becomes stable. Only local oscillations are still taking place and they originate at reduced (red) centers. There are many such centers while there were only a few blue centers earlier in the same system. We observe a reducing wavefront (RW) growing steadily as a red disk in the oxidized, blue background (plates 4-10). Within this disk, oscillations continue and a blue center appears, very nearly superimposed on the original red center (plates 7,8). The reducing wavefront (red moving into blue) continues its propagation, followed now by an oxidizing wavefront (OW, blue moving into red) (plates 7-10). One observes a new property of the Zhabotinskii waves: the speeds of the reducing and oxidizing wavefronts are not equal. Because the speed of the OW is larger, the OW catches up with the reducing wavefront (plates 9-11). In this process, the OW declerates until its speed is almost equal and disappears (plate 11). In many cases, a few more oscillations take place at the location of the red center which reappears, propagating a new reducing wave followed by another oxidizing wave (plates 9-12). In the end, the system is entirely oxidized.

$[MA]_0$ has a strong effect on the speed of the reducing wavefronts while it does not seem to change the speed of the oxidizing ones (Fig. 5). At high $[MA]_0$, both wavefronts have the same speed. At lower $[MA]_0$, the speed of the reducing wavefront drops rapidly. (Note that, in Fig. 5, only the average speed is given, without indication of standard deviations. In the case of decelerating OW's only the maximum speed is indicated.) Further information on the propagation of OW's and RW's is given in Fig. 6. The radius of the wavefronts is plotted as a function of time. RW designates the reducing wavefront which moves into a blue, oxidized region, leaving a red, reduced region in its wake. The opposite is true for an OW. Along the abcissae, we have indicated the red and blue regions by BLUE (or B) and RED. The bottom part of Fig. 6 corresponds to a system at high $[MA]_0$: the reducing wavefront follows the oxidizing one with equal speed. The middle part of Fig. 6 represents an intermediate case: $[MA]_0$ becomes low enough to produce a speed difference between the OW and RW. However, the oxidizing wavefront does not decelerate because of a timely bulk reduction which leaves most of the system in a reduced state. The results at the top of Fig. 6 are qualitatively similar to those in Fig. 4. A typical case of deceleration of the OW and elimination of the RW is illustrated.

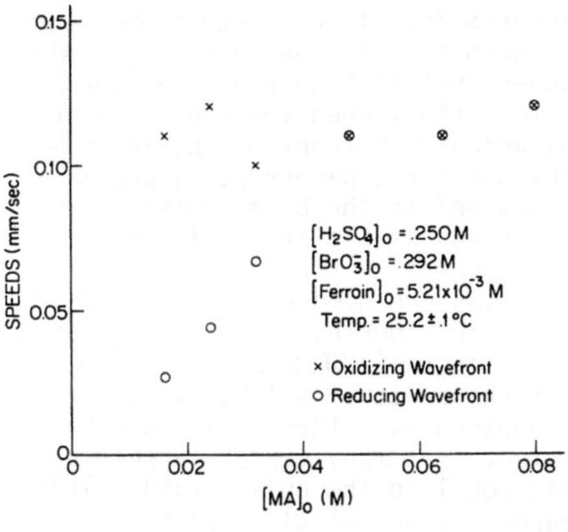

◄ Fig. 5 Speeds of the oxidizing (x) and reducing (o) wavefronts as a function of the initial concentration of malonic acid, $[MA]_0$. Error limits omitted for clarity.

Fig. 6 (below) Time dependence of the radii of oxidizing (OW's) and reducing (RW's) wavefronts in three systems. $[MA]_0$ decreases from bottom to top. (See Table 1 for conditions.)

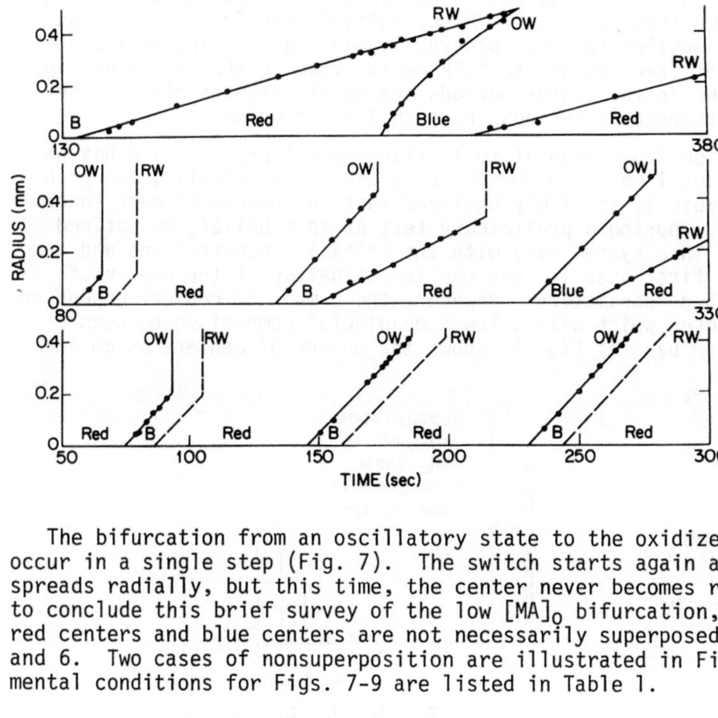

The bifurcation from an oscillatory state to the oxidized steady-state may also occur in a single step (Fig. 7). The switch starts again at local centers and it spreads radially, but this time, the center never becomes reduced again. Finally, to conclude this brief survey of the low $[MA]_0$ bifurcation, we must clarify that the red centers and blue centers are not necessarily superposed, as they are in Figs. 4 and 6. Two cases of nonsuperposition are illustrated in Figs. 8 and 9. The experimental conditions for Figs. 7-9 are listed in Table 1.

3. From Temporal to Spatio-temporal Organization: The Centers

After adequate homogenization of the distributed system, the centers of propagating waves are punctual. For this reason, their formation is often attributed to the presence of (punctual) dust or bubbles [15], [16], [19]. However, in absence of special care, the desynchronization appears as broad patches and thin filaments which initiate spirals rather than circular waves. These patches and filaments argue against the importance of dust in the desynchronization process. Also, we have recently verified that scratches in the Petri dish and CO_2 bubbles have little, if any, effect on the initiation of wave formation. This confirms earlier reports by Zhabotinskii [28]. Nevertheless, from time to time, one observes a center which is clearly due to dust and there is no doubt that dust may sometimes start a wave. Although the manipulations are not entirely dust-free, maximum care has been exercised to avoid dust contamination in the experiments.

Let us return to Fig. 2 and discuss some characteristics of the centers observed there. The blue center at 11 o'clock appears 132 seconds after the last synchronization (by stirring). It has a period of 214 seconds, not much shorter than the bulk period (296 s). This center illustrates the fact that, when the center period is only slightly shorter than the bulk period, the unique wavefront may be annihilated before the reapparition of the center. Here, the first wavefront is annihilated by the first fast accelerating wave 145 seconds before the blue center reappears (plage 5). During this time interval, there is no visible trace of the existence of a center at that location (plate 4). This process is repeated later. Because Tcenter is only 72% of Tbulk, several bulk periods are needed before multiple wavefronts with a given wavelength appear at this center. During this waiting time, we must refer to a virtual rather than an actual wavelength ($\lambda \sim$Tcenter x minimum speed). It should be understood that the period of the centers is taken as the average time interval between consecutive center appearances. These time intervals show some variation. Fig. 2 also illustrates the fact that many centers appear at the time of bulk oxidation and have a period almost equal to the bulk period. This is true of the centers appearing in plate 8 but barely or not at all apparent in plate 3 of Fig. 2. In Fig. 3, the system has a shorter bulk period (93.2 s). The number of centers is greater than in Fig. 2. Also, multiple wavefronts, and thus actual rather than virtual wavelengths are observed in many cases. The period of several centers is a smaller fraction of the bulk period than in the system of Fig. 2. Note that the variations in the center periods are mainly responsible for the differences in wavelengths; speed variations are much less important.

The spontaneous transition from temporal to spatio-temporal organization has not been studied in any depth until now. In the United States in particular, the problem has been neglected because it is widely believed that in absence of dust there are no spontaneous centers. During a preliminary test of this belief, we noticed that the number of centers in a system vary with the initial concentrations and temperature of the system. A first step is thus the investigation of the number of centers as a function of these parameters. However, the number of centers depend on time and a time must be chosen which will allow a meaningful comparison between various systems. The bottom part of Fig. 10 shows the number of centers which are

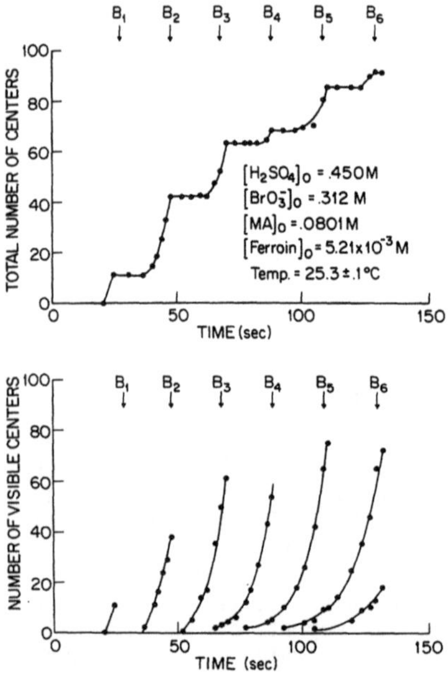

Fig. 10 Conditions listed above apply to Fig. 10 and Fig. 12 (bottom).

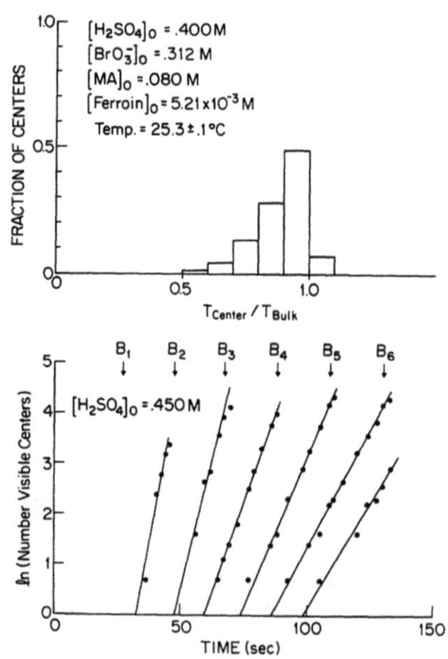

Fig. 12 See text.

visible as a function of time. The arrows B_1 to B_6 indicate the times at which successive bulk oxidations occur. The number of centers grows until the time of bulk oxidation when almost all centers "disappear." They reappear shortly thereafter and grow again. The process is periodic. Figure 11 shows the actual system before, during, and after the third bulk oxidation. In plate 1, 49 centers are "visible," due to the presence of at least one propagating wavefront. In plate 3, only 6 centers are still "visible." But the earlier centers are of course still present and they are again "visible" before the next bulk oxidation.

We may thus conclude that the meaningful variable is the total number of centers formed since the last synchronization rather than the number actually visible at any time. This variable is shown in the top part of Fig. 10. Note that the total number of centers increases with time but that the rate of formation of new centers is null immediately after the bulk oxidation. Returning to the number of visible centers, we see in Fig. 12 (bottom) that this number is an exponential function of time. The periodicity is also evidenced. However, with the increase in the number of bulk oxidations, the slope decreases. Exponentiality, periodicity, and decreasing slopes are observed very generally. These phenomena are a manifestation of the underlying frequency distribution of the center periods. An example of such a distribution is given in Fig. 12 (top) which shows the fraction of centers for each value of the ratio Tcenter / Tbulk in a given system. Note that the majority of centers have a period between 90 and 100% of the bulk period. (The centers listed with a period greater than the bulk period represent an artefact: the bulk period used in Fig. 12 (top) is an average value. The actual bulk period may be longer than the average one in some parts of the dish. In these parts, the center period itself may then be longer than the average bulk period.) Further work on the frequency distribution of center periods as a function of initial concentrations and temperature is in progress.

We will now mention briefly some results on the total number of centers in systems with various initial concentrations or temperature. All systems are compared at the time of their third bulk oxidation after the last synchronization. In Fig. 13, we have plotted the total number of centers as a function of the angular bulk frequency (2π / Tbulk). Each point represents a system with a different set of initial concentrations or temperature. In parts 1 and 2, only the initial concentrations of sulfuric acid (H_2SO_4) are varied. The difference in slopes for part 1 and 2 is due to different concentrations of bromate ions ($[BrO_3^-]_0$ = 0.312 M in part 1 and 0.250 M in part 2). $[MA]_0$, $[Ferroin]_0$ and temperature are identical in both series (see Table 1). The slopes are 150 ± 24 s and 99 ± 22 s. In part 3, the concentrations of sodium bromate ($[BrO_3^-]_0$) are varied and a slope of 190 ± 31 s is found for the conditions listed in Table 1. Finally, in part 4, the temperature is varied while the initial concentrations remain constant. When the total number of

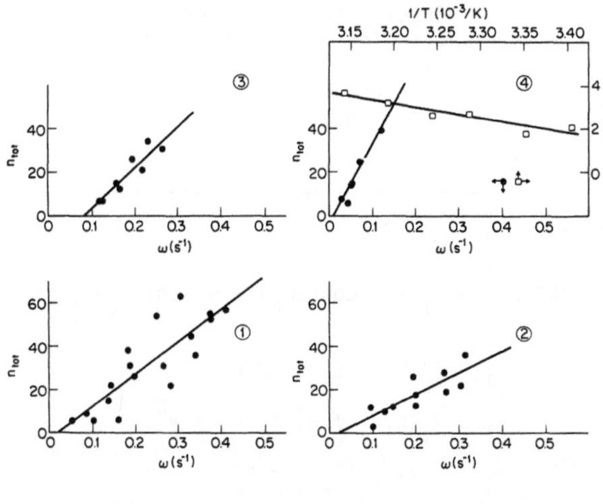

Fig. 13
part 1: Variations in $[H_2SO_4]_0$ with $[BrO_3^-]_0$ = 0.312 M

part 2: Variations in $[H_2SO_4]_0$ with $[BrO_3^-]_0$ = 0.250 M

part 3: Variations in $[BrO_3^-]_0$

part 4: Variations in Temperature

See Table 1 for other conditions

centers is plotted as a function of the bulk angular frequency, the slope is 376 ± 43 s. The same results have also been plotted as the logarithm of the number of centers versus the inverse of the temperature. The result can be expressed in terms of an "activation energy" (E_a),

$$n_{\text{total at } B_3} = n_o \exp(-E_a/RT)$$

with E_a = 55 kJ/mol and R, the gas constant. One must keep in mind, however, that the value of E_a varies from bulk oxidation to bulk oxidation and is mostly of qualitative interest at this point. A knowledge of the frequency distribution of center periods at various temperatures will probably lead to a better understanding of the temperature effect. This research is still in progress and we will only make a few comments here. We think that the temperature dependence (in part 4) indicates a correlation between the noise (or fluctuation) level and the number of centers in the system. It eliminates any naive, one-to-one correlation between dust and number of centers. The experiments in which concentrations are varied show that a high bulk frequency favors the formation of centers. However, the different slopes obtained in parts 1, 2, and 3 of Fig. 13 indicate that another factor must be involved, such as the amplitude of the oscillations. We plan to test this hypothesis in the near future.

B. Some Theoretical Results

The Bautin model is the Ising model of chemical oscillations. We cannot describe this system in detail here, but several previous articles [2], [10], may be consulted. Here, we will simply introduce some new results such as the asymmetric version of the Bautin system and mention some previous results of particular interest. One of them is the saddle-node transition from limit cycle to bistability (or monostability in the asymmetric model). There will be a short note on frequency modulation, double periodicity and interrupted oscillations in a non-autonomous version of the Bautin model. Finally, we will summarize our results concerning chemical waves in a distributed Bautin system. We will show that some restrictions introduced in 1973 [13] can and should be removed entirely; this model for chemical waves includes diffusion and ignores the two-wave theory.

1. Symmetric Bautin Model [2,10]

The rate equations are expressed here in terms of relative concentrations $X \equiv X'-X_o$, $Y \equiv Y'-Y_o$, where X' and Y' are the actual concentrations of intermediates. We have

$$\frac{dX}{dt} = aY + E_1X - X^3 - XY^2 \qquad \frac{dY}{dt} = -aX + E_2Y - Y^3 - YX^2 \tag{2}$$

If $E_1 < X_o^2 - (a^2/4X_o^2)$ and $E_2 < Y_o^2 - (a^2/4Y_o^2)$, the actual concentrations will always be positive, as they should. At $X = 0$, $Y = 0$, a singularity exists for any value of a, E_1, E_2. These parameters define several dynamic regions (Fig. 14). If $|E_2 - E_1|<|2a|$, we are in region C or E and the system oscillates. For $(E_2 + E_1)>0$, in region C, the oscillations are sustained and independent of the initial values of X and Y. In the asymptotic limit, we have a limit cycle given by

$$\lim_{t\to\infty} X^2(t) = \frac{\cos^2(C_o - az_2t)}{(\alpha_1 + \beta_1\cos(2C_o - 2az_2t) + \gamma_1\sin(2C_o - 2az_2t) + \exp(-(E_2 + E_1)t)((\cos^2C_o/X^2(0)) - \alpha_1 - \beta_1\cos(2C_o) - \gamma_1\sin(2C_o)))}$$

with $C_o \equiv \tan^{-1}((\tan\phi(0) - z_1)/z_2)$ $\qquad \alpha_1 \equiv 2/(E_2 + E_1) \equiv 2/\mu^2$

$z_1 \equiv (E_2 - E_1)/2a$ $\qquad \beta_1 \equiv z_1(z_1E_1 + a)/(E_2E_1 + a^2)$

$z_2 \equiv (1 - z_1^2)^{1/2} = \omega/a$ $\qquad \gamma_1 \equiv z_1z_2E_1/(E_2E_1 + a^2)$

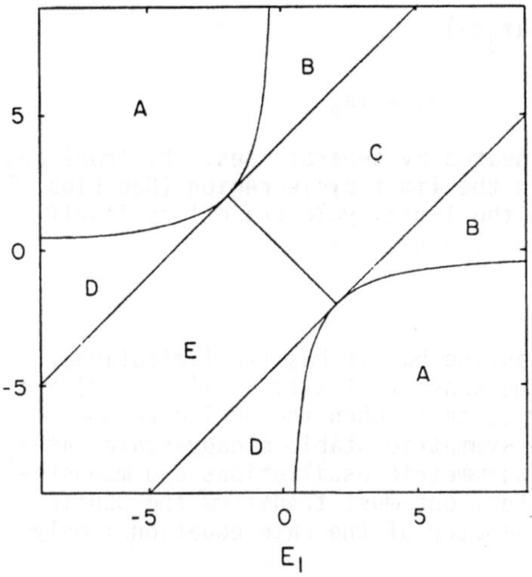

Fig. 14 Symmetric Bautin. Stability regions in parameter space (a=2)

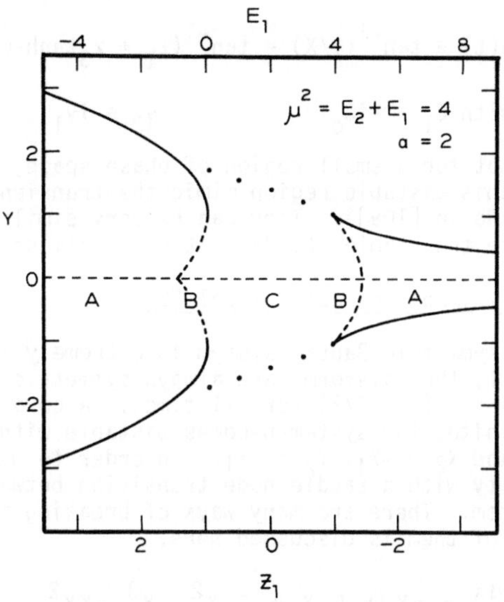

Fig. 15 Symmetric Bautin. Bifurcation diagram

In general, for nonlinear oscillations, the phase is not a linear function of time.

$$\phi(t) \equiv \tan^{-1}(Y/X) = \tan^{-1}(z_1 + z_2\tan(C_o - az_2t))$$

It has been determined that this limit cycle is an ellipse [10c]. The period and amplitude depend on the parameters a, E_1, E_2.

$$T = 4\pi/\left(4a^2 - (E_2 - E_1)^2\right)^{1/2} \tag{3}$$

$$Y^2_{max} = \left(\mu^2(\mu^4 \pm 2(a^2 - \omega^2)^{1/2}\mu^2 + 4a^2)\right) / \left(2(\mu^4 + 4a^2)\right)$$

with the upper sign for $z_1 > 0$. X^2_{max} is given by the same expression but with the lower sign for $z_1 = 0$.
When $z_1 = 0$, $\omega = a$ and $X^2_{max} = Y^2_{max} = \mu^2/2$. When $|z_1|$ varies from 0 to 1, the oscillations change from sinusoidal to relaxational. For $|z_1| = 1$, on the C - B boundary, the period is infinite. For $|z_1| > 1$, the system is bistable with five singularities in region B and three in region A. Fig. 15 is a bifurcation diagram which shows the values taken by Y in various steady-states when z_1 is varied. In region A, we have two stable nodes and one saddle point. In region B, two new singularities appear. We have two stable nodes, one unstable node and two saddle points. The unstable steady-states are represented by dashed lines, the stable states by solid lines. The dots in region C represent the amplitude ($\pm$) Y_{max} of the limit cycle oscillations.

Since the symmetric Bautin system is solvable in closed form, we also know the transients in the bistable system. This might be the only case of an analytic bistable system in chemical instabilities. In the bistable region with 5 singularities (region B), we have

$$X^2(t) = \frac{\cosh^2(C_1 + az_3t)}{\begin{array}{l}(\alpha_1 + \beta_1\cosh(2C_1 + 2az_3t) + \gamma_3\sinh(2C_1 + 2az_3t) \\ + \exp(-(E_2 + E_1)t)((\cosh^2 C_1/X^2(0)) - \alpha_1 - \beta_1\cosh(2C_1) - \gamma_3\sinh(2C_1)))\end{array}}$$

$$\phi(t) \equiv \tan^{-1}(Y/X) = \tan^{-1}(z_1 + z_3\tanh(C_1 + az_3t))$$

with $C_1 \equiv -iC_o$ $\gamma_3 \equiv i\gamma_1$ $z_3 \equiv iz_2$

Except for a small region of phase space, delineated by separatrices, the transients in this bistable region mimic the transients in the limit cycle region (See Figs. 1 and 9a in [10a]). They can be very similar to the limit cycle trajectory itself since the stable steady-states are almost on the limit cycle path.

2. Asymmetric Bautin Model [2]

The symmetric Bautin system is extremely interesting but it has two limitations. First, the waveforms are always symmetric in the sense that $X(t) = -X(t + T/2)$ and $Y(t) = -Y(t + T/2)$ for all times. A corollary is that, when the period becomes infinite, the system becomes bistable with two symmetric stable steady-states at X_1, Y_1 and $X_2 = -X_1$, $Y_2 = -Y_1$. In order to model asymmetric oscillations and monostability with a saddle-node transition between them, one must transform the Bautin system. There are many ways of breaking the symmetry of the rate equations; only one of them is discussed here.

$$\frac{dX}{dt} = aY + E_1X + C_1X^2 - X^3 - XY^2 \qquad \frac{dY}{dt} = -aX + E_2Y - Y^3 - YX^2$$

This will be called the C_1-asymmetric Bautin sysem. In Fig. 16 we show the changes due to an increase in C_1 from 0 (top) to 0.8 (bottom). The middle of Fig. 16 corresponds to $C_1 = 0.4$. The waveforms $X(t)$ are shown on the extreme right; the phase plane trajectories are in the middle section. On the extreme left, the nullclines are shown in the phase plane. When these nullclines intersect, we have a steady-state, by definition. For $C_1 = 0$ and 0.4, there is only one steady state

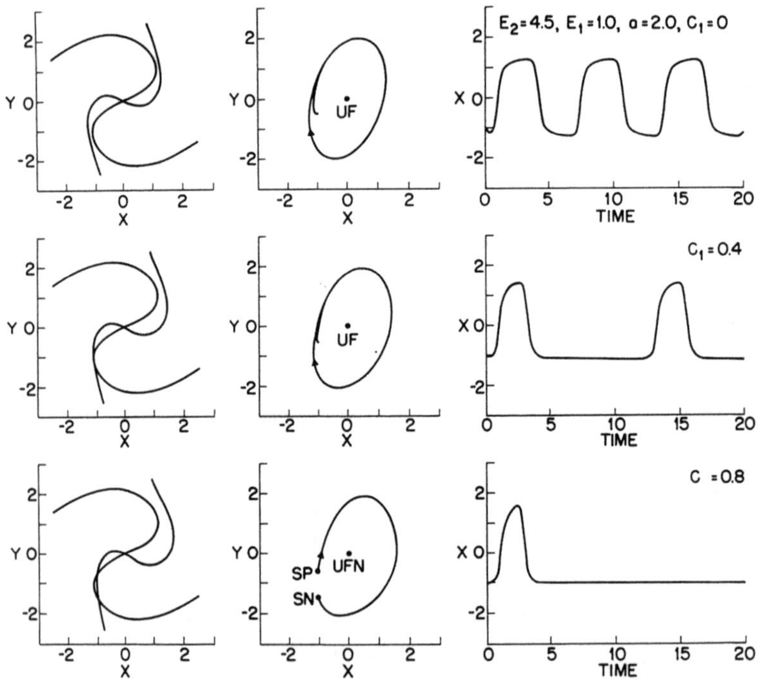

Fig. 16 C_1-asymmetric Bautin model. From top to bottom: C_1 = 0, 0.4, 0.8

(unstable focus) at X=Y=0. For $C_1 = 0.8$, the nullclines intersect three times. Only one of these three steady-states is stable; the system is monostable. As in the case of the symmetric Bautin system, in the infinite period limit, we have a saddle-node transition from oscillations to a stable steady-state. Note the similarity between the oscillatory trajectory and waveform for $C_1 = 0.4$ and the monostable case for $C_1 = 0.8$. When such a system is encountered in the laboratory, it is not a trivial problem to distinguish a case of oscillations with a long period from a monostable case. This type of difficulty seems to occur with the Zhabotinskii system in the so-called "excitable" region [21]. Note also that the saddle-node transition occurring between $C_1 = 0.4$ and $C_1 = 0.8$ corresponds to a transition between full blown oscillations and a stable steady-sate. A sudden start (or end) of finite amplitude oscillations is also observed in the experimental Zhabotinskii system (see Fig. 8 in [2]).

3. Nonautonomous Bautin Model [10c]

In continuously stirred tank reactors (CSTR), oscillatory systems may show complex oscillations which may or may not be of the chaotic type. In glycolysis, double periodicity has been known for a long time [29]. In the Zhabotinskii system, there are reports of interrupted oscillations [30], and more recently of possible chaos [31]. It must be emphasized that no such complex oscillations have been observed in the batch reactor for the Zhabotinskii system. The period of oscillations may be quite irregular [2] but the waveforms remain simple. In 1975, we showed that double periodicities and interrupted oscillations can be modelled with a nonautonomous variation of the symmetric Bautin system. These phenomena can be reproduced by allowing periodic coefficients in the rate equations. In the simplest analytic case, there is only frequency modulation. The constant parameters of (2) are replaced by

$$E_2(t) = e_2 + \varepsilon z_1 \sin\omega_p t \qquad E_1(t) = e_1 - \varepsilon z_1 \sin\omega_p t \qquad a(t) = a_o + \varepsilon \sin\omega_p t$$

This can be solved for the phase

$$\phi(t) \equiv \tan^{-1}(Y/X) = \tan^{-1}(z_1 + z_2 \tan(C_o - z_2 \int_o^t a(t)dt))$$

and, if $e_2 = e_1$, one can also solve for X and Y

$$Y(t) = e_1^{1/2} \sin(C_o - z_2 \int_o^t a(t)dt) \qquad X(t) = e_1^{1/2} \cos(C_o - z_2 \int_o^t a(t)dt)$$

If $e_1 \neq e_2$, there is frequency and amplitude modulation, with auto-crossing of the trajectory in phase plane. For more details, one may consult [10c]. Fig. 3 of [10c] gives the results for double periodicity and interrupted oscillations.

4. Distributed Bautin Model [10a], [13], [14]

This model was proposed in 1973 as a one-wave [14] interpretation of the Zhabotinskii waves. It is in the spirit of a simpler but remarkably good model introduced by ZHABOTINSKII himself in 1970 [4] when he first reported the two-dimensional waves. When our model was first developed, we thought that some restrictions were required concerning the diffusion coefficients of the various compounds. These restrictions have now been removed, as they were never really required. The distributed system is described by:

$$\text{(a)} \quad \frac{\partial X}{\partial t} = aY + E_1 X - X^3 - XY^2 + D_x \nabla^2 X \qquad (4)$$

$$\text{(b)} \quad \frac{\partial Y}{\partial t} = - aX + E_2 Y - Y^3 - YX^2 + D_y \nabla^2 Y$$

$$\text{(c)} \quad \frac{\partial E_2}{\partial t} = D_{E2} \nabla^2 E_2$$

Eqs. (4a, 4b) are in agreement with the tradition of dissipative structures [8]; they include chemical reaction and diffusion. Eq. (4c) however, introduces a new feature in order to take some experimental facts into account. Many centers of propagating waves oscillate with a period shorter than the bulk period (Fig. 12). We assume that this shorter period is due to an instantaneous, local perturbation in some factor affecting the bulk period (e.g., concentrations, temperature, etc.). As a first approximation, one can model the formation of centers by introducing perturbations in the Bautin parameters a, E_1, or E_2, since they control the period (3) and they represent products of parametric concentrations and temperature-dependent rate constants [10a]. These perturbations correspond to instantaneous, pulse-like, local variations in the concentration E_2. They propagate by diffusion to the entire system (4c). Due to a misunderstanding, numerous reviews [15b], [17], [32] mention the distributed Bautin model as one that does not involve diffusion! The waves obtained have been called kinematic waves, pseudo-waves, or WFN phase waves [14]. This is an error: in the early model, D_X and D_Y were null in order to save computer time, but D_{E2} was finite. We have now simulated the waves with equal and unequal diffusion of all compounds. Work is in progress in which we use also temperature perturbations, heat conduction, and differences in heat production [26] to model the different speeds for the OW's and RW's observed at low $[MA]_0$.

When even the simplest version of the distributed Bautin system (4) is integrated numerically, all features of the Zhabotinskii waves at high $[MA]_0$ (including the spirals) are reproduced qualitatively. Figs. 2 and 3 in [13] should be compared with the Zhabotinskii model [4].

Conclusion

In September 1978, we wrote [11b] that "in spite of the extreme ease with which (the Zhabotinskii) waves are obtained, the experimental work is lagging behind the theoretical work on this subject." With the new experimental results above, the outlook in this field is radically changed. We are now faced with a great wealth of new experimental results which raise numerous questions on which further theoretical work is required. These new observations increase considerably the interest in the Zhabotinskii system as an example of chemical synergetics.

Table 1 Initial concentrations, temperature, and bulk period

Figure	$[BrO_3^-]_0$/M	$[H_2SO_4]_0$/M	$[MA]_0$/M	Temp./°C	Tbulk /s
2	0.317	0.150	0.080	25.0	296.0
3	0.229	0.325	0.080	24.9	93.2
4	0.358	0.200	0.024	room T	$\to \infty$
6 top	0.292	0.250	0.016	25.2	$\to \infty$
6 middle	0.292	0.250	0.032	25.2	109.2
6 bottom	0.292	0.250	0.048	25.2	89.6
7	0.250	0.250	0.008	25.1	$\to \infty$
8	0.358	0.200	0.024	24.5	$\to \infty$
9	0.250	0.350	0.024	24.9	$\to \infty$
11	0.312	0.500	0.080	25.3	16.8
13 (1)	0.312	0.200-0.500	0.080	25.2	15-121
13 (2)	0.250	0.300-0.550	0.080	25.2	13-66
13 (3)	0.187-0.312	0.400	0.080	25.2	24-52
13 (4)	0.312	0.200	0.080	20.5-45.0	88-204

$[Ferroin]_0 = 5.21 \times 10^{-3}$ M

Acknowledgments

I am grateful to the Chemistry Department at the University of Michigan and especially to Professor T. M. Dunn for their hospitality and support.

References

1 A.M. Zhabotinskii, A.N. Zaikin, M.D. Korzukhin, G.P. Kreitser, Kinet. and Cat. 12, 516-521 (1971)

2 M-L. Smoes, "Period of homogeneous oscillations in the ferroin-catalyzed Zhabotinskii system" to appear in J. Chem. Phys. (tentative issue: Dec. 1, 1979)

3 H. Haken, "Synergetics: An Introduction" 2nd ed. Springer Series in Synergetics (Springer Verlag, N.Y.) (1979)

4 A.N. Zaikin and A.M. Zhabotinskii, Nature London 225, 535-537 (1970)

5 D. Edelson, R.M. Noyes and R.J. Field, Int. J. Chem. Kinet. 11, 155-164 (1979) Note: 1. The mechanism proposed here is for the cerium-catalyzed Zhabotinskii system. 2. See also a critique of this mechanism in A.B. Rovinskii and A.M. Zhabotinskii, Theoret. and Exp. Chem. 14 (2) 142-150 (1978) Translation from Russian.

6a R.J. Field and R.M. Noyes, J. Chem. Phys. 60, 1877-1884 (1974)
b R.J. Field, J. Chem. Phys. 63 (6), 2289-2296 (1975)
c K. Showalter, R.M. Noyes and K. Bar-Eli, J. Chem. Phys. 69 (6), 2514-2524 (1978)

7 "Synergetics: Far from Equilibrium" Eds. A. Pacault and C. Vidal (Springer-Verlag, N.Y.) See article by A. Pacault, pp. 128-146 (1979)

8 G.Nicolis and I. Prigogine, "Self-Organization in Nonequilibrium Systems. From Dissipative Structures to Order through Fluctuations" Wiley-Interscience (1977)

9 A.A. Andronov, A.A. Vitt and S.E. Khaikin, "Theory of Oscillators" Addison-Wesley (1966) Note: The Bautin system is discussed on p. 336-340. A saddle-node transition is illustrated in Fig. 318

10a M-L Smoes, Ph.D. Thesis, University of Colorado (1973)
b J. Dreitlein and M-L. Smoes, J. Theor. Biol. 46, 559-572 (1974) Note: Fig. 1 is incorrect. See references [10a] and [10c] for correct waveforms
c M-L. Smoes, "Proceedings of the International Conference on Nonlinear Oscillations (1975) Abhandlungen der Akademie der Wissenschaften der DDR." p. 385-390

11a M-L. Smoes, Bull. Amer. Phys. Soc. 23 (4) 534 (1978)
b M-L. Smoes, Abstracts of Papers, 176th ACS National Meeting, Miami Beach, Florida, Sept. 1978

12 M-L. Smoes, Abstracts of Papers, 175th ACS National Meeting, Anaheim, California, March 1978. Also, paper in preparation

13 M-L. Smoes and J. Dreitlein, J. Chem. Phys. 59, 6277-6285 (1973)

14 M-L. Smoes, Bull. Amer. Phys. Soc. 24 (3), 477 (1979) See Abstracts KQ12 and KQ14

15a A.T. Winfree, Science, 175, 634-636 (1972)
b A.T. Winfree, Lecture Notes in Biomathematics 2, 241-260 (1974) (Mathematical Problems in Biology, Victoria Conference, Ed. P. van den Driessche, Springer-Verlag, N.Y.) Note: Winfree's comments on [13] are incorrect and misleading. There is no way of classifying [13] in terms of the two-wave theory since it is essentially a denial of such a theory.)

16 E.J. Reusser and R.J. Field, JACS 101, 1063-1071 (1979). This paper is an example of the increasing problems encountered by the two-wave theory.

17 R.M. Noyes and R.J. Field, Ann. Rev. Phys. Chem. 25, 95-119 (1974). The note in [15b] applies here too.

18 See references [10a], [13], [14]

19 R.J. Field and R.M. Noyes, Acc. Chem. Res. 10, 214-221 (1977)

20 "Theoretical Chemistry. Advances and Perspectives: Periodicities in Chemistry and Biology" eds. H. Eyring and D. Henderson (Academic Press) vol. 4 (1978)

21 Our oscillatory systems are apparently different from those used by Winfree, Field and Noyes [16]. These authors claim that their systems are excitable and nonoscillatory. However, it should be noted that repeated bulk oxidations do occur in the "nonoscillatory" systems used by these authors. This might indicate the presence of bulk oscillations with long and irregular period. See also R.J. Field and R.M. Noyes, Faraday Chem. Soc. 9, 21 (1974) and M-L. Smoes, ibid. p. 85

22 H.E. Stanley "Introduction to Phase Transitions and Critical Phenomena." Oxford Univ. Press, N.Y. (1971)

23 After carrying an experimental check, we are satisfied that scratches have no significant effect on the number of centers of waves.

24 D. Thoenes, Nature Phys. Sc. 243, 18-20 (1973)

25 When the center period is longer than the bulk period, the wave is moving inwards, toward the center. Such a wave could be observed only at the time of bulk oxidation and is thus much more difficult to detect. See [10a], [13]

26 E. Körös, M. Orban, and Zs. Nagy, Nature Phys. Sc. 242, 30-31 (1973)

27 See reference 14, abstract KQ14 for three different definitions of the expression "phase waves"

28 A.M. Zhabotinskii and A.N. Zaikin, J. Theor. Biol. 40, 45-61 (1973)

29 B. Hess and A. Boiteux, in Järnefelt J., Ed. "Regulatory Functions of Biological Membranes" Elsevier, Amsterdam (1968) p. 148

30 A.N. Zaikin and A.M. Zhabotinskii in "Biological and Biochemical Oscillators" eds. B. Chance, E.K. Pye, A.K. Ghosh, and B. Hess (Academic Press) See Fig. 2, p. 83 (1973)

31 O.E. Rössler and K. Wegmann, Nature, 271, 89-90 (1978)

32 A.T. Winfree, Faraday Symp. Chem. Soc. 9, 38-46 (1974)

Propagating Waves and Target Patterns in Chemical Systems

P.C. Fife

Mathematics Department, University of Arizona
Tucson, AZ 85721, USA

1. Introduction

The discovery of propagating waves of various types in chemical reagents has provoked a great deal of research, during the last ten years, into the phenomenology and the underlying mechanisms for such wavelike activity. The research has been performed by natural scientists and mathematicians alike. Most of it has been experimental, but much computer simulation and mathematical analysis has also been done. Chemical wave activity is believed to be prevalent in biological organisms, but the most readily accessible reagent for laboratory study is that discovered by Belousov and Žabotinskiĭ. This mixture has oscillatory or excitable kinetics, depending on the concentrations of the various chemicals in the solution. Both of these regimes have at least two natural time scales: During one period of an oscillation or during one excited "excursion", most of the variation in the concentration of the reactants occurs within a brief interval of time. The time scale associated with this brief spurt of activity is much shorter than that associated with the slow variation which occurs before and after. This is well known from experiment and computation, and is evident from scaling analyses of model kinetic equations performed in [1] and elsewhere. Spatial structures are also prevalent in unstirred layers of this reagent ([23]; [2], [22], [24], and references therein). Target patterns (expanding concentric circular waves) are among the most prevalent of these structures. Here again, disparate space and time scales are evident from computer simulation of propagating waves [3].

It is natural, therefore, to use multiple scaling techniques when attempting to reduce wave and pattern phenomena to mathematical analysis. The study of propagation phenomena in excitable media such as biological membranes has indeed profited from this use [4-8]. In general reaction-diffusion settings, these methods were pursued in some detail in [9,10,26]. Nevertheless, their full implications have yet to be determined. In particular, their use in analyzing target patterns has been neglected. The purpose of the present paper is to explain some basic principles in the analysis of wave fronts by scaling techniques, and to discuss the application of these principles to the task of modeling target patterns.

We begin with a brief account, in sec. 2, of the fundamental theory of wave fronts for scalar nonlinear diffusion equations, as these are the component parts of the more complex wave phenomena to be examined in later sections. A scaling method to reduce the study of more complex propagating fronts to that of scalar fronts is elaborated in sec. 3. In 1974, Winfree [11] suggested that at least two broad categories of chemical waves exist: phase and trigger waves (see also [3]). We can see such a distinction very clearly in the context of wave fronts associated with relaxation oscillatory, (or even excitable) kinetics; this distinction is explained and discussed at some length in sec. 4.

Besides treating the existence and properties of chemical structures, one may wish to inquire how they might arise in a reagent in the first place. Two such mechanisms are detailed in sec. 5. Sec. 6 takes up the problem of modeling target patterns. It is found that the theory of sharp wave fronts, as developed in the preceding section, is very useful here. These patterns are constructed from a series of propagating circular trigger-type wave fronts, generated periodically at one point. At first glance, the idea is conceptually clear; but some nontrivial difficulties arise when a mathematical analysis is performed. For example, it is found that self-sustaining targets of this type cannot exist unless at least three reacting components are present. The roles of, and distinction between, oscillatory, excitable, and bistable kinetics are discussed in Sec. 7. This is relevant because the Z-reagent (the one mentioned above) can exist in these three regimes, and developing target patterns exist which turn a quiescent medium (with excitable kinetics) into an oscillating one.

Other writers ([12,13], and especially Kopell and Howard [14]) have studied target patterns within the context of λ-ω systems. An approach via Padé approximants is in [25]. These approaches are entirely different from that discussed here. The reader interested in the Z-reagent can profit from the books by Žabotinskiǐ [15] and by Tyson [16]. Some of the material in this paper was presented from a different point of view, and in more mathematical detail, in [17]. Some of the results announced here represent joint work with R. Smock; some others are joint with J. Tyson. I wish to thank M. Marek for brining to my attention papers [4] and [7].

2. Scalar fronts

Here we review some basic facts about wave front solutions $u = U(x-ct)$ of scalar nonlinear diffusion equations

$$u_t = u_{xx} + f(u) \tag{1}$$

where f has two zeros: $f(U_1) = f(U_2) = 0$. Under fairly general circumstances, there exist fronts satisfying

$$U(-\infty) = U_1,\ U(\infty) = U_2. \tag{2}$$

Two important cases arise in applications:

(a) $f'(u_1) < 0$, $f'(U_2) < 0$, and f has only one intermediate zero between U_1 and U_2;

(b) $f(u) \neq 0$ for u in the interval between U_1 and U_2.

In case (a), there exists a unique velocity c and a profile $U(z)$, unique up to shifts in z, satisfying (2), such that $U(x-ct)$ satisfies (1) [18]. This front is very stable; if it is perturbed by any bounded function whose bound does not surpass a certain known constant, then the resulting solution of (1) evolves, as $t \to \infty$, back to the same front (possibly shifted by a certain amount) [19].

In case (b), on the other hand, there exists a whole range of possible front velocities [20]. For example, suppose $f \geq 0$ and $U_1 > U_2$. Then there is a positive minimal speed c^* such that for any value c, $c^* \leq c < \infty$, there exists a unique (except for shifts) wave front solution $U(z)$, satisfying (2). These fronts are stable to small perturbations which are zero except in a finite interval, but they are certainly not stable to the same extent as those of type (a).

In any case, the front moves in such a direction that for each x, $u(x,t)=U(x-ct)$ approaches the "dominant state" as $t \to \infty$. The dominant state is defined to be the constant U_1 if $\int_{U_1}^{U_2} f(s)ds < 0$, and U_2 if this integral > 0. If the integral equals zero, then $c = 0$ and the front is stationary.

3. Decoupling and free boundary problems

In typical singular perturbation problems, a complex system may be reduced to several simpler ones by rescaling and exploiting the smallness of some parameter. The simpler problems may govern the solution in different parts of its domain of definition; thus there may be boundary layers versus regions of relatively slow variation.

Analogous situations arise in reaction-diffusion problems [9,10]. To illustrate this, we consider a system of $n = n_1 + n_2$ reacting and diffusing components, n_1 of them "fast" and the others not fast. Let $u = (u_1, \dots, u_{n_1})$ be the vector of fast components, and v the others. The system of RD equations in one space variable is of the form

$$U_t = \alpha D_1 U_{xx} + kf(u,v), \tag{3a}$$

$$_t = D_2 v_{xx} + g(u,V). \tag{3b}$$

Here D_i are (diffusion) matrices; $k \gg 1$ is a parameter expressing the fact that reactions affecting the concentration u are fast; and the parameter $\alpha \ll k$ is inserted to account for the possibility that the diffusion rate of u is small or large. The lowest order approximation, in regions where u_t and αu_{xx} are not large, is obtained by setting the coefficient of k equal to zero.:

$$f(u,v) = 0. \tag{4}$$

We assume that this equation can be solved for u in a nonunique manner: there are at least two functions $h_+, h_- : \mathbb{R}^{n_2} \to \mathbb{R}^{n_1}$, such that (4) holds when

$$u = h_{\pm}(v). \tag{5}$$

In other words, $f(h_+(v),v) \equiv f(h_-(v),v) \equiv v$.

We imagine that the x-t plane is partitioned into two parts $\Omega_{\pm}$, in which (5) holds (approximately) with the corresponding sign. Under certain circumstances, such a partitioning is possible, with sharp wave fronts forming the boundaries between the two domains.

To investigate this further, we scale x and t differently in the layer between Ω_+ and Ω_-. The scaling will be chosen so as to eliminate the parameters in (3a). Setting

$$\tau = kt, \quad \zeta = \sqrt{\frac{k}{\alpha}}\, x,$$

we reduce (3a) to

$$u_\tau = D_1 U_{\zeta\zeta} + f(u,v). \tag{6}$$

If we assume that v(x,t) varies smoothly across the transition zone between Ω_+ and Ω_-, then within this narrow zone, v may be treated as constant. In this case it is reasonable to suppose that (6), like (1), has a traveling front solution connecting the two known zeros of f, namely $h_-(v)$ and $h_+(v)$ (if $n_1 = 1$, the theory is governed by the considerations in sec. 2). Let us assume this is true, and that this front is a higher-dimensional analogue of the scalar front of type (a) in sec. 2. That is, we assume the velocity c and profile are uniquely determined from the parameter v in (6). Thus $u = U(\zeta - c(v)\tau\, ; v)$. This type of reasoning was used in [4], [9], and later papers.

In the original variables, $u = U(\sqrt{\frac{k}{\alpha}}(x - \sqrt{\alpha k}\, ct))$, revealing that the actual velocity is of the order $\sqrt{\alpha k}$, and the width of the front is of the order $\sqrt{\frac{\alpha}{k}} \ll 1$. Now let $x = y(t)$ denote the position of one such front. Knowing its velocity, we we may write

$$\frac{dy}{dt} = \sqrt{\alpha k}\; c(v(y,t)). \tag{7}$$

Suppose there is only one such front, and Ω_- lies to the left of it, with Ω_+ to its right. Then to lowest order, we have found that when v is known, u is determined by (5) in $\Omega_\pm$, and that the boundary between Ω_+ and Ω_- moves according to (7). Thus, it appears that we have uncoupled u from v.

This is true in that the problem reduces to one for v alone; but it is a free boundary problem. We must solve (3b) with the replacement (5) made, the "+" holding for $x > y(t)$, the "-" for $x < y(t)$, and the boundary $y(t)$ governed by a nonlinear differential equation (7).

If $D_2 = 0$ or is small enough to be negligible, the problem simplifies considerably.

In more general situations, the boundary between Ω_+ and Ω_- could consist of several wave fronts. And, of course, the extension of this reasoning to higher space dimensions is clear.

4. Phase and Trigger fronts.

We refer to the setting in sec. 3 with $n_1 = n_2 = 1$; corresponding phenomena in higher dimensions remain to be explored. So now u and v are scalar functions. The additional complication we impose is that $h_\pm$ are not defined for all values of v; rather their graphs lie on a nullcline of f as shown:

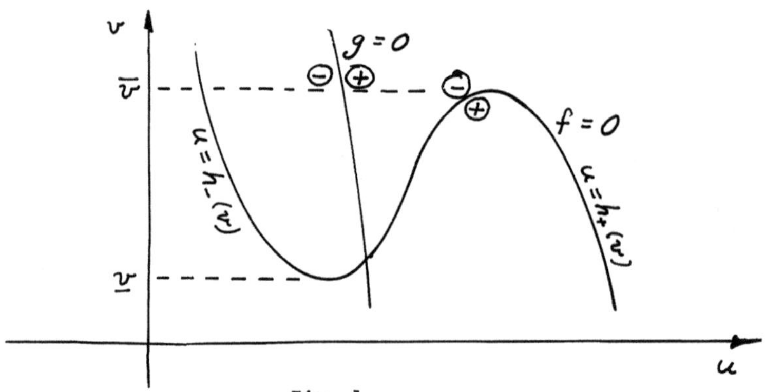

Fig. 1

For values of v in the interval $\underline{v} < v < \bar{v}$, $f(u,v)$ has the features of the function in (1) in case (a), provided $f_u(h_\pm(v),v) \neq 0$.

We assume, mainly for simplicity, that $D_2 = 0$ in (3b), and that $g(h_-(v),v) \equiv G_-(v) < 0$, $g(h_+(v),v) \equiv G_2(v) > 0$. If there is a single front with trajectory $y(t)$ and v varying continuously across it, then

$$\frac{\partial v}{\partial t} = \begin{cases} G_-(v) < 0 , & x < y(t) , \\ G_+(v) > 0 , & x > y(t) . \end{cases} \tag{8}$$

If $\underline{v} < v < \bar{v}$, $y(t)$ is governed by (7), because the function $c(v)$ in that equation comes from consideration of the bistable case ((a), sec. 2). Furthermore, if v is initially continuous at the front (as we assume), it must remain so. For otherwise, as the front passes a fixed value of x, v would change discontinuously in time, meaning that v_t would have a δ-function behavior. This is contradicted by (8): the right side, though discontinuous, is bounded.

Now suppose that $c = y' > 0$: the front is advancing into the region where $u \sim h_+(v)$. This will be the case when v is near $\underline{v}$, for then h_+ is the dominant state, according to the definition in sec. 2. Then the values of $v(x,t)$ ahead of the front $(x > y)$ determine the motion of the front: $y' = c(v(y,t))$.

It may happen, however, that $v(y(t)t)$ attains the minimal value $\underline{v}$ at some time t_0. At that point, v is prohibited from any further decrease: there can exist no front with $v < \underline{v}$. We must therefore have, at $t = t_0$,
$0 = \frac{d}{dt} v(y(t),t) = v_x(y(t) + 0,t)y'(t) + v_t = v_x{}^+ y' + G_+(\underline{v})$, where $v_x{}^+$ is the x-derivative of v at the front evaluated from the right. Hence

$$\frac{dy}{dt} = - G_+(\underline{v})/v_x{}^+ . \tag{9}$$

This relation replaces (7) when v attains the value $\underline{v}$, and in fact continues to hold as long as $v(y,t) = \underline{v}$.

We may inquire whether a front can exist when $v = \underline{v}$, for at this value of v, we are not in the bistable case (a). However, we are in case (b), as $f(u,\underline{v})$ is of one sign for $h_-(\underline{v}) < u < h_+(\underline{v})$, and fronts exist in this case as well. In fact, as we have seen, their velocity is arbitrary, subject only to a minimal value c^*. This means that one with velocity given by (9) is indeed possible.

We therefore have two types of propagation laws for fronts: (7) and (9). These types correspond to "trigger" and "phase" waves, in the terminology of Winfree [11]. To summarize, trigger fronts occur when $\underline{v} < v < \overline{v}$; their speed is determined by (7), where the function c comes from a law for scalar fronts, and is produced by the combination of diffusion and reaction. Phase fronts, on the other hand, occur when $v = \underline{v}$ or $\overline{v}$; their speeds are totally unaffected by either diffusion or the reaction term f. Rather (from (9)) they depend on the distribution of values of v ahead of the front (specifically, on $v_x{}^+$). In this sense, the motion of phase fronts is determined by the initial values of v.

5. The generation of fronts

We ask now, by what process may fronts be formed, if they do not originally exist. There are two such processes:

(i) (See [9].) Referring back to Fig. 1, suppose that, at time $t = 0$, the pair of functions $(u(x,0),v(x,0))$ map the x-axis on to a curve Γ in the (u-v) plane as shown:

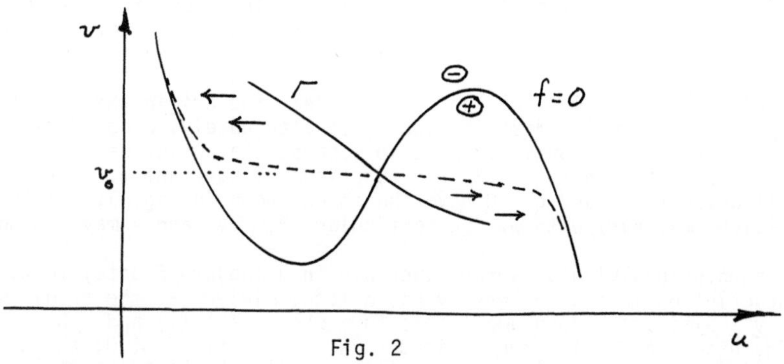

Fig. 2

Initially, the second term in (3a) is negligible compared to the last term, and we have $u_t \sim kf(u,v)$. This means that as time increases, u rapidly (since $k >> 1$) changes in such a manner that the phase-plane image is drawn from its initial position Γ to one or both of the stable descending branches $u = h_{\pm}(v)$. If Γ intersects the ascending intermediate branch (as shown) at some value $v = v_0$, then the evolved image curve is split between the two stable branches, as shown by the dotted line. At some point the x-variation in u will be abrupt enough that the first term on the right of (3a) is no longer negligible; and at this point, a sharp wave front in u is formed, with v varying continuously across the front, attaining the value v_0 at the front itself.

(ii) (See [10].) Suppose, on the other hand, that Γ does not intersect the the intermediate branch. Then the image is attracted entirely to only one of the stable branches, say h_-. After this happens, a slower process takes place, in which v evolves according to (3b) with u replaced by $h_-(v)$. Again for simplicity, assume $D_2 = 0$, so $v_t = G_-(v) < 0$. Eventually, for some $x = x_0$, v will attain a minimal value of $\underline{v}$. Further decrease of v causes the image to leave the branch $u = h_-(v)$ for x in a neighborhood of x_0. Then by the process described in (i) above, that part of the image curve is rapidly attracted to the other stable branch $u = h_+(v)$. This localized attraction to h_+ causes a pair of fronts, facing oppositely, to be formed near $x = x_0$. For each such front, h_+ will be the dominant state, so the fronts will move apart, increasing the interval on which $u \sim h_+(v)$.

6. Target patterns with trigger fronts

These are a series of concentric circular chemical waves, expanding outward, new ones regularly being generated at the center (usually called a "leading center" [21]). Such patterns have been observed in various forms of the Z-reagent ([23]; [2], [22], and references therein). Some of the patterns observed are associated with externally imposed heterogeneities at the center. We shall indicate in (i) below how such targets may be modelled. Self-sustaining target patterns, not dependent upon external stimuli, may also be modelled by the techniques discussed above; see (ii) below. Such models, based on sharp trigger fronts, require at least three reacting chemicals. We exclude consideration of phase fronts, as they would not account for the regular circular patterns observed [2].

All of the models we describe involve the same two basic phenomena: (a) spontaneous generation of wave fronts at the leading centers as described in sec. 5 (ii), and (b) their subsequent motion, according to the rules brought out in sec. 4. The generation process in 5 (ii) was for pairs of diverging fronts moving in one space dimension; its two-dimensional analog is the spontaneous appearance of a small circular front which spreads outward. Since the fronts are very narrow in our analysis, they appear locally as plane waves. Therefore it suffices to treat the problem in a one-dimensional framework, which we shall do. Alternatively, the variable x could be interpreted as distance to the origin in a configuration with radial symmetry.

(i) Imposed heterogeneities. We suppose that near the origin there is a substance with prescribed density distribution w(x); alternately, w could represent an imposed temperature distribution. We also suppose that w influences the reaction process, so that f and y in (3) are functions of u,v, and w. For each value of w, the nullcurves $f = 0$ and $g = 0$ have the shape shown in Fig. 1; but their relative positions may vary with w. In particular, $h_{\pm}$, $\overline{v}$, and $\underline{v}$ may depend on w.

At $x = 0$, fronts involving an abrupt increase in u (upjump fronts) form, according to the description in sec. 5, when v has a local minimum at the origin which decreases to $\underline{v}$. Similarly, downjump fronts form when $u = h_+(v)$ and v has a maximum which increases to $\overline{v}$. The result is that when x is fixed at 0, the trajectory $(u(o,t),v(o,t))$ is that of a relaxation oscillator with the kinetics of Fig. 1.

So the nullcurve $g = 0$ must be placed as shown, to ensure that the kinetics are oscillatory.

We wish to model targets made up of expanding trigger fronts; these correspond to the circular patterns observed in experiments; because phase waves, being essentially influenced by initial conditions, would only be circular by accident. This means their paths are described by (7), and

$$\underline{v}(w(x)) < v(x,t) < \overline{v}(w(x)) \ , \qquad (10)$$

at least at the front. The exception is at $x = 0$, where the generation takes place and equality holds. Our models have (10) everywhere except at the origin.

The pattern must be periodic in time. The period T is set by the period of the relaxation oscillatory motion followed by the solution at $x = 0$, $w = w(0)$. This is presumed known.

The mathematical analysis of the above conceptual model consists in determining the trajectories of the expanding fronts, and the function $v(x,t)$, so that all the above evolutionary laws and constraints are fulfilled. It is convenient to express the fronts' motion by the functions $\tau^{\pm}(x)$. Here $\tau^{+}(x)$ is the time at which some specific upjump front reaches position x, and $\tau^{-}(x)$ is the time for the next succeeding downjump. The next upjump is then at time $\tau^{+} + T$. The equations to be satisfied are:

$$\frac{d}{dx}\tau^{\pm}(x) = \pm c^{-1}(v(x,\tau^{\pm}(x)), w(x)) \ ,$$

$$\frac{\partial v}{\partial t} = \begin{cases} G_{+}(v,w(x)), & \tau^{+}(x) < t < \tau^{-}(X) \ , \\ G_{-}(v,w(x), & \tau^{-}(x) < t < \tau^{+}(x) + T \end{cases}$$

(it sufficies to determine v in the interval $\tau^{+} < t < \tau^{+} + T$). The periodicity constraint on v is that $v(x,\tau^{+}(v)) = v(x,\tau^{+}(x) + T)$, and constraint (10) is also to be fulfilled. In addition, we must require that $v(0,t)$ be the values of v corresponding to the relaxation oscillator at the center, and that as $x \to \infty$, the $\tau^{\pm}(x)$ approach linear functions (corresponding to a plane wave train).

In general, it is difficult and rather delicate to determine whether there exist functions $\tau^{\pm}$, v satisfying all the above.

Tyson [1] has produced a scaled version of a set of reaction-diffusion equations (the "Oregonator") realistically modeling the Belousov-Žabotinskiĭ reaction. He and the author (in preparation) have obtained conditions under which the required functions do exist for Tyson's equations so that the above modeling procedure is valid.

(ii) Self-sustaining target patterns may be modelled by retaining the function $w(x)$, but supposing it to obey a third reaction-diffusion equation coupled to the first two. Thus the distribution of w is obtained as part of the model, not imposed by external conditions. This type of model is explored in some detail in [17]. Again, the dynamics of the wave front and the functions v and w involve complicated mathematics, but some reasonable simplifications are possible.

With both kinds of patterns, discussed above, the presence of a third "species" w is essential. To see why this is so, suppose there were only two reacting species u and v, and that a target pattern existed. Then the source functions f and g in (3) are independent of position x, and have the properties shown in Fig. 1. The period T of the pattern is set at the center and is (approximately) the period of the relaxation oscillator associated with that figure. At a fixed position away from the center, the solution is still periodic with period T. During a cycle, it traverses portions of the branches $h_{\pm}(v)$, skipping from one to the other by sharp downjump and upjump fronts represented by straight horizontal trajectories in Fig. 1. The only way the same period T can be maintained is for these jumps to occur at $v = \underline{v}$ and $v = \overline{v}$; otherwise the period would be smaller. Far away from the center, the two kinds of fronts must travel with the same speed. It would be purely accidental, however, if the trigger front speeds at $\underline{v}$ and $\overline{v}$ were the same; at least one of the fronts will therefore in general be a phase front. By design and in accordance with experimental observations, we have excluded consideration of phase fronts.

7. Targets in excitable media

With nullcurves as depicted in Fig. 1, the kinetic equations

$$u_t = kf(u,v), \quad v_t = g(u,v)$$

have stable relaxation oscillatory solutions. When the g nullcurve is shifted to one of the positions in Figs. 3 and 4, however, the kinetics become excitable or bistable.

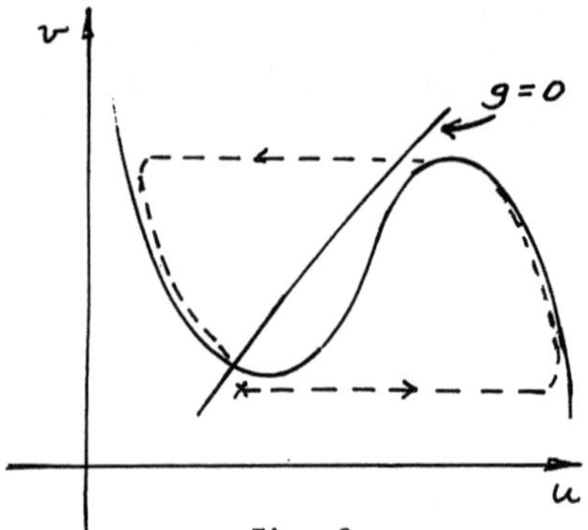

Fig. 3

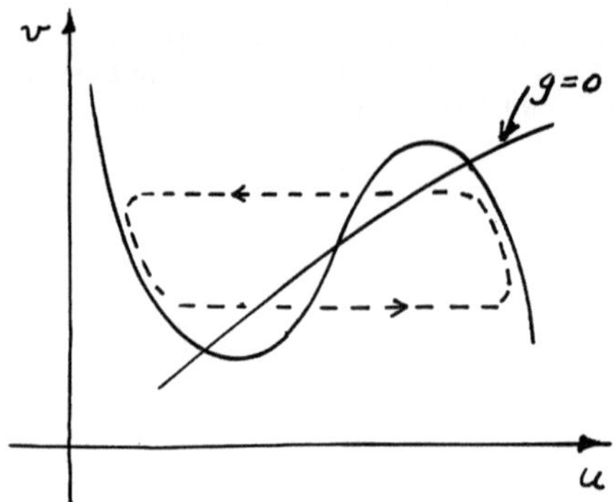

Fig. 4

For example in Fig. 3, there is one stable rest state (the unique intersection point); but when this state is perturbed downward a small amount, the solution mades a large excursion (dotted line) before returning to the rest state.

In modeling target patterns in sec. 6, we needed the kinetics to be oscillatory at the origin; but away from the origin, the configurations in Figs. 3 and 4 are not excluded. As $x \to \infty$ the target develops into a regular wave train, so of course the kinetics must support such a train there. Excitable and bistable (as well as oscillatory) kinetics do support wave trains. For example, the phase plane image (orbit) of a train at a fixed value of x is shown by the dotted loop in Fig. 4. Therefore there is no contradiction involved in having a periodic target pattern emerge in an excitable medium.

This is apparently what often happens with the Z-reagent, which can exist in an excitable, as well as oscillatory, regime. The initial formation of a target pattern in such a medium involves a wave train entering a quiescant region at a stable rest state. It may be difficult to visualize this process of excitation into a periodic state, but several mechanisms are available for accomplishing it (R. Smock, in preparation).

Research supported by the National Science Foundation and the Mathematics Research Center, Madison. Results presented at a conference on Synergetics, Bielefeld, September, 1979.

Literature

1. J. Tyson 1979, Oscillations, bistability, and echo waves in models of the Belousov-Zhabotinskii reaction. Ann. N.Y. Acad. Sci. 36, 279-295.

2. A. T. Winfree 1978, Stably rotating patterns of reaction and diffusion, Theor. Chem., Vol. 4, Academic Press, New York, pp. 1-51.

3. D. J. Reusser and R. J. Field 1979, The transition from phase waves to trigger waves in a model of the Zhabotinskii reaction, J. Amer. Chem. Soc. 101, 1063-1071.

4. L. A. Ostrovskiĭ and V. G. Jahno 1975, The formation of pulses in an excitable medium. Biofizika 20, 489-493.

5. R. Casten, H. Cohen, and P. Lagerstrom 1975, Perturbation analysis of an approximation to Hodgkin-Huxley theory. Quart. Appl. Math. 32, 365-402.

6. J. P. Keener 1979, Waves in excitable media, preprint.

7. V. G. Jahno 1975, On a model for leading centers, Biofizika 20, 669-674. See also G. M. Žislin, V. G. Jahno, and Ju. K. Gol'cova 1976, Biofizika 21, 692-697; Ju. K. Gol'cova, G. M. Žislin, and V. G. Jahno 1976, Biofizika 21, 893-897; V. G. Jahno 1977, Biofizika 22, 876-881.

8. G. Carpenter 1978, Bursting phenomena in excitable membranes, SIAM J. Appl. Math., to appear.

9. P. C. Fife 1976, Pattern formation in reacting and diffusing systems, J. Chem. Phys. 64, 854-864.

10. P. C. Fife 1976, Singular perturbation and wave front techniques in reaction-diffusion problems, in: SIAM-AMS Proceedings, Symposium on Asymptotic Methods and Singlar Perturbations, New York, 23-49.

11. A. T. Winfree 1974, Wavelike activity in biological and chemical media, in: Lecture Notes in Biomathematics (Ed. P. van den Driessche), Springer-Verlag, Berlin.

12. J. M. Greenberg 1977, Axisymmetric time-periodic solutions to λ-ω systems, to appear.

13. P. Ortoleva and J. Ross 1974, On a variety of wave phenomena in chemical and biochemical oscillations, J. Chem. Phys. 60, 5090-5107.

14. N. Kopell and L. N. Howard 1979, Target patterns and horseshoes from a perturbed central force problem: some temporally periodic solutions to reaction diffusion equations. preprint.

15. A. M. Žabotinskiĭ 1974, Concentration Oscillations (Russian) Nauka, Moscow.

16. J. J. Tyson 1976, The Belousov-Zhabotinskii Reaction, Lecture Notes in Biomathematics No. 10, Springer, New York.

17. P. C. Fife 1979, Wave fronts and target patterns, in: Applications of Nonlinear Analysis in the Physical Sciences, Pitman Publishing, London, to appear.

18. Ya. I. Kanel' 1962, On the stabilization of solutions of the Cauchy problem for the equations arising in the theory of combustion, Mat. Sbornik 59, 245-288.

19. P. C. Fife and J. B. McLeod 1977, The approach of solutions of nonlinear diffusion equations to traveling front solutions, Arch. Rational Mech. Anal. 65, 335-361. Also: Bull. Amer. Math. Soc. 81, 1075-1078 (1975).

20. A. N. Kolmogorov, I. G. Petrov skiǐ, and N. S. Piskunov 1937, A study of the equation of diffusion with increase in the quantity of matter, and its application to a biological problem, Bjul. Moskovskovo Gos. Univ. 17, 1-72.

21. A. M. Zhabotinsky and A. N. Zaikin 1973, Autowave processes in a distributed chemical system, J. Theor. Biol. 40, 45-61.

22. M. Marek and J. Juda 1979, Controlled generation of reaction-diffusion waves, Sci. Papers of Prague Inst. Chem. Technol. Ser. K, to appear.

23. A. N. Zaikin and A. M. Zhabotinsky 1970, Concentration wave propagation in two-dimensional liquid-phase self-oscillating system, Nature 225, 535-537.

24. M. L. Smoes, these preceedings.

25. P. Ortoleva 1978, Dynamic Padé approximants in the theory of periodic and chaotic chemical center waves, J. Chem. Phys. 69, 300-307.

26. P. Ortoleva and J. Ross 1975, Theory of propagation of discontinuities in kinetic systems with multiple time scales: fronts, front multiplicity, and pulses, J. Chem. Phys. 63, 3398-3408.

On the Consistency of the Mathematical Models of Chemical Reactions

L. Arnold

Fachbereich Mathematik, Forschungsschwerpunkt Dynamische Systeme, Universität Bremen
D-2800 Bremen 33, Fed. Rep. of Germany

1. Introduction

There are two main principles according to which chemical reactions in a spatial domain are modeled:

(i) *global* description (i.e. without diffusion, spatially homogeneous or 'well-stirred' case) versus *local* description (i.e. including diffusion, spatially inhomogeneous case),

(ii) *deterministic* description (macroscopic, phenomenological, in terms of concentrations) versus *stochastic* description (on the level of numbers of particles, taking into account internal fluctuations).

The combination of these two principles gives rise to essentially four mathematical models, see e.g. HAKEN [7] and NICOLIS and PRIGOGINE [15].

This paper briefly describes these models and deals with the question of whether and in what mathematical sense those models are consistent.

Since we want to be conceptual rather than aim at highest generality, we restrict ourselves to the case of one single reactant X whose concentration is not kept constant, to one-dimensional volume $\Omega \subset \mathbb{R}^1$ and to a reaction scheme of the following type:

$$\sum_{j=1}^{r} \alpha_{ij}A_j + \beta_i X \underset{k_i'}{\overset{k_i}{\rightleftarrows}} \sum_{j=1}^{r} \alpha_{ij}'A_j + (\beta_i+1)X,$$

$$\alpha_{ij},\ \alpha_{ij}',\ \beta_i \in \mathbb{N} = \{0,1,2,\ldots\};\ k_i, k_i' \geq 0,\ i=1,\ldots,s. \tag{1}$$

2. Mathematical Models of Chemical Reactions

2.1 Global deterministic model

One can read-off from the reaction scheme (1) the gain term

$$\lambda(x) = \sum_{i=1}^{s} \alpha_i k_i x^{\beta_i} = \sum_{i=0}^{p} a_i x^i,\ a_i \geq 0,$$

and the loss term

$$\mu(x) = \sum_{i=1}^{s} \alpha_i' k_i' x^{\beta_i+1} = \sum_{i=1}^{q} b_i x^i, \ b_i \geq o \ .$$

Thus, the evolution of the concentration $\varphi = \varphi(t)$, $t \in [o,\infty)$, of the reactant X is described by the nonlinear ODE (kinetic equation)

$$\frac{d\varphi\ (t)}{dt} = f(\varphi(t)), \ \varphi(o) = \varphi_o = o \ , \tag{2}$$

$$f(x) = \lambda(x) - \mu(x) = \sum_{i=o}^{d} c_i x^i, \ c_o \geq o, c_d \neq o \ .$$

If $c_d < o$ then (2) has a non-negative solution existing for all times t which we assume from now on.

2.2 Local deterministic model

Now besides reaction transport of matter via diffusion is taken into account. If $\varphi = \varphi(r,t)$, $r \in \Omega \subset \mathbb{R}^1$, $t \in [o,\infty)$, is the concentration, $D > o$ the diffusion coefficient,

$\Delta = \frac{\partial^2}{\partial r^2}$ the Laplacian, then

$$\frac{\partial\varphi}{\partial t}(r,t) = D\ \Delta\ \varphi(r,t) + f(\varphi(r,t)) \tag{3}$$

(equation of reaction and diffusion, see e.g. FIFE [6]). The domain Ω can be bounded or unbounded, and φ is subjected to boundary and initial conditions. For the problem of existence and uniqueness of solutions and invariant sets we refer to KUIPER [8] and AMANN [1], for asymptotic behavior and steady states see FIFE [5] and [6].

For later use we state the following existence and uniqueness result which can be obtained from KUIPER [8]:
Take $\Omega = (o,1)$, assume $c_d < o$, and pick a $\rho > o$ to the right of the biggest zero of $f(x) = o$. Let the initial condition $\varphi(r,o) = \varphi_o(r)$ satisfy $o \leqq \varphi_o(r) \leqq \rho$, $r \in [o,1]$, $\varphi_o \in H^2_{bc}$, the closure in $H^2(o,1)$ of those functions in $C^2[o,1]$ which satisfy the boundary conditions. Choose as boundary conditions
$\varphi(o,t) = \varphi(1,t) = o$ or
$\partial\varphi/\partial r(o,t) - \alpha\varphi(o,t) = o$, $\partial\varphi/\partial r(1,t) + \beta\varphi(1,t) = o$ $(\alpha,\beta \geqq o)$.
Then there exists a unique global solution φ of (3) satisfying the initial and boundary conditions such that
$\varphi \in C^1([o,\infty), L_2(o,1)) \cap C([o,\infty), H^2(o,1))$
and
$o \leqq \varphi(r,t) \leqq \rho$ for all $r \in [o,1]$ and all $t \in [o,\infty)$.

2.3 Global stochastic model

We now look at the total number of particles $X(t)$ of reactant X in a volume of length L at time t. This function is modeled as a Markov jump process, in particular as a birth and death process with state space $\mathbb{N}$ and transition probabilities

$$P(X(t+s) = k+1 \mid X(s) = k) = p_{k,k+1}(t) = \lambda_k t + o(t),\ k \in \mathbb{N},$$

$$P(X(t+s) = k-1 \mid X(s) = k) = p_{k,k-1}(t) = \mu_k t + o(t),\ k-1 \in \mathbb{N},$$

where the birth rates λ_k and the death rates μ_k are given by

$$\lambda_k = L(a_o+a_1\frac{k}{L}+a_2\frac{k(k-1)}{L^2}+\ldots+ a_p\frac{k(k-1)\ldots(k-p+1)}{L^p})$$

and (4)

$$\mu_k = L(b_1\frac{k}{L}+b_2\frac{k(k-1)}{L^2}+\ldots+ b_q\frac{k(k-1)\ldots(k-q+1)}{L^q}) .$$

Introducing the functions

$$\lambda_L(x) = \sum_{i=o}^{p} a_i\ x(x-\frac{1}{L})\ldots(x-\frac{i-1}{L}),$$

$$\mu_L(x) = \sum_{i=1}^{q} b_i\ x(x-\frac{1}{L})\ldots(x-\frac{i-1}{L}), \tag{5}$$

we can write

$$\lambda_k = L\ \lambda_L(k/L),\ \mu_k = L\mu_L(k/L) . \tag{6}$$

Observe that for $L \to \infty$ and each x

$$\lambda_L(x) = \lambda(x) + O(1/L),\ \mu_L(x) = \mu(x) + O(1/L) .$$

The transition rates λ_k and μ_k together with an initial distribution uniquely determine the stochastic process $X(t)$ provided that either the reaction is linear or $c_q < o$. In particular, the probabilities $p_k(t) = P(X(t) = k)$ are the unique solution of the so-called *Master equation* (or KOLMOGOROV's second equation)

$$\dot{p}_k(t) = \lambda_{k-1}p_{k-1}(t) + \mu_{k+1}p_{k+1}(t) - (\lambda_k+\mu_k)p_k(t),\ k \in \mathbb{N} .$$

for initial distribution $p_k(o) = p_k^o,\ \sum_{k\in\mathbb{N}} p_k^o = 1$.

2.4 Local stochastic model

Now the total volume of length L is divided into N cells of equal size $\ell = L/N$, where adjacent cells are connected by diffusion while in each cell reaction goes on. If $X_j(t)$ denotes the number of particles of reactant X in the j-th cell at time t, then

$$X(t) = (X_1(t),\ldots,X_N(t))$$

is modeled as a Markov jump process with state space $\mathbb{N}^N$ and the following transition intensities:

$$q_{k,k+e_j} = \lambda_{k_j}, \quad q_{k,k-e_j} = \mu_{k_j}, \quad j=1,\ldots,N, \quad \text{(reaction)}$$

$$q_{k,k+e_{j+1}-e_j} = (D^*/2)k_j, \quad j=1,\ldots,N-1, \qquad (7)$$

$$\text{(diffusion)}$$

$$q_{k,k+e_{j-1}-e_j} = (D^*/2)k_j, \quad j=2,\ldots,N,$$

$q_{k,k+m} = o$, otherwise. Here
$k = (k_1,\ldots,k_j,\ldots,k_N)$, $k\pm e_j$, $k+e_{j\pm 1}-e_j$, $k+m \in \mathbb{N}^N$, e_j is the j-th unit vector in $\mathbb{R}^N$, and

$$\lambda_{k_j} = \ell\lambda_\ell(k_j/\ell), \quad \mu_{k_j} = \ell\mu_\ell(k_j/\ell), \qquad (8)$$

and $D^* > o$ are the birth rate, death rate, and diffusion parameter, resp., and $\lambda_\ell(x)$ and $\mu_\ell(x)$ are the functions defined by (5).

We also have to fix boundary conditions. Saying nothing additional amounts to reflection at the boundary (zero flux boundary conditions). The coupling of the system to reservoirs at its boundaries with prescribed fixed (e.g. zero) particle concentrations can be modeled by adding a cell at each boundary with a fixed number of particles and by coupling the new cell to its neighbor by diffusion with intensities as given above.

The transition intensities as fixed above together with an initial distribution on $\mathbb{N}^N$ uniquely determine the stochastic process $X(t)$ provided that reaction is either linear or $c_d < o$. In particular, the probabilities $p_k(t) = P(X(t)=k)$, $k \in \mathbb{N}^N$, are the unique solution of the *multivariate Master equation* (written down e.g. for zero flux boundary conditions)

$$\dot{p}_k(t) = \sum_{j=1}^{N} (\lambda_{k_j-1}\, p_{k-e_j}(t) + \mu_{k_j+1}\, p_{k+e_j}(t) - (\lambda_{k_j}+\mu_{k_j})p_k(t))$$

$$+ \frac{D^*}{2} \sum_{j=2}^{N} ((k_j+1)p_{k+e_j-e_{j-1}}(t) - k_j p_k(t))$$

$$+ \frac{D^*}{2} \sum_{j=1}^{N-1} ((k_j+1)p_{k+e_j-e_{j+1}}(t) - k_j p_k(t))$$

with initial distribution $(p_k^o)_{k\in\mathbb{N}^N}$.

3. Relations between the models

Since the stochastic (global or local) model is considered more detailed than the deterministic one, it should be possible to recover from the master equations the deterministic equations if the volume L tends to infinity (thermodynamic limit). On the other hand, one should be able to recover the global (stochastic or deterministic) description from the local one by letting the diffusion become dominant compared to reaction.

The subject of this section is to make the above statements precise and show in what sense the models are consistent.

3.1 Relation between the global stochastic and deterministic models

This question has been completely settled by KURTZ ([10], [11], [13]) by proving the following results:

Theorem 1. Given the set-up of sections 2.1 and 2.3. Then we have:

(i) *Law of large numbers*: Suppose $\lim_{L\to\infty} X(o)/L = \varphi_o$ in probability, then for each finite T and $\delta > o$

$$\lim_{L\to\infty} P\left(\sup_{o\leq t\leq T} \left| \frac{X(t)}{L} - \varphi(t) \right| > \delta\right) = o,$$

(ii) *Central limit theorem*: Suppose that in addition

$$\lim_{L\to\infty} \sqrt{L}\left(\frac{X(o)}{L}-\varphi_o\right) = y_o \quad \text{in probability, then}$$

$$\lim_{L\to\infty} \sqrt{L}\left(\frac{X(t)}{L} - \varphi(t)\right) = y(t) \quad \text{weakly}$$

(i.e. on the level of the corresponding probability measures in function space), where $y(t)$ is a Gaussian diffusion process which is the solution of the linear stochastic differential equation

$$dy(t) = \frac{d(\lambda-\mu)}{dx}(\varphi(t))y(t)dt + ((\lambda+\mu)(\varphi(t))^{1/2}dW_t,$$

$$y(o) = y_o.$$

(iii) *Diffusion approximation*: There is a probability space on which

$$\sup_{o\leq t\leq T} \left| \frac{X(t)}{L} - u(t) \right| \leq K_T \frac{\log L}{L},$$

where K_T is a random variable depending on T and $u(t)$ is a diffusion process identical in distribution to the solution of the nonlinear stochastic differential equation

$$du(t) = (\lambda-\mu)(u(t))dt + \frac{1}{\sqrt{L}}((\lambda+\mu)(u(t)))^{1/2}dW_t,$$

$$u(o) = X(o)/L.$$

Remark: The approximations in Theorem 1 (i) and (ii) can also be made sample-wise (at least at the expense of restriction to linear reaction, see KURTZ [13]) with the result

$$\begin{aligned} X(t)/L &= \varphi(t) + O(1/\sqrt{L}), \\ &= \varphi(t)+(1/\sqrt{L})y(t) + O(\log L/L), \\ &= u(t) + O(\log L/L) \end{aligned}$$

so that it turns out that the error of the Gaussian and of the diffusion approximation have the same order of magnitude.

3.2 Relation between the local stochastic and deterministic models

As it turns out, the presence of diffusion forces us to rescale the length so that on a new r-axis the system is defined on the inverval [O,R] with new cell size h given by

$$h = R\,\ell/L = R/N$$

where necessarily $h \to 0$. Depending on whether equation (3) is considered on a finite or infinite interval, $R = \text{const}$ or $R \to \infty$. For simplicity we assume $R = 1$, so that $h = \ell/L = 1/N$.

The N-vector process $X(t)/\ell$ corresponds to a step function on [O,1] defined by

$$x(r,t) = X_j(t)/\ell \quad \text{on} \quad ((j-1)/N, j/N],\ j=1,\dots,N.$$

We will compare $x(\cdot,t) = x(t)$ with the solution $\varphi(t)$ of (3) as elements of $L_2(0,1)$, where the parameters L,ℓ and D^* in the stochastic model are moving such that $L \to \infty$ and $\ell/L = 1/N \to 0$. The parameter D in (3) is considered fixed.

Theorem 2 (ARNOLD and THEODOSOPULU [3]).

Given the set-up of sections 2.2 and 2.4. Take fixed or zero flux boundary conditions in both models.
Suppose $L \to \infty$, $N \to \infty$ such that

(i) $\lim \| x(o)-\varphi_o \|_{L_2} = 0$,

(ii) $D^*/2 = DN^2$ (entailing $D^* \to \infty$),

(iii) $N^2/\ell = L^2/\ell^3 \to 0$ (entailing $\ell \to \infty$).

Then we have the *law of large numbers*: For each finite T and $\delta > o$

$$\lim \sup_{o \leq t \leq T} P(\|x(t)-\varphi(t)\|_{L_2} > \delta) = 0 .$$

Remark: Condition (ii) is needed to end up with the Laplacian. Condition (iii) requires that the cell size ℓ has to increase to infinity quite fast, e.g. like $\ell = L^{2/3+\epsilon}$, $o < \epsilon < 1/3$. On the other hand, if $L^2/\ell^3 \to \infty$ then the law of large numbers breaks down, so (iii) cannot be relaxed if one aimes at convergence in L_2-norm. However, (iii) can be dismissed if one is content with a weaker kind of convergence, e.g.

$$\int_0^1 \psi(r)x(r,t)dr \to \int_0^1 \psi(r)\varphi(r,t)dr \quad \text{in probability}$$

for all $\psi \in C^1[o,1]$.

Theorem 3 (ARNOLD and KOTELENEZ [2]).

Suppose that in addition to the assumptions of Theorem 2

$$\lim \|\sqrt{L}(x(o)-\varphi_o) - y_o\| = o \text{ in probability.}$$

Then the following *central limit theorem* holds:

$$\lim \sqrt{L}(x(t)-\varphi(t)) = y(t) \quad \text{weakly}$$

(i.e. on the level of the corresponding probability measures in the space of L_2-valued functions of time t), where $y(t)$ is a Gaussian diffusion process with values in L_2 which is the solution of the linear stochastic partial differential equation

$$dy(t) = \frac{d}{dx}(\lambda-\mu)(\varphi(t))y(t)dt + D\Delta y(t)dt + G(\varphi(t))dW_t,$$

$$y(o) = y_o,$$

where

$$T = GG^* = -D\frac{\partial}{\partial r}\varphi(t)\frac{\partial}{\partial r} + (\lambda+\mu)(\varphi(t)) : H^1 \to H^{-1},$$

$T \geq o$, $G^* : H^1 \to L_2$, $G : L_2 \to H^{-1}$, and W_t is

the Wiener process with values in L_2 so that W_1 is the Wiener measure in the abstract Wiener space (H^1, L_2, i) in the sense of GROSS and KUO [9].

3.3 Relation between the local and global deterministic models

We now look at the solution $\varphi(r,t)$ of (3) on [o,1] with the boundary conditions given in 2.2 for $D \to \infty$. We expect that immediately after start a strong diffusion straightens out any spatial inhomogeneity so that $\varphi(r,t)$ is approximately equal to a constant $\psi(t)$ with respect to r, and $\psi(t)$ solves the global kinetic equation (2).

Note first that the only boundary conditions in the class considered fulfilled for a non-zero straight line are the zero flux conditions. Therefore, only under zero flux conditions there is a chance of recovering the global law by "turning on" diffusion.

Theorem 4 Given the solution $\varphi(r,t)$ of (3) on [o,1] with zero flux boundary conditions

$$\frac{\partial\varphi(o,t)}{\partial r} = o, \quad \frac{\partial\varphi(1,t)}{\partial r} = o$$

and initial condition φ_o. Let $\psi(t)$ be the solution of (2) with initial condition

$$\psi_o = \int_o^1 \varphi_o(r)dr.$$

Then for all $\delta > o$ and finite $T > o$

$$\lim_{D\to\infty} \sup_{\delta\leq t\leq T} \| \varphi(r,t)-\psi(t) \|_{L_2} = o .$$

Proof. The operator $D\,\Delta$ generates the analytic contraction semigroup $T(t)$ on L_2 given by

$$T(t)x(r) = c_o + \frac{1}{\sqrt{2}}\sum_{n=1}^{\infty} c_n\, e^{-n^2\pi^2 Dt}\cos n\pi r$$

where $c_o = \int_o^1 x(r)dr$, $c_n = \frac{1}{\sqrt{2}}\int_o^1 x(r)\cos n\pi r\, dr$, $x \in L_2$

(see CURTAIN and PRITCHARD [4], p. 46). We have $T(t)c = c$ for any spatial constant and $\| T(t) \| = 1$. We have $\lim_{D\to\infty} T(t) = \Pi$ uniformly for $\delta \leq t \leq T$, where Π is the projection onto the one-dimensional subspace generated by 1 defined by

$$\Pi x(r) \equiv \int_o^1 x(r)dr = c_o.$$

Writing both equations in their semigroup form taking into account $\Pi\varphi_o = \psi_o$, we obtain

$$\varphi(t) - \psi(t) = (T(t) - \Pi)\varphi_o + \int_o^t T(t-s)(f(\varphi(s)) - f(\psi(s)))ds.$$

For any interval [o,T] and given φ_o there is a sup-topology sphere S_ρ of radius ρ such that $\varphi(r,t) \in S_\rho$ and $\psi(t) \in S_\rho$ for all $r \in [o,1]$, $t \in [o,T]$. In S_ρ f is Lipschitz with constant L_ρ. Therefore

$$\| \varphi(t)-\psi(t) \| \leqq \| T(t)-\Pi \| \, \| \varphi_o \| + L_\rho \int_o^t \| \varphi(s)-\psi(s) \| \, ds.$$

Putting $\| T(t) - \Pi \| \, \| \varphi_o \| = \epsilon(t)$, the Bellman-Gronwall lemma yields

$$\| \varphi(t)-\psi(t) \| \leqq \epsilon(t) + L_\rho \int_o^t \exp(L_\rho(t-s))\epsilon(s)ds$$

from which the result follows. □

3.4 Relation between the local and global stochastic models

We again assume zero flux boundary conditions for the local stochastic model (i.e. reflection at the boundary) and consider the case $D^* \to \infty$, L and ℓ fixed. The total number of particles

$$\overline{X}(t) = \sum_{j=1}^{N} X_j(t)$$

is not a Markov process anymore, but we expect that for big D^* $\overline{X}(t)$ is close in a certain sense to the process pertaining to the global stochastic description. Attempts have been made in this direction e.g. by MALEK-MANSOUR [14], but to our knowledge no rigorous proof has been given.

Theorem 5 Let $X(t)$ be the stochastic process belonging to the local stochastic model with transition intensities given by (7) and (8), let $\overline{X}(t)$ be the total number of particles in the local model. Then

$$\lim_{D^* \to \infty} \overline{X}(t) = Y(t) \quad \text{weakly} \qquad (9)$$

(i.e. on the level of the corresponding probability measures in function space), where $Y(t)$ is the stochastic process belonging to the global stochastic model with transition intensities given by (6) startet with the distribution of $\overline{X}(o)$.

Proof. We use a semigroup approximation method developed by KURTZ [12]. Without restricting the generality we can assume that our process is being stopped at first exit time from a set

$$\mathbb{N}_\rho^N = \{ k \in \mathbb{N}^N : \overline{k} = \sum_1^N k_j \leqq \rho \}$$

with arbitrarily big ρ. Thus, we can restrict our whole investigation to this set.

The infinitesimal generator A_{D^*} of the Markov process $X(t)$ is defined on the Banach space $B(\mathbb{N}_\rho^N)$ of all bounded functions on $\mathbb{N}_\rho^N$ by

$$A_{D^*} = D^* L_1 + L_2$$

where

$$L_1 f(k) = \frac{1}{2} \sum_{j=1}^{N-1} (f(k+e_{j+1}-e_j)-f(k))k_j + \frac{1}{2} \sum_{j=2}^{N} (f(k+e_{j-1}-e_j)-f(k))k_j$$

and

$$L_2 f(k) = \sum_{j=1}^{N} ((f(k+e_j)-f(k))\lambda_{k_j} + (f(k-e_j)-f(k))\mu_{k_j})$$

with λ_{k_j}, μ_{k_j} given by (8), and for $k \in \mathbb{N}_\rho^N$.

The infinitesimal generator B of the Markov process $Y(t)$ is defined on the Banach space $B(\mathbb{N}_\rho)$ of all bounded functions on $\mathbb{N}_\rho$ by

$$Bf(j) = (f(j+1) - f(j))\lambda_j + (f(j-1) - f(j))\mu_j, \; j \in \mathbb{N}_\rho,$$

with λ_j, μ_j given by (6).

Let M be the set of all $f \in B(\mathbb{N}_\rho)$ such that there is a sequence $f_{D^*} \in B(\mathbb{N}_\rho^N)$ with

$$\lim_{D^*\to\infty} \sup_k \mid f_{D^*}(k) - f(\bar{k}) \mid = o \tag{10}$$

and

$$\lim_{D^*\to\infty} \sup_k \mid A_{D^*} f_{D^*}(k) - Bf(\bar{k}) \mid = o . \tag{11}$$

We will prove that $M = B(\mathbb{N}_\rho)$. Since $\lambda - B$ is invertible for some $\lambda > o$, a theorem of KURTZ ([12], S. 630-631) assures that under these conditions (9) holds.

We proceed as in PAPANICOLAOU [16] and put for any $f \in B(\mathbb{N}_\rho)$

$$f_{D^*}(k) = f(\bar{k}) + \frac{1}{D^*} g(k), \; k \in \mathbb{N}_\rho^N ,$$

with g to be determined. Certainly (10) is true for any g.

Furthermore,

$$A_{D^*} f_{D^*}(k) = D^* L_1 f(\bar{k}) + L_1 g(k) + L_2 f(\bar{k}) + \frac{1}{D^*} L_2 g(k).$$

Because $\overline{k+e_{j\pm1}-e_j} = \bar{k}$ we have $L_1 f(\bar{k}) = o$.

Thus (11) will be satisfied if g can be chosen such that

$$L_1 g(k) = - L_2 f(\bar{k}) + Bf(\bar{k}).$$

Observe that L_1 is the infinitesimal generator of the Markov process on $\mathbb{N}_\rho^N$ modeling pure diffusion of intensity 1 with reflection at the boundary. This process is ergodic with stationary distribution P_o being the multinomial distribution

$$p_k = \frac{m!}{k_1!k_2! \, \ldots \, k_N!} N^{-m} , \text{ if } \bar{k}=m, \; k \in \mathbb{N}_\rho^N ,$$

where $m \leq \rho$ is the initial number of particles in the system (which is conserved), and $p_k = 0$ otherwise (see VAN DEN BROECK, HORSTHEMKE and MALEK-MANSOUR [17]).

Therefore $L_1 g = u$ has a solution provided the solvability condition

$$\int u(x)P_o(dx) = \sum_k u(k)p_k = 0$$

holds for our particular u. This is the only thing that remains to be checked.

We have

$$\sum_k (-L_2 f(\bar{k}) + Bf(\bar{k}))p_k =$$

$$= (f(m+1)-f(m))(\lambda_m - \sum_k (\sum_{j=1}^N \lambda_{k_j})p_k) + (f(m-1)-f(m))(\mu_m - \sum_k (\sum_{j=1}^N \mu_{k_j})p_k).$$

For the multinomial distribution of m particles into $N = L/\ell$ cells

$$\sum_k (\sum_{j=1}^N \lambda_{k_j})p_k = \sum_k (\sum_{j=1}^N \ell\lambda_\ell(k_j/\ell))p_k$$

$$= N\ell \sum_{j=o}^m \lambda_\ell(k_j/\ell)\left(\frac{1}{N}\right)^j \left(1-\frac{1}{N}\right)^{m-j} \binom{m}{j}$$

$$= L\lambda_L(m/L) = \lambda_m,$$

similarly for the μ-terms, so that indeed the solvability condition is satisfied. □

4. Summary

We summarize the relations between the four models in the following scheme:

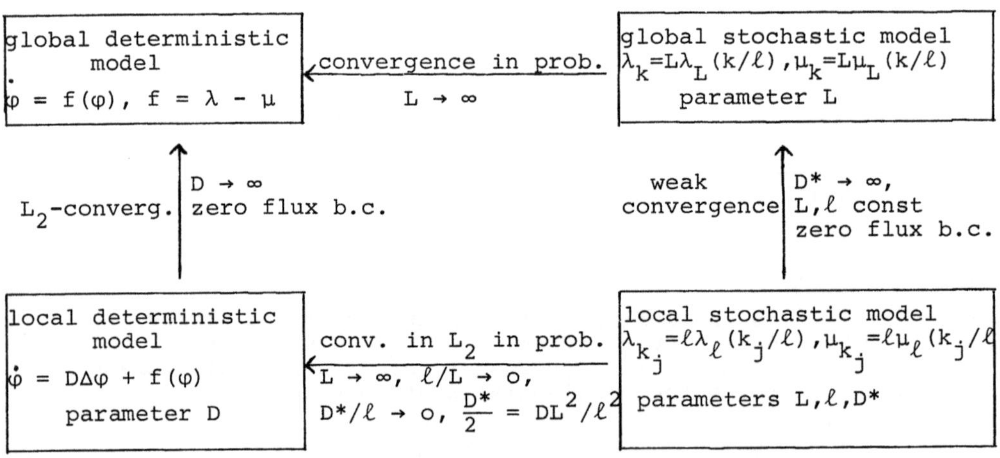

REFERENCES

1 Amann, H.: Invariant Sets and Existence Theorems for Semilinear Parabolic and Elliptic Systems. J. Math. Anal. Appl. 65 (1978), 432 - 467.

2 Arnold, L., and P. Kotelenez: Central Limit Theorem for the Stochastic Model of Chemical Reactions with Diffusion (in preparation).

3 Arnold, L., and M. Theodosopulu: Deterministic Limit of the Stochastic Model of Chemical Reactions with Diffusion. Adv. Appl. Prob. (1980).

4 Curtain, R. F., and A. J. Pritchard: Infinite Dimensional Linear Systems Theory. Lecture Notes in Control and Information Sciences 8. Springer, Berlin - Heidelberg - New York 1978.

5 Fife, P. C.: Asymptotic States for Equations of Reaction and Diffusion. Bull. Amer. Math. Soc. 84 (1978), 693 - 726.

6 Fife, P. C.: Mathematical Aspects of Reacting and Diffusing Systems. Lecture Notes in Biomathematics 28. Springer, Berlin-Heidelberg- New York 1979.

7 Haken, H.: Synergetics. Springer, Berlin-Heidelberg-New York 1978 (second edition).

8 Kuiper, H. J.: Existence and Comparison Theorems for Nonlinear Diffusion Systems. J. Math. Anal. Appl. 60 (1977), 166 - 181.

9 Kuo, H.-H.: Gaussian Measures in Banach Spaces. Lecture Notes in Mathematics 463. Springer, Berlin-Heidelberg-New York 1975.

10 Kurtz, T.: Solutions of Ordinary Differential Equations as Limits of Pure Jump Markov Processes. J. Appl. Prob. 7 (1970), 49 - 58.

11 Kurtz, T.: Limit Theorem for Sequences of Jump Markov Processes Approximating Ordinary Differential Processes. J. Appl. Prob. 8 (1971), 344 - 356.

12 Kurtz, T.: Semigroups of Conditioned Shifts and Approximation of Markov Processes. Annals of Prob. 3 (1975), 618 - 642.

13 Kurtz, T.: Strong Approximation Theorems for Density Dependent Markov Chains. Stoch. Proc. and their Appl. 6 (1978), 223 - 240.

14 Malek-Mansour, M.: Fluctuation et Transition de Phase de Non-equilibre dans les Systemes Chimiques. These. Université Libre de Bruxelles 1979.

15 Nicolis, G., and I. Prigogine: Selforganization in Nonequilibrium Systems. Wiley, New York 1977.

16 Papanicolaou, G. C.: Asymptotic Analysis of Stochastic Equations. In: Rosenblatt, M. (ed.): Studies in Probability Theory. Studies in Mathematics Vol. 18. The Mathematical Association of America 1978.

17 Van den Broeck, C., W. Horsthemke, and M. Malek-Mansour: On the Diffusion Operator of the Multivariate Master Equation. Physica 89 A (1977), 339 - 352.

Note: After this paper was finished H. G. Othmer brought to our attention the paper by E. Conway, D. Hoff and J. Smoller, Large Time Behavior of Solutions of Systems of Nonlinear Reaction-Diffusion Equations (SIAM J. Appl. Math. 35 (1978), 1-16) in which results similar to Theorem 4 are proved.

The Critical Behavior of Nonequilibrium Transitions in Reacting Diffusing Systems

A. Nitzan

Department of Chemistry, Tel Aviv University
Tel Aviv, Israel

The term "synergetics" which is the title of the present meeting has been invented by Haken to describe the field of non-equilibrium cooperative phenomena; the idea being that such phenomena in largely different systems should be amenable to analysis within a common mathematical formalism. With this in mind, considerable work has been done within the past decade [1] in an effort to reduce the non-linear equations of motion (EOM's) describing the dynamics of systems of interest,to as few as possible common forms. This of course cannot be done under the most general circumstances. However, for systems near the critical point of their non-equilibrium transitions (e.g. the system described by Fig.1), multiple time and length scales procedures were used to reduce the EOM's to the corresponding time dependent Ginzburg Landau (TDGL) equations for the appropriate order parameter characterizing the transition. Once such a TDGL equation is derived, the powerful machinery developed for equilibrium critical phenomena can be used to make definite predictions about the critical behavior of the given system. In particular the idea of universality implies that non-equilibrium critical phenomena will fall into a few universality classes with systems belonging to the same class showing the same type of critical behavior (e.g. the same critical exponents).

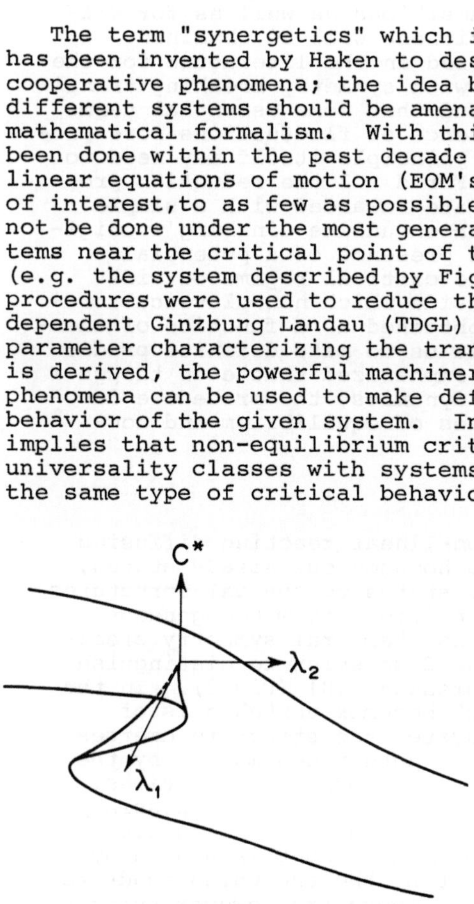

Fig.1. The critical point for a cusp catastrophe. C* is a steady state variable, λ_1 and λ_2 are externally controlled constraints.

The existence of a TDGL equation results from the separation of time and length scales between the critical mode(s) and the rest of the modes, in the vicinity of the critical point. The fast non-critical modes follow the dynamics of the slow, critical ones or, in the pictorial language of Haken [1a], are "slaved" by them, the later thus play the role of the order parameters of the transition. This slaving principle, or adiabatic following as it became to be known in other contexts, has been used to extract TDGL equations for critical pointsassociated with hydrodynamic instabilities [2a], the laser threshold [2b], chemical instabilities [2c,3-5] and more. Some problems remained: it appeared initially [2c] that the multiple time and length scale expansion techniques which are used in the rigorous reduction procedure did not work for transitions between homogeneous steady states or for transitions to multidimensional (d>1) structures. It was subsequently shown [3,4] that these problems may be overcome by taking proper account of the critical conditions which insure that the system is indeed near a critical point. Other problems arise in the presence of fluctuations and are still not fully resolved.

In this talk I will first discuss the critical conditions which are necessary ingredients into any reduction procedure near the critical point. I will then describe a reduction procedure based on scaling ideas recently introduced by Mori.[6] Some typical generalized TDGL equations for homogeneous transitions as well as for different kinds of symmetry breaking transitions will be obtained. I will argue that some ad-hoc forms proposed in the literature for the Ginzburg Landau Hamiltonian associated with symmetry breaking transitions may have been too simple minded and that some essential terms may have been disregarded. In the presence of fluctuations more problems arise: for example, the slaving principle itself may need to be modified for dimensionality $d \leq 2$. For $d > 2$ the reduction procedure may be still carried out and we are rewarded with a simple derivation of the Ginzburg criterion, which unlike Ginzburg's original derivation [7] does not rely on the presence of a potential. This criterion estimates the size of the critical regime outside which mean field theory is valid and inside which non-classical critical behavior is predicted. This is obtained as a function of chemical and transport parameters and enable us to make definite predictions for given systems. It should be emphasized that even though I use the language of reacting-diffusing dynamics, the procedure described is equally useful for other types of equilibrium and non-equilibrium transitions.

Transitions in Chemical Systems

The simplest transitions observed in non-linear reacting diffusing systems are (a) transitions between two homogeneous steady states, (b) transitions from homogeneous steady states to spatial structures (spatial symmetry breaking) and (c) transitions from homogeneous steady states to homogeneous oscillations (temporal symmetry breaking). It is important for the following discussion to distinguish between two types of spatial symmetry breaking [8] (Fig.2). In the first type (Fig.2a) the first wave which becomes unstable is of finite wavelength. In a large enough system the structure emerges with a characteristic wavelength which is determined mostly by the system's dynamics and only marginally by the boundary conditions. We call these *intrinsic symmetry breaking transitions*. A possibly realistic system is provided by the illuminated thermodiffusive system [8] where a mixture of two components A and B is heated by light which is absorbed only by component A, the one which tends to thermodiffuse towards the hotter region. There are however systems characterized by stability diagrams of the type shown in Fig.2b where the first wave to become unstable has infinite wavelength. Spatial structure will emerge only far enough from the transition point where the longest wave compatible with the size of the system has become unstable. The structure in these so-called *extrinsic symmetry breaking transitions* thus depends on the boundary (external) conditions in an essential way. In a large system such transitions behave in some respect like homogeneous transitions but the zero initial slope of $\mathrm{Re}\gamma_0$ vs. k^2 (γ_0 is the relevant eigenvalue of the stability matrix and Re denotes real part) leads to some peculiarities in the resulting TDGL as will be discussed below. I note in passing that any

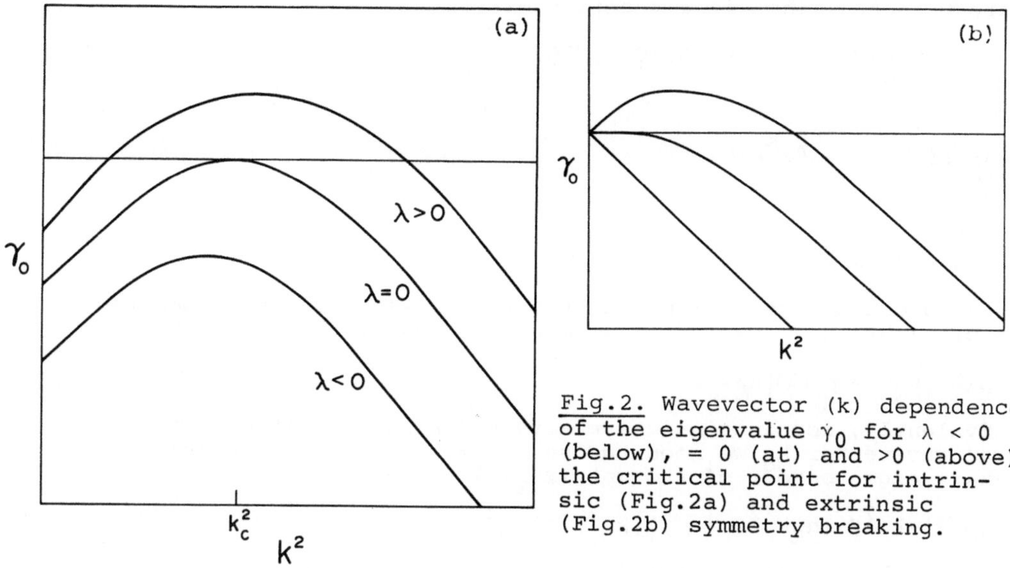

Fig.2. Wavevector (k) dependence of the eigenvalue γ_0 for $\lambda < 0$ (below), $= 0$ (at) and >0 (above) the critical point for intrinsic (Fig.2a) and extrinsic (Fig.2b) symmetry breaking.

transition occurring in a reacting two component system $A \rightleftharpoons B$ with global conservation ($\int d^3r(A+B) = \text{const}$) belongs to this class. [8]

Critical Conditions [3,4]

Figure 1 shows the configuration space of a system characterized by a cusp catastrophy. The critical point is defined by specifying the values of the two external (control) parameters λ_1 and λ_2 and the concentrations $\underset{\sim}{C} = (C_1, C_2 \ldots)$. Transitions which occur during the variation of λ_1 (say), will be discontinuous (hard) unless λ_2 is kept at its critical value. In the later case the transition is continuous (soft). From the experimental point of view only soft transition points are achievable because in the hard transition cases fluctuations will lead to transitions before the actual bifurcation point has been reached. [9]

A proper reduction of the EOM's of any given system may be achieved only if the proper conditions insuring that the critical point is indeed being approached are satisfied. To express these conditions in a suitable mathematical form consider a general reaction-diffusion scheme

$$\frac{\partial \underset{\sim}{C}}{\partial t} = \underset{\approx}{D}\nabla^2 \underset{\sim}{C} + \underset{\sim}{F}(\underset{\sim}{C}, \underset{\sim}{\lambda}) \qquad (1)$$

where $\underset{\sim}{C}$ is a vector of concentrations, λ are control parameters, $\underset{\approx}{D}$ is the diffusion matrix and $\underset{\sim}{F}$ is a vector of reaction rates. It will be convenient to take the critical point as the origin for the $\underset{\sim}{\lambda}$ variables. Also, for any given $\underset{\sim}{\lambda}$ the solution $\underset{\sim}{C}^\circ(\underset{\sim}{\lambda})$ of the homogeneous steady state equations $\underset{\sim}{F}(\underset{\sim}{C}^\circ(\lambda), \underset{\sim}{\lambda}) = 0$ is taken as the origin of $\underset{\sim}{C}$ ($\underset{\sim}{C} - \underset{\sim}{C}^\circ(\underset{\sim}{\lambda}) \rightarrow \underset{\sim}{C}$). The reaction dynamics represented by $\underset{\sim}{F}(\underset{\sim}{C}, \underset{\sim}{\lambda})$ may be then expanded in powers of $\underset{\sim}{C}$

$$\underset{\sim}{F}(\underset{\sim}{C}, \underset{\sim}{\lambda}) = \underset{\approx}{\Omega}(\underset{\sim}{\lambda})\underset{\sim}{C} + \underset{\approx}{N}(\underset{\sim}{\lambda}) : \underset{\sim}{C}\underset{\sim}{C} + \ldots \qquad (2)$$

where $\underset{\approx}{\Omega}$ is the linearized reaction matrix

$$\underset{\approx}{\Omega}(\lambda) = \underset{\approx}{\Omega}(\underset{\sim}{C}^{\circ}(\lambda),\lambda) = (\partial \underset{\sim}{F}(\underset{\sim}{C},\underset{\sim}{\lambda})/\partial \underset{\sim}{C})_{\underset{\sim}{C}^{\circ}(\lambda)} \tag{3}$$

and where

$$(\underset{\approx}{A}:\underset{\sim}{b}\underset{\sim}{c}) = \sum_i \sum_j A_{ij}^{k} b_i c_j$$

with

$$N_{ij}^{k} = (\partial F_k/\partial C_i \partial C_j)_{\underset{\sim}{C}^{\circ}(\lambda)} \tag{4}$$

For homogeneous transitions it is further useful to introduce the set of right $|\alpha\underset{\sim}{\lambda}>$ and left $<\alpha\underset{\sim}{\lambda}|$ eigenvectors of the matrix $\underset{\approx}{\Omega}(\lambda)$

$$\underset{\sim}{\Omega}(\underset{\sim}{\lambda})|\alpha\underset{\sim}{\lambda}> = \gamma_{\alpha}(\underset{\sim}{\lambda})|\alpha\underset{\sim}{\lambda}> \tag{5}$$

Equivalently, for intrinsic symmetry breaking transitions with critical wavevector $\underset{\sim}{k}_c$ we need the set of right $|\alpha\underset{\sim}{\lambda}k_c^2>$ and left $<\alpha\lambda k_c^2|$ eigenvectors of the matrix $\underset{\approx}{\Omega}(\lambda) - k_c^2\underset{\approx}{D}$

$$(\underset{\approx}{\Omega}(\lambda)-k_c^2\underset{\approx}{D})|\alpha\underset{\sim}{\lambda}k_c^2> = \gamma_{\alpha}(\underset{\sim}{\lambda},k_c^2)|\alpha\underset{\sim}{\lambda}k_c^2> \tag{6}$$

Consider the simplest situation where a single root $\alpha = o$ vanishes at the critical point

$$\gamma_{\alpha}(\underset{\sim}{\lambda}) \text{ or } \gamma_{\alpha}(\underset{\sim}{\lambda},k_c^2) \xrightarrow[\lambda \to 0]{} 0 \tag{7}$$

It may be shown that a necessary condition for this transition to be soft is that

$$N_{00}^{0}(\underset{\sim}{\lambda}) \xrightarrow[\underset{\sim}{\lambda} \to 0]{} 0 \tag{8}$$

where for homogeneous transitions

$$N_{\alpha'\alpha''}^{\alpha}(\underset{\sim}{\lambda}) = <\alpha\underset{\sim}{\lambda}|\underset{\approx}{N}:|\alpha\ \underset{\sim}{\lambda}>|\alpha\ \underset{\sim}{\lambda}> \tag{9a}$$

and equivalently for intrinsic symmetry breaking

$$N_{\alpha'\alpha''}^{\alpha}(\underset{\sim}{\lambda}) = <\alpha\lambda k_c^2|\underset{\approx}{N}:|\alpha'\underset{\sim}{\lambda}k_c^2>|\alpha''\underset{\sim}{\lambda}k_c^2> \tag{9b}$$

In Eqs. (9)

$$<a|N:|b>|c> \equiv \sum_i \sum_j \sum_k (<a|)_i(|b>)_j(|c>)_k N_{jk}^{i} \tag{10}$$

I call the condition expressed by (8) the *critical condition*. This condition has to hold for soft homogeneous transitions or for multi-dimensional symmetry breaking transitions. However, for transitions to one dimensional spatial structure or to a limit cycle this condition is not necessary. [4]

Finally it should be mentioned that for a symmetry breaking transition the relation

$$D_{00}(\lambda,k_c^2) \equiv <0\underset{\sim}{\lambda}k_c^2|\underset{\approx}{D}|0\underset{\sim}{\lambda}k_c^2> \xrightarrow{\lambda \to 0} 0 \tag{11}$$

is also satisfied [2c,4]

Scaling and Reduction of the EOM's [4]

The scaling of variables and control parameters, as well as of the space and time coordinates in the EOM's near the critical point follows from the expected divergence of the space and time scales of interest as the critical point is approached, i.e. as $\underset{\sim}{\lambda}$ and $\underset{\sim}{c}$ approach zero under the conditions (7) (8) and (11). To demonstrate the scaling idea consider the simplest case of a transition between two homogeneous states in a system characterized by two variables (C_1, C_2) with quadratic nonlinearity. We may assume that a transformation $(C_1, C_2) \rightarrow (W_0, W_1)$ has been made where the W variables diagonalize the linear dynamics (W_0 corresponds to the root which vanishes at the critical point). In this representation the EOM's are

$$W_0 = D_{00}(\lambda)\nabla^2 W_0 + D_{01}(\lambda)\nabla^2 W_1 + \gamma_0(\lambda) W_0 + N^0_{00}(\lambda) W_0{}^2 + N^0_{01}(\lambda) W_0 W_1 + N^0_{11}(\lambda) W_1{}^2 \tag{12a}$$

$$W_1 = D_{10}(\lambda)\nabla^2 W_0 + D_{11}(\lambda)\nabla^2 W_1 + \gamma_1(\lambda) W_1 + N^1_{00}(\lambda) W_0{}^2 + N^1_{01}(\lambda) W_0 W_1 + N^1_{11}(\lambda) W_1{}^2 \tag{12b}$$

In these equations $D_{\alpha\beta}(\lambda) \equiv \langle\alpha\lambda|\underset{\sim}{D}|\beta\lambda\rangle$. I have taken all control parameters excluding one (λ) to be fixed at their critical values.

Near the critical point

$$\gamma_0(\lambda) = O(\lambda) \quad ; \quad \gamma_1(\lambda) = \gamma_1(0) + O(\lambda) \tag{13a}$$

$$D_{\alpha\beta}(\lambda) = D_{\alpha\beta}(0) + O(\lambda) \tag{13b}$$

$$N^{\alpha}_{\alpha'\alpha''}(\lambda) = N^{\alpha}_{\alpha'\alpha''}(0) + O(\lambda) \text{ if } \alpha' \alpha'' \text{ not all } 0 \tag{13c}$$

$$N^0_{00}(\lambda) = O(\lambda) \tag{13d}$$

Introduce now the following change in variables

$$r = Lr' \tag{14a}$$

$$t = L^z t' \tag{14b}$$

$$\lambda = L^{-y}\lambda' \tag{14c}$$

$$\gamma_0 = L^{-y}\gamma_0' \tag{14d}$$

$$N^0_{00} = L^{-m} N^{0'}_{00} \tag{14e}$$

$$W_\alpha(r,t,\lambda) = L^{-x_\alpha} W'_\alpha(r',t',\lambda') \tag{14f}$$

The scaling parameter L goes to infinity at the critical point. Equation (14a) is in fact the definition of L where r' is taken to be of order 1. The exponents in Eqs. (14b-f) will be chosen so that also t', λ', γ_0', $N^{0'}_{00}$ and W'_α are of order 1. Inserting Eqs. (14) into Eqs. (12) we obtain

$$\dot{W}_0' = L^{z-2} D_{00}\nabla'^2 W_0' + L^{z-y}\gamma_0' W_0' + L^{z-x_1} N^{0'}_{01} W_0' W_1' + L^{z-m-x_0} N^{0'}_{00} W_0'^2 + L^{z+x_0-2-x_1} D_{01}\nabla'^2 W_1' + L^{z+x_0-2x_1} N^0_{11} W_1'^2 \tag{15a}$$

$$L^{-z}\dot{W}_1' = \gamma_1 W_1' + L^{x_1-2x_0} N^1_{00} W_0'^2 + L^{x_1-x_0-2} D_{10}\nabla'^2 W_0' + L^{-2}\nabla'^2 W_1'$$

$$+ L^{-x_0} N^1_{01} W_0' W_1' + L^{-x_1} N^1_{11} W_1'^2 \qquad (15b)$$

The basic ansatz of the scaling theory is that when the critical point is approach, i.e. when $L \to \infty$, only those terms remain in Eqs. (15) which leave these equations invariant. The other terms vanish as $L \to \infty$ and are irrelevant as far as the critical point is concerned. In practice, one tries to keep as many terms as possible in the invariant equations. This leads to the choice of exponents

$$x_0 = 1 \ , \ z = y = x_1 = 2 \qquad (16)$$

With this choice Eq. (15b) yields the adiabatic following or slaving equation

$$W_1' = - \frac{N_{00}'}{\gamma_1} W_0'^2 \qquad (17)$$

which, together with the terms persisting in Eq. (15a) yields the equation for the *order parameter* W_0

$$\dot{W}_0 = D_{00}\nabla^2 W_0 + \gamma_0 W_0 - \frac{N^0_{01} N^1_{00}}{\gamma_1} W_0{}^3 \ (+ N^0_{00} W_0{}^2) \qquad (18)$$

In Eq. (18) the primes denoting scaled variables have been dropped. The bracketed term in (18) may appear or not, depending on the value of the exponent m in (14e). This in turn is determined by the dependence of N^0_{00} on λ as implied by the particular problem considered. (For a further discussion of this point see [3].

Finally it should be noted that the assumed invariance of Eq. (18) to distance from the critical point implies the scaling property of W

$$W_\alpha(r\ t\ \lambda) = L^{-x_\alpha} W(L^{-1}r, L^{-2}t, L^2\lambda) \qquad (19)$$

with $x_0 = 1$ and $x_1 = 2$.

The execution of this procedure for the critical point of a symmetry breaking transition is similar, however some differences arise: first, relation (11) now holds so that D_{00} has to be scaled also (usually $D_{00}(\lambda k_c{}^2) \xrightarrow{\lambda \to 0} O(\lambda)$); secondly and more important, only the length scale associated with deviations from the critical structure is scaled: expanding

$$\mathfrak{g} = \frac{1}{V} \Sigma\ \Sigma\ e^{i\vec{k}_c{}^I \cdot \vec{r}} \ |\alpha\lambda k_c{}^2 > W_{I\alpha}(\lambda, \vec{r}) \qquad (20)$$

where I denotes a particular direction on the critical shell, it is the $\vec{r}$ argument in W which is scaled, but not that in the exponential function. With these modifications the procedure is carried out essentially as before [4] yielding the set of equations for the coupled order parameters $W_J \equiv W_{J0}$

$$\frac{\partial W_J}{\partial t} = \gamma_0 W_J + 4\langle 00k_c^2|\underset{\approx}{D}(\underset{\approx}{\Omega}_o - \underset{\approx}{D}k_c^{\,2})^{-1}\underset{\approx}{D}|00k_c^2\rangle(\vec{k}_c^{\,J}\cdot\vec{\nabla})^2 W_J$$
$$-2i\langle 00k_c^2|\underset{\approx}{D}(\underset{\approx}{\Omega}_o - \underset{\approx}{D}k_c^{\,2})^{-1}|v_1\rangle \sum_I \vec{k}_c^{\,J}\cdot\vec{\nabla}\,(W_I W_{I(J)})$$
$$-4i\langle 00k_c^2|\underset{\approx}{N}_2:|00k_c^2\rangle|v_2\rangle \sum_I W_{I(J)}\vec{k}_c^{\,I}\cdot\vec{\nabla}\, W_I \tag{21}$$
$$-2\sum_{II'I''}\delta(\vec{k}_c^I + \vec{k}_c^{I'} + \vec{k}_c^{I''}, \vec{k}_c^J)\langle 00k_c^2|\underset{\approx}{N}_2:|00k_c^2\rangle|v_3^{II'}\rangle W_I W_{I'} W_{I''}$$

where $\vec{I}(J)$ is a direction defined such that $\vec{I} + \vec{I}(J) = \vec{J}$ where $\vec{I}$, $\vec{I}(J)$ and $\vec{J}$ are unit vectors in the corresponding directions, $\delta(a,b)$ is the Kronecker delta and where

$$|v_1\rangle = \underset{\approx}{N}_2:|00k_c^2\rangle|00k_c^2\rangle$$

$$|v_2\rangle = (\underset{\approx}{\Omega}_o - \underset{\approx}{D}k_c^{\,2})^{-1}\underset{\approx}{D}|00k_c^2\rangle$$

$$|v_3^{II'}\rangle = [\underset{\approx}{\Omega}_o - \underset{\approx}{D}(\vec{k}_c^{\,I} + \vec{k}_c^{\,I'})^2]^{-1}|v_1\rangle\ . \tag{22}$$

The matrix $\underset{\approx}{\Omega}_o - \underset{\approx}{D}k_c^2$ is singular, however the inverse $(\underset{\approx}{\Omega}_o - \underset{\approx}{D}k_c^{\,2})^{-1}$ operates only on vectors which are orthogonal to $|00k_c^2\rangle$, so that (21) and (22) are well defined.

TDGL equations for coupled order parameters has been obtained also for symmetry breaking equilibrium transitions. The interesting new feature of the result (21) is the appearance of diffusion terms which are quadratic in the order parameter. This result casts doubts on TDGL equations for symmetry breaking transitions written heuristically without such terms. It should be pointed out however that these terms do not appear in cases of transition to one dimensional structures. [4]

In the case of extrinsic symmetry breaking transitions (Fig.2b), $k_c = 0$ and the situation is similar in this respect to homogeneous transitions. Unlike the later, (11) is satisfied (with k_c=0) so that D_{00} has to be scaled as in the intrinsic case. The resulting TDGL is

$$\frac{\partial W}{\partial t} = D_{00}\nabla^2 W + 2\sum_{\alpha\neq 0}\frac{N^0_{0\alpha}\,N^\alpha_{00}}{\gamma_\alpha}W^3 - \sum_{\alpha\neq 0}\frac{D_{0\alpha}\,N^\alpha_{00}}{\gamma_\alpha}\nabla^2(W^2) + (N^0_{00}\,W_0^2) \tag{24}$$

where the last bracketed term may or may not appear as discussed above for the homogeneous transitions case. Note that also here nonlinear diffusion terms appear.

Scaling of Fluctuations and Ginzburg Criteria [3,4]

Fluctuations are taken into account phenomenologically by supplementing the deterministic terms in (1) by delta correlated Gaussian stochastic source terms

$$\frac{\partial \underset{\sim}{c}}{\partial t} = \underset{\approx}{D}\nabla^2 \underset{\sim}{c} + \underset{\sim}{F}(\underset{\sim}{c},\underset{\sim}{\lambda}) + \underset{\sim}{f}(\vec{r},t) \tag{25}$$

with

$$\underset{\sim}{f}(\vec{r},t) = \underset{\sim}{f}_D(\vec{r},t) + \underset{\sim}{f}_R(\vec{r},t) \tag{26}$$

f_D corresponds to fluctuations associated with the conservative (diffusion) processes while f_R represents the nonconservative (reaction) random noise. [10] In the present simple model

$$<\underset{\sim}{f}_D \underset{\sim}{f}_R> = <\underset{\sim}{f}_D> = <\underset{\sim}{f}_R> = 0 \tag{27}$$

$$<f_D(\vec{r}_1,t_1) f_D(\vec{r}_2,t_2)> = - \underset{\approx}{S}\nabla_1{}^2 \delta(\vec{r}_1-\vec{r}_2)\delta(t_1-t_2) \tag{28}$$

$$<\underset{\sim}{f}_R(\vec{r}_1,t_1)\underset{\sim}{f}_R(\vec{r}_2,t_2)> = \underset{\approx}{Q}\delta(\vec{r}_1-\vec{r}_2)\delta(t_1-t_2) \tag{29}$$

where

$$Q_{ij} = \sum_R \nu_{iR}\nu_{jR} \frac{M_i M_j}{A_0} (r^\circ_{fR}+r^\circ_{bR}) \tag{30}$$

and (for diagonal diffusion matrix)

$$S_{ij} = \frac{2D_j\, C^\circ_j\, M_j}{A_0} \delta_{ij} \tag{31}$$

In (30) and (31) A_0 is the Avogadro number, M_i is the molecular weight of component i, ν_{iR} is the stoichiometric coefficient of component i in the reaction R, C°_j is the steady state concentration (in mass/unit volume) of component j and D_j - its diffusion coefficient. Finally, r°_{fR} and r°_{bR} are the forward and backwards rates (in mole/(unit time)) associated with the reaction R. The sum in (27) is taken over all chemical reactions in the systems.

Starting from (25) the scaling procedure may be repeated essentially as before. The scaling of the random term, resulting from its time and space dependence yields important information about the role of fluctuations in critical phenomena. As an example, return to equation (12), now supplemented by random source terms

$$\dot{W}_0 = D_{00}\nabla^2 W_0 + D_{01}\nabla^2 W_1 + \gamma_0 W_0 + N^0_{00}W_0{}^2 + N^0_{01}W_0W_1 + N^0_{11}W_1{}^2 + f_0(\vec{r},t) \tag{32a}$$

$$\dot{W}_1 = D_{10}\nabla^2 W_0 + D_{11}\nabla^2 W_1 + \gamma_1 W_1 + N^1_{00}W_0{}^2 + N^1_{01}W_0W_1 + N^1_{11}W_1{}^2 + f_1(\vec{r},t) \tag{32b}$$

The scaling (14), (16) now leads to

$$\gamma_1 W_1' + N^1_{00}W_0'^2 + L^{-\frac{1}{2}d} f_{1D} + L^{-\frac{1}{2}(d-2)} f_{1R} = 0 \tag{33}$$

where d is the dimensionality of the system. This leads, for $L \to \infty$, to the slaving equation (17) only for $d > 2$. *For* $d \leq 2$ *slaving is noisy.*

Assuming $d > 2$ and using (17) we further obtain (disregarding the primes which denote scaled variables)

$$\dot{W}_0 = D_{00}\nabla^2 W_0 + \gamma_0 W_0 - \frac{N^0_{01}\,N'_{00}}{\gamma_1} W_0{}^3 \quad (+N^0_{00}W_0{}^2) + L^{(d-4)/2} f_{0R} \tag{34}$$

The scaled fluctuating term now appears with a multiplicative factor which depends on the distance L^{-1} from the critical point and on the dimensionality d. If the scaling is chosen such that all the deterministic terms in (34) are of order unity, the condition for validity of Eq. (18) near the critical point is

$$L^{-(d-4)}\; Q_{00} \ll 1 \tag{35}$$

For $d > 4$ this inequality is satisfied even at the critical point. For $d < 4$ the noise becomes dominant within the critical region where (35) ceases to hold, and mean field theory (i.e. Eq. (18) for the averaged order parameter) breaks down. The inequality (35) may be considered as a reduced version of a Ginzburg criterion [7] which, for equilibrium critical phenomena provides a measure to the size of the critical region. I proceed to show that (35) is indeed equivalent to the Ginzburg criterion.

Consider the stochastic equation for a single order parameter

$$\frac{\partial W}{\partial t} = D\nabla^2 W + F(W) + f(\vec{r},t) \tag{36}$$

where

$$F(W) = \partial U(W)/\partial W \tag{37}$$

$$U(W) = \tfrac{1}{2}\lambda W^2 + \tfrac{1}{2}\nu W^4 \tag{38}$$

and where $f(\vec{r},t)$ is a nonconservative noise term satisfying

$$\langle f \rangle = 0$$

$$\langle f(\vec{r}_1 t_1) f(\vec{r}_2 t_2)\rangle = Q\delta(\vec{r}-\vec{r}_1)\,\delta(t-t_1) \tag{39}$$

Eq. (36) corresponds to the equilibrium probability distribution

$$P[W(r)] = \frac{1}{Z}\exp\Big[-\frac{1}{Q}\int dr\big(U[W(r)] + \frac{1}{2}D(\nabla W)^2\big)\Big] \tag{40}$$

where Z is a normalizing factor. For this system, based on equilibrium considerations involving the free energy functional (40), Ginzburg has derived the following criterion for the validity of mean field theory [7]

$$\frac{QK_d \nu N}{(2\pi)^d (\frac{1}{2}D)^2}\left(\frac{\frac{1}{2}D}{\lambda}\right)^{(4-d)/2} \ll 1 \tag{41}$$

where N is a number of order unity and K_d is the surface area of a one dimensional unit sphere. If $d < 4$, mean field theory breaks down in the close vicinity of the critical point ($\lambda \to 0$).

In order to compare this result to that obtained from the scaling transformation it is convenient to recast it in a dimensionless form. To this end introduce a characteristic time τ (e.g. typical reaction time), characteristic mean value of the order parameter $\bar{W}$ (e.g. the total concentration of a given component at the critical point) and the characteristic length $(D\tau)^{1/2}$. The definitions

$$\tilde{W} = W/\bar{W} \; ; \; \tilde{t} = t/\tau \; ; \; \tilde{r} = r/(D\tau)^{1/2}$$

$$\tilde{\nabla} = (D\;)^{1/2}\nabla \; ; \; \tilde{\lambda} = \lambda\tau \; ; \; \tilde{\nu} = \frac{1}{2}\tau\,\bar{W}^2\nu \tag{42}$$

lead to a transformed form of (36)

$$\frac{\partial\tilde{W}}{\partial\tilde{t}} = \tilde{\nabla}^2\,\tilde{W} + \tilde{\lambda}\tilde{W} + 2\tilde{\nu}\tilde{W}^3 + \tilde{f}(\tilde{r},\tilde{t}) \tag{43}$$

where

$$\langle\tilde{f}(\tilde{\vec{r}}_1,\tilde{t}_1)\;\tilde{f}(\tilde{\vec{r}}_2,\tilde{t}_2)\rangle = \tilde{Q}\delta(\tilde{\vec{r}}-\tilde{\vec{r}}_1)\,\delta(\tilde{t}-\tilde{t}_1) \tag{44}$$

and where

$$Q = \frac{\tau}{\bar{W}^2\,(D\tau)^{d/2}}\;Q \tag{45}$$

In terms of the redefined variables the Ginzburg criterion (41) takes the form

$$\frac{4K_d\tilde{\nu}N}{(2\pi)^d\;2^{(4-d)/2}}\;\tilde{Q}\;\tilde{\lambda}^{(d-4)/2} \;<<\; 1 \tag{46}$$

If the conversion (42) to dimensionless quantities was done with characteristic values of τ and $\bar{W}$ we have

$$\frac{4K_d\tilde{\nu}N}{(2\pi)^d\;2^{(4-d)/2}} \;\sim\; O(1) \tag{47}$$

Ginzburg criterion is therefore

$$\tilde{Q}\tilde{\lambda}^{(4-d)/2} \;<<\; 1 \tag{48}$$

Finally, return to the dimensionless Eq. (43) and introduce the scaling

$$\tilde{\lambda} = L^{-2}\tilde{\lambda}'$$

$$\tilde{t} = L^{2}\tilde{t}'$$

$$\tilde{W} = L^{-1}\tilde{W}'$$

$$\tilde{r} = L^{-1}\tilde{r}'$$

$$\tilde{\nu} = \tilde{\nu}'$$

$$\tilde{Q} = \tilde{Q}' \tag{49}$$

From the discussion following (14) we may conclude that $\tilde{\lambda}'$, $\tilde{t}'$, $\tilde{W}'$ and $\tilde{r}'$ are of order 1. Introducing (49) into (43) taking also (44) into account leads to

$$\frac{\partial \tilde{W}'}{\partial t'} = \tilde{\nabla}'^2 \tilde{W}' + \tilde{\lambda}' \tilde{W}' + 2\tilde{\nu}' \tilde{W}'^3 + L^{(4-d)/2} \tilde{f}(\tilde{r}', \tilde{t}') \tag{50}$$

In (50) all the deterministic terms are of order unity, therefore fluctuations are negligible (and mean field theory holds) provided

$$(L^{2-d/2})^2 \tilde{Q} \ll 1 \tag{51}$$

and since $L^2 \sim \tilde{\lambda}^{-1}$ we obtain again the result (48).

This procedure provides a powerful method for deriving Ginzburg criteria for the validity of mean field theory in different equilibrium and non-equilibrium systems. In particular, unlike the original Ginzburg derivation the existence of a potential (37) is not required.

It is interesting to note that for the case of homogeneous transitions considered so far, the conservative fluctuations do not play any role near the critical point. This is seen in the analysis which leads to (34) and is intuitively expected: as the correlation range increases, the role of fluctuations associated with the diffusion process decreases as the system of interest becomes smoother in space. This will not be the case for intrinsic symmetry breaking transitions where the critical structure prevails at the critical point. Indeed for this case it may be shown [4] that (35) is replaced by the inequality

$$L^{-(d-4)} (Q_{00} + k_c^2 S_{00}) \ll 1 \tag{52}$$

To conclude, I list below the results for the Ginzburg criteria obtained for the different critical points discussed in this article (g and δ are characteristic concentration and diffusion coefficient respectively).

Type	TDGL Equation	Ginzburg Criterion	Critical Dimensionality
Homogeneous Transitions	(18)*	$\lambda^{(4-d)/2} \gtrsim \frac{M}{A_0} \frac{(\delta\tau)^{-d/2}}{g}$	4
Intrinsic Symmetry Breaking	(21),(22)	$\lambda^{(4-d)/2} \gtrsim \frac{M}{A_0} \frac{(\delta\tau)^{-d/2}}{g} \times (1+\delta\tau k_c{}^2)$	4
Extrinsic Symmetry Breaking	(24)	$\lambda^{(6-d)/2} \gtrsim \frac{M}{A_0} \frac{(\delta\tau)^{-d/2}}{g}$	6

*For the general result obtained in the case of many chemical components see ref. [3,4].

A closer look at the Ginzburg criteria summarized in this Table [3] reveals that non-classical behavior may in principle be observed in non-equilibrium critical pointsassociated with chemical instabilities. This is in contrast to other non-equilibrium transitions (e.g. the laser threshold [2a] and the Bernard instability [2b] that were investigated by similar methods. In these cases mean field theory has been found to be valid for all conceivable values of the control parameters. On the other hand, chemical systems offer a much wider range of parameters: reaction rates vary over more than ten orders of magnitude and concentrations and diffusion coefficients may also vary over a few orders of magnitude. Thus for chemical instabilities the critical regime may be observed in principle for reasonable choices of chemical and transport parameters. We should bear in mind however that in presently known systems like the Belousov-Jabotinsky reaction [11] or the photo-thermochemical instability in the NO_2/N_2O_4 system [12] the reactions are too slow or the diffusion coefficients too large to allow, with the experimental resolution in the corresponding control parameters, a controlled close approach to the non-classical critical regime.

Conclusions

The present study has resulted in a few conclusions which are summarized below.

(1) Critical conditions are necessary for a proper reduction of the equations of motion to obtain the TDGL equation for the order parameter near the critical point.

(2) Reduction was achieved by scaling. The scaling reflects the critical slowing down and the divergence of the correlation length.

(3) Nonconventional TDGL equations containing non-linear transport terms are obtained for symmetry breaking transitions.

(4) The scaling used is "mean field scaling" which breaks down in the close vicinity of the critical point. The procedure provides its own limits of validity in the form of Ginzburg criteria obtained from the scaling of the noise.

(5) For $d \leq 2$ slaving is noisy. Careless neglect of apparently irrelevant variables may lead to erroneous results.

(6) Chemical transitions offer enough variety of parameters to make the observation of non-mean field critical behavior possible in principle (in contrast to the laser and the hydrodynamic transitions).

(7) All presently known chemical transitions are predicted to behave according to mean field theory.

The procedure described in the present paper provides a general and powerful tool for a quick analysis of critical behavior of non-equilibrium systems. As an example, homogeneous transitions in ionic systems in the presence of an electric field were recently investigated [5], with the result that the critical dimensionality becomes 3 for a finite electric field strength. As the field intensity grows from zero a typical cross-over behavior should be observed. Similarly, more systems may be investigated and their experimental feasibility and significance predicted using these methods. Further application of the theory to tricritical and more complicated phenomena is of course possible and will be carried out in the future.

Acknowledgement

This research was supported by the US-Israel Binational Science Foundation, Jerusalem, Israel. I am grateful to Dr. P. M. Rentzepis and to Bell Telephone Laboratories for their hospitality during the time when the manuscript was written.

References

1. H. Haken, *Synergetics, an Introduction*, (Springer, Berlin 1977).
2. a) R. Graham, Phys. Rev. A10, 1762 (1974); R. Graham and A. Pleiner, Phys. Fluids 18 130 (1975); J. Swift and P. C. Hohenberg, Phys. Rev. A15 315 (1977).
   b) R. Graham in Fluctuations, instabilities and Phase Transitions, edited by T. Riste (Plenum, N.Y., 1976).
   c) Y. Kuramoto and T. Tsuzuki, Prog. Theor. Phys. 52 1399 (1974) and 54 687 (1975); A. Wunderlin and H. Haken, Z. Physik B21 393 (1975); H. Haken, Z. Physik B29, 61 (1978) and references therein.
3. A. Nitzan, Phys. Rev. A, 17, 1513 (1978) and in *Synergetics, a Workshop* edited by H. Haken (Springer 1977).
4. A. Nitzan and P. Ortholeva, "Scaling and Ginzburg Criteria for Critical Bifurcations in Nonequilibrium Reacting Systems", Phys. Rev. A, in press.
5. M. DelleDonne, S. Schmidt, A. Nitzan and P. Ortoleva, "Scaling and Crossover in Nonequilibrium Reaction Diffusion Phenomena in Presence of Electric Field", to be published.
6. H. Mori, Prog. Theor. Phys. 52, 433 (1974) and 53, 1617 (1975); H. Mori and K. J. McNeil, Prog. Theor. Phys. 57, 770 (1977).
7. V. L. Ginzburg, Fizika Tverdogo Tela 2 2031 (1960).
8. A. Nitzan, P. Ortholeva and J. Ross, J. Chem. Phys. 60, 3134 (1974).
9. A. Nitzan, P. Ortoleva, J. Deutch and J. Ross, J. Chem. Phys., 61 1056 (1974).
10. C. W. Gardiner, J. Stat. Phys. 15 451 (1976); S. Grossman, J. Chem. Phys. 65, 2007 (1976).
11. R. M. Noyes and R. J. Field, Acc. Chem. Res. 10 215; 273 (1977).
12. C. L. Creel and J. Ross, J. Chem. Phys. 65 3779 (1976).

Part V

Turbulence and Chaos

Diffusion-Induced Chemical Turbulence

Y. Kuramoto

Department of Physics, University of Kyoto
Kyoto 606, Japan

1. Introduction

It is almost a common belief today that there exist such macroscopic phenomena that are certainly governed by *deterministic chaos*. Many of the chaotic phenomena appear as *spatio-temporal chaos*, and their adequate mathematical modeling often calls for a set of partial differential equations. In the past, the studies of spatio-temporal chaos have largely been concerned with a variety of turbulent phenomena in fluid systems [1,2] obeying the Navier-Stokes equation. Recent studies made it clear that a much simpler class of partial differential equations called reaction-diffusion equations is also capable of showing complicated space-time behavior. This was demonstrated for the Brussels model [3-5], the Rashevsky-Turing model [6], and also for a still wider class of systems consisting of diffusion-coupled oscillators [3-10]. In view of the surprising richness of patterns which reaction-diffusion equations can exhibit, it is quite natural to expect that chemical turbulence of diffusion-induced type could hardly be confined to the kinds discovered so far. The purpose of the present paper is to explore the possibility of a novel class of chemical turbulence, in particular, the one associated with the wavefront behavior of some chemical waves in media with the spatial dimension *two*.

Figure 1 shows schematically some chemical waves traveling in two-dimensional media. The first one (Fig.1a) is a *pulse* or also called a trigger wave. This kind of waves may actually be observed in various forms in excitable media [11] such as the Belousov-Zhabotinsky reaction system [12] and nerve axon.

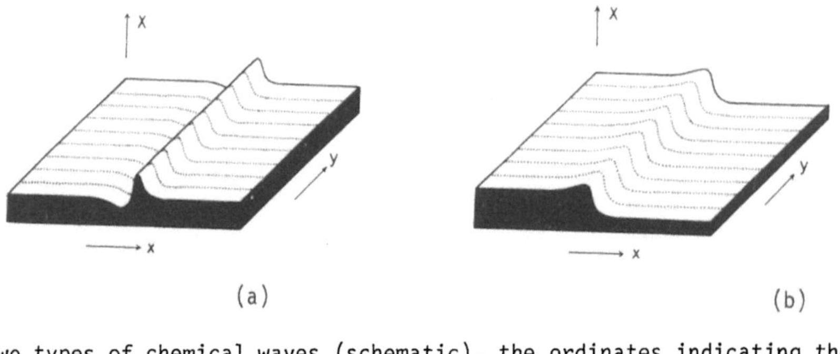

Fig.1 Two types of chemical waves (schematic), the ordinates indicating the concentration of some species. (a) Pulse; (b) Front

The second one (Fig.1b) is called a *front* (in a restricted sense) which is expected to appear in systems possessing multistability [13,14], although no one has ever succeeded in visually realizing it in chemical experiments. A feature common to pulses and fronts is that the medium is quiescent in most part while a sharp spatial variation of concentrations appears only in a narrow region in space. Despite the term *front* representing a certain type of chemical waves, we shall also use below the term *wavefront* or simply *front* in a slightly different meaning, namely, for representing (loosely) the contour on which a sharp concentration change occurs for either type of waves.

The waves in Fig.1 are laterally uniform. A simple question comes to mind, however: Is a laterally uniform wavefront always stable? In thermodynamic systems such as coexisting liquid and gas, the interface has surface tension, which will always act as a flattening force on the interface according to the principle of minimum free energy. This kind of rule does not seem to apply, however, when chemically reacting systems are under consideration because they are in general far from thermal equilibrium. In fact, a soluble example of reaction-diffusion equation to be studied later proves to have *negative surface tension*, implying the instability of a uniform wavefront. It will also be shown later that the instability of this kind leads to turbulence after a number of bifurcations.

In general, whenever a chaotic behavior has first made its appearance after some bifurcations, it will still preserve a rather simple character in some sense. Thus one might reasonably expect that it could be reduced to a known prototype of chaos found so far e.g. for some three-component systems of coupled ordinary differential equations [15]. If one is fortunate enough, it might even be reducible to a one-dimensional discrete chaos such as the one obtained from the logistic type difference equation [16,17]. It turns out that such a reduction is actually possible for our wavefront chaos.

2. Stability of a Uniform Wavefront

To begin with, we try to establish the stability condition for a chemical wave which is laterally uniform [18,19]. Let $(X_1, X_2, \cdots X_n) \equiv X$ be a concentration vector, and we study a general reaction-diffusion equation in the form

$$\partial_t X = F(X) + D(\partial_x^2 + \partial_y^2)X \tag{1}$$

in a two-dimensional medium with the coordinates x and y. D is a diffusion matrix with non-negative elements. We assume that the system is sufficiently extended in both the x and y directions. Let $X_s(x-vt)$ be a solution of (1) representing a steadily propagating chemical wave of the type of either Fig.1a or Fig.1b; periodic wave trains are excluded from our consideration for the reason found later. The wavefront has been chosen to be parallel to the y axis, and the propagation velocity has been denoted by v.

According to the standard linear stability analysis, we introduce a small deviation u around X_s by

$$X(x, y, t) = X_s(z) + u(z, y, t) \;, \tag{2}$$

where z is the moving coordinate, $z \equiv x-vt$. Since our interest is in the behavior of a laterally *nonuniform* disturbance or y-dependent u, it is reasonable to require $X_s(z)$ to be stable at least with respect to any laterally *uniform* disturbance u(z,t). This stability requirement may be formulated as follows. We substitute (2) without y-dependence of u into (1), linearize it in u, and get

$$\partial_t u(z, t) = \Gamma u(z, t) \;, \tag{3}$$

where Γ is the linear operator

$$\Gamma = \Gamma_0 + D\partial_z^2 + v\partial_z \;, \tag{4}$$

$$(\Gamma_0)_{ij} = \partial F_i(X_s)/\partial X_{sj} \;. \tag{5}$$

Equation (3) generates various eigenvectors $u_\ell(z)$ with the eigenvalues λ_ℓ, where the time-dependence of the eigenmodes has been assumed as $\exp(\lambda_\ell t)$. Thus the stability of X_s means the inequality

$$\mathrm{Re}\lambda_\ell \leq 0 \tag{6}$$

to hold for any ℓ. It is well known that the stability problem of wave propagation in autonomous dissipative systems necessarily involves one special kind of disturbance whose eigenvalue is exactly zero. In the present case, this corresponds to an infinitesimal spatial translation of $X_S(z)$ along the z direction. We take this neutral mode to be the ℓ=0 mode and write

$$\lambda_0 = 0 \ , \qquad u_0(z) = \partial_z X_s(z) \ . \tag{7}$$

In contrast to the translational mode, the other disturbance deforms the wave structure itself. Thus we assume it to be decaying, or $\mathrm{Re}\lambda_\ell < 0$ ($\ell \neq 0$). To be precise, we assume further that the zero-eigenvalue is isolated, which is expected to be true as far as the steady solutions of the type of Fig.1 are concerned, but may not be true for the case of periodic wave trains.

We now come back to the more general case, i.e. the case of y-dependent u. In particular, this dependence is assumed to be sufficiently slow. Let us seek the eigenvectors in the form

$$u(z, y \ , \ t) = u(z, \ t)\exp(iQy) \ , \tag{8}$$

where Q is a sufficiently small wavenumber. As before, we get the linearized equation

$$\partial_t u = (\Gamma - DQ^2)u \ . \tag{9}$$

Note that the existence of the term DQ^2, which is sufficiently small by assumption, makes the only difference of (9) from (3). Thus it is appropriate to make a perturbation expansion of the eigenvalue $\lambda_{\ell Q}$ of (9) in powers of Q^2 around λ_ℓ:

$$\lambda_{\ell Q} = \lambda_\ell - \alpha_\ell Q^2 - \gamma_\ell Q^4 - \cdots \ . \tag{10}$$

The formal expression for the expansion coefficients above may readily be found from an elementary perturbation theory [19]. For vanishing Q we have already assumed the stability of X_S. The problem now is whether its stability property is affected by allowing for nonvanishing but sufficiently small Q. The answer is clear. Namely, the change in the stability is possible, but it is only with respect to the translational mode because its unperturbed eigenvalue λ_0 is identically zero. Thus the sign of the first expansion coefficient α in the dispersion equation for the translational branch

$$\lambda_{0Q} = -\alpha Q^2 - \gamma Q^4 - \cdots \tag{11}$$

becomes decisive of the stability of X_S. Namely, X_S is stable if α is positive, and vice versa. From the perturbation theory [19], α is found to be

$$\alpha = \int_{-\infty}^{\infty} {}^t u_0^\dagger(z) D u_0(z) dz \ , \tag{12}$$

where $u_0^\dagger$ is the null-eigenvector of the adjoint operator of Γ, and the superscript t means the transpose. For reference, the mentioned adjoint operator $\Gamma^\dagger$ is given by

$$\Gamma^\dagger = {}^t\Gamma_0 + {}^t D d_z^2 - v d_z \ . \tag{13}$$

The higher coefficients may also be calculated. In particular, γ has the form

$$\gamma = \sum_{\ell \neq 0} \lambda_\ell^{-1} \int_{-\infty}^{\infty} {}^t u_0^\dagger(z) D u_\ell(z) dz \int_{-\infty}^{\infty} {}^t u_\ell^\dagger(z) D u_0(z) dz \ . \tag{14}$$

In addition to the stability criterion, the above stability analysis tells us something more. In fact, the linear dispersion equation (11) implies the form of the evolution equation to be obeyed by a small amplitude translational disturbance. Since the translational disturbance is the same thing as the positional disturbance of the wavefront, it is appropriate to introduce a positional or phase disturbance

ψ in such a way that the x coordinate of the front for given y is the sum of ψ and the systematically changing part vt (see Fig.2). Thus the quantity ψ is expected to obey the equation

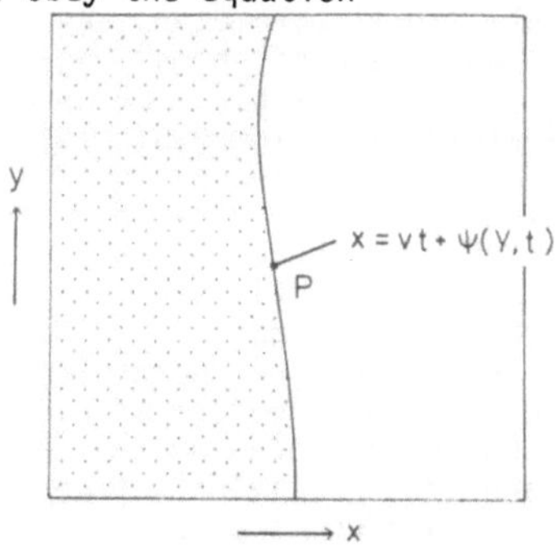

Fig.2

$$\partial_t\psi = \alpha\partial_y^2\psi - \gamma\partial_y^4\psi + \cdots , \tag{15}$$

which has the form consistent with (11). The quantity α in (15) now appears as the diffusion coefficient of the front disturbance, and the study of its sign is the main subject of the following section. As to the sign of γ, however, we shall assume its positivity throughout.

3. An Example Showing Negative α

Before trying to extend (15) to nonlinear regime, we will stop here to examine the possibility of α becoming negative. By looking at the formula (12), the following fact is noticed. As a special case, let D be a diagonal matrix with the equal (positive) diagonal elements d. Then, we immediately get the equality $\alpha=d$ on account of the assumed orthonormality condition for the eigenvectors. This fact shows that the difference in diffusion constants is a necessary condition for the occurrence of the front instability. Further information on the sign of α is contained in the null-eigenvectors u_0 and $u_0^\dagger$, hence we are led to the study of a concrete model. The rest of this section is a brief summary of the results obtained in [19].

As a simplest nontrivial model which permits the analytical calculation of α, we consider the piecewise linear Bonhoeffer-van der Pol model including diffusion with the spatial dimension one:

$$\begin{aligned} \partial_t X &= -X + H(X-a) - Y + D_X\partial_x^2 X , \\ \partial_t Y &= bX - cY + D_Y\partial_x^2 Y . \end{aligned} \tag{16}$$

Here H is the Heaviside step function, and the parameters a, b, c, D_X and D_Y are assumed to be positive. The studies by McKean [20], and Rinzel and Keller [21] correspond to the case $c=D_Y=0$. Then model then represents the piecewise linear FitzHugh-Nagumo nerve conduction equation [22,23]. Rinzel and Keller obtained analytic forms of some pulse solutions, and discussed their stability. Winfree studied the two-dimensional extension of (16) for the case of equal diffusion constants and without cY term again, and obtained spiral waves by numerical experiment [11]. These works will serve as a sufficient evidence that the present model is undoubtedly nontrivial in spite of its simple appearance.

A brief remark on the local property of the system (16) is now given. Let us drop the diffusion for the moment. Figure 3 shows the nullclines $d_tX=0$ and $d_tY=0$. The first nullcline is an idealization of a sigmoidal manifold, and the second one is a straight line. Depending on the values of the parameters, one may have one or three fixed points. If we have only one fixed point, the system is considered to represent a monostable excitable system and, by including diffusion, trigger waves may be excited under appropriate initial conditions. In the case of three fixed points, the system is bistable, and the solution of expanding domain may exist. These two types of waves are the things we have just been concerned with, hence our model seems to be appropriate for our purpose. In fact, one may obtain analytically the solution representing a single pulse or a single front, thanks to the piecewise linear character of (16). Its stability is strongly suggested from the numerical calculation of the eigenvalues. The null-eigenvectors, and hence α may also be given analytically, but in an implicit form. Actually, it turns out that the determination of α involves so complicated transcendental equations, that one is hardly able to extract any useful information without going to some limiting case, in particular, the case where some of the parameters are vanishingly small.

An interesting extreme case is provided by

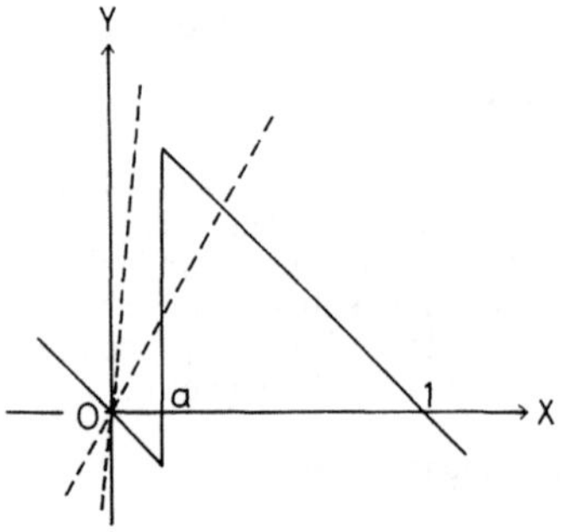

Fig.3 The nullclines of the model (16) without diffusion (Solid line: $d_tX=0$; Broken lines: $d_tY=0$ for monostable and bistable cases)

$$b = \epsilon \tilde{b}, \quad c = \epsilon \tilde{c}, \quad D_X = \epsilon^{1/2}\tilde{D}_X , \tag{17}$$

where ϵ is a sufficiently small parameter, and $\tilde{b}$, $\tilde{c}$, $\tilde{D}_X$ as well as a and D_Y are of normal magnitude. The fact that b and c are small means that our sigmoidal manifold is a slow manifold [24]. Calculation shows that the increasing slowness of the sigmoidal manifold means the increasing difficulty for α to become negative. However, if we let D_X be sufficiently small simultaneously, then we can make α negative again. Thus the decreasing b and c, and the decreasing D_X have the opposite effects on the sign of α. These effects may be counter-balanced by the scaling choice (17), and the transition between a positive and negative α becomes possible.

The calculated wave profiles of the pulse and front solutions are illustrated in Fig.4, where the infinitely sharp wavefronts are the result of letting $\epsilon \to 0$. For both the pulse and front,

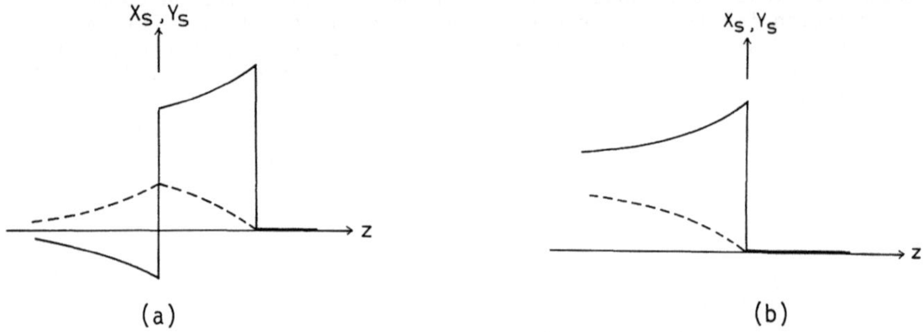

Fig.4 Analytically obtained wavepatterns for the extreme case (17). (a) Pulse; (b) Front (Solid lines: X_s; Broken lines: Y_s)

the expression for α is found to be identical, taking the form

$$\alpha = \epsilon^{1/2}\tilde{D}_X\{1 - \frac{\tilde{b}}{(1-2a)^2}(\frac{D_Y}{\tilde{D}_X})^2\} . \tag{18}$$

It is obvious from the expression above that α may become either negative or positive.

4. Qualitative Interpretation of Front Instability

Although we could demonstrate analytically the possibility of α becoming negative, it is still desirable that the origin of the front instability will be understood in qualitative terms, which is now described briefly. The system considered in the preceding section may be viewed as a system of an (potentially) activating substance X and inhibiting one Y with D_Y much larger than D_X. One may recall here that the study of activator-inhibitor systems with (relatively) fast diffusion rate of the inhibitor has long been a subject of numerous works ever since the pioneering works by Rashevsky [25] and Turing [26]. Indeed such systems often show spontaneous spatial differentiation, and hence they have served as convenient reaction-diffusion models of morphogenesis [27-30]. As is easily checked by the linear stability analysis, for the above kind of symmetry-breaking instability to occur, the system must satisfy the condition that the two nullclines $d_tX=0$ and $d_tY=0$ have the same sign in their tangents at their intersection point (see Fig.5). (Although this condition is already a constituent of the frequently used definition of an activator-inhibitor system, one may still call a system *potentially* an activator-inhibitor system even if the above condition is not satisfied like the case b in Fig.5 and the system (16) .) It is obvious from Fig.3 that our previous example does not satisfy this condition, and naturally we did not get any symmetry-breaking instability, obtaining instead propagating waves as far as the media with the spatial dimension one is concerned. Nevertheless, an important fact to be noticed is that the system's potential

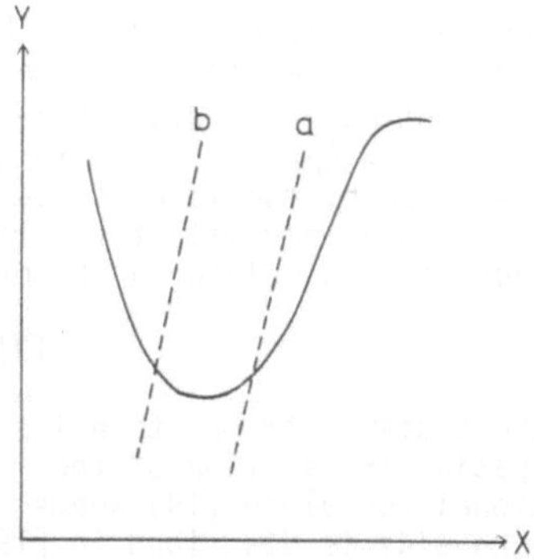

Fig.5 A schematic nullclines of an activator-inhibitor system. The Turing type instability is possible for the case a, but not for b

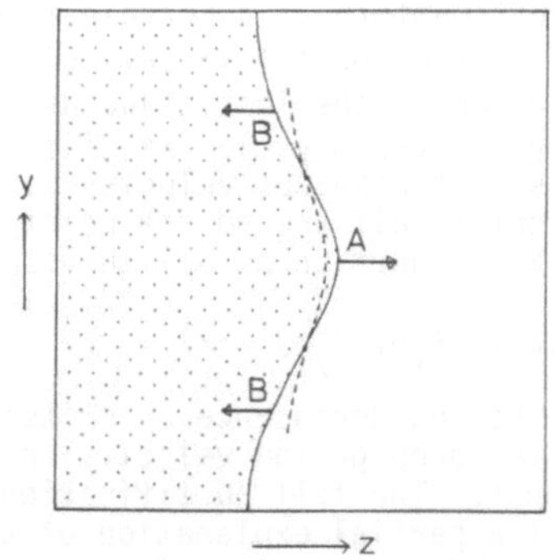

Fig.6 An intuitive picture of the front instability. The shaded region is the activated domain. The arrows indicate the motion of the front due to the cross-inhibiting effect

nature of activation and inhibition combined with the fast diffusion of the inhibiting substance causes another kind of spatial instability, namely a front instability, once many identical one-dimensional waves have been laterally coupled through diffusion. Such a system is therefore of double characters, namely, cross-excitory and cross-inhibitory ones. The former character is dominant along the direction of propagation, while the latter along the lateral direction causing spatial instability.

Let a system (X,Y) be a bistable system for the sake of definiteness, and let $D_Y \gg D_X$. An intuitive picture of what is brought about by the cross-inhibition along the lateral direction is as follows. Suppose that the activated domain with a uniform wavefront is expanding. In the activated region, the concentration of the activating substance is higher, which implies the inhibitor, induced by the activator, also has the higher concentration there. Suppose that a part of the wavefront has been pushed foward to produce a nonuniformity (Fig.6), and we observe what follows. Since the inhibitor is rapidly diffusing, it will immediately flow out of the peninsula, thereby forming its own wavefront somewhat smoother than the activator's front. In any case, both substances of course have positive diffusion rates, so that the front nonuniformity will have more or less a tendency to be flattened. An important countereffect should not be overlooked, however. At the promontory A, the activator will come to experience the scarcity in the inhibitor concentration because the latter has diffused out as a result of the front deformation. This implies that the autocatalytic production of the activator is even promoted, thereby increasing the local propagation velocity. What happens in the neighboring regions B is the exact converse. There the activator will come to experience the richer concentration of the inhibitor than before, implying the decrease in the local propagation velocity. It is evident that the front nonuniformity is strengthened by this kind of cross-inhibition process. The above-described two tendencies conflict with each other, making the wavefront stable or unstable.

5. Nonlinear Evolution Equation for the Wavefront

So far the matter of our main concern has been the stability of a laterally uniform wavefront, and we tried to clarify its mathematical and physical aspects. We are now ready to attack the next problem: How does the wavefront evolve after instability? This study necessarily involves the derivation of a nonlinear evolution equation governing the wavefront, which might sound like a little hard task. Actually, however, this turns out unexpectedly simple provided the wavefront is only weakly unstable (or stable). The method developed below is slightly phenomenological, although a more sophisticated perturbation theory may also be developed [19].

Let the front pattern at a certain moment be like the one in Fig.2 as before, but the amplitude of the spatial variation of ψ being no longer confined to the linear regime. For given y, the instantaneous velocity of the front observed along the direction x should be $v+\partial_t\psi(y,t)$. At this point, the following assumption seems to be natural: This instantaneous velocity is completely determined by the local shape of the wavefront itself around the point P of interest. This is equivalent to claiming that $\partial_t\psi$ is a function of all the spatial derivatives of ψ calculated at P, or

$$\partial_t\psi = \Omega(\partial_y\psi,\ \partial_y^2\psi,\ \partial_y^3\psi,\ \cdots)\ . \tag{19}$$

Note that $\partial_t\psi$ has no dependence on ψ itself, because the system's translational symmetry leaves the propagation velocity unchanged by a spatial translation of the front, $\psi\to\psi+\text{const}$. The full justification of the functional postulate (19) seems difficult, yet a partial explanation of the underlying physics is described in [19], where it is stated that a kind of *local equilibrium assumption* or Haken's *slaving principle* [31] is certainly behind.

The problem now is how to make the form of Ω explicit. There are at least two obvious properties satisfied by Ω. The first is almost trivial and is expressed as $\Omega(0,0,\cdots)=0$. This follows from the fact that the propagation velocity along the x direction is v, or $\partial_t\psi=0$, if the wavefront is a straight line and parallel to the y axis. The second property is given by $\Omega(\partial_y^{2n}\psi,\ -\partial_y^{2n-1}\psi)=\Omega(\partial_y^{2n}\psi,\ \partial_y^{2n-1}\psi)$, namely, Ω is invariant under the simultaneous inversion of the sign of all the odd-order derivatives. This property comes from the fact that even if we have replaced the front pattern by its mirror image as shown in Fig.7, the propagation velocity along x at P should not be changed due to the isotropy of the system. The above two properties may briefly be written as

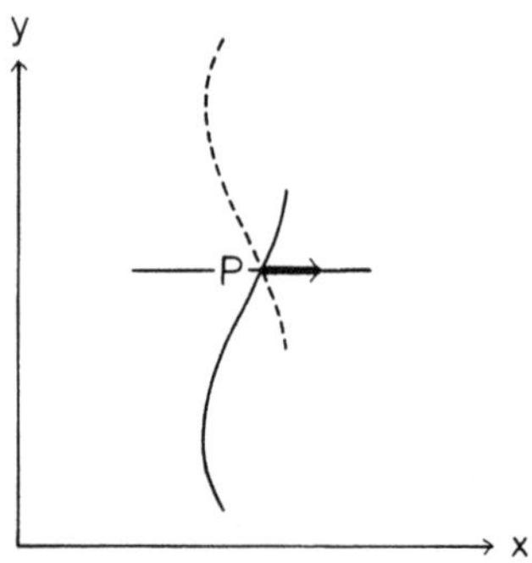

Fig.7 The replacement of a wavefront (solid line) by its mirror image (broken line) with respect to the horizontal line passing through P leaves the propagation velocity (thick arrow) along the x direction unchanged

$$\begin{aligned}&(\mathrm{I})\quad \Omega(0) = 0\ ,\\ &(\mathrm{II})\quad \Omega(-\partial_y) = \Omega(\partial_y)\ .\end{aligned} \tag{20}$$

The next step is the formal expansion of Ω with respect to all its arguments. The resulting expansion series may conveniently be rearranged in powers of the formal expansion parameter ∂_y. No zeroth order term in ∂_y appears due to the property (I). Terms like $\partial_y\psi$, $\partial_y^3\psi$, $\partial_y^2\psi\partial_y\psi$, that is, terms containing odd numbers of ∂_y are also absent on account of the property (II). One may therefore write generally,

$$\partial_t\psi = \{\alpha\partial_y^2\psi+\beta(\partial_y\psi)^2\}+\{-\gamma\partial_y^4\psi+\cdots\}+O((\partial_y)^6)\ . \tag{21}$$

In each order in ∂_y, we have several different types of terms. The constants α, β, γ and so on appear as phenomenological constants. As far as α, β and γ are concerned, however, their determination is rather easy in the following manner. When the nonlinearity is neglected, (21) must coincide with the linear equation (15), hence α and γ in (21) should be identical with the quantities given by (12) and (14), respectively. The constant β may be determined by considering another special case where the wavefront is a straight line but only a little non-parallel

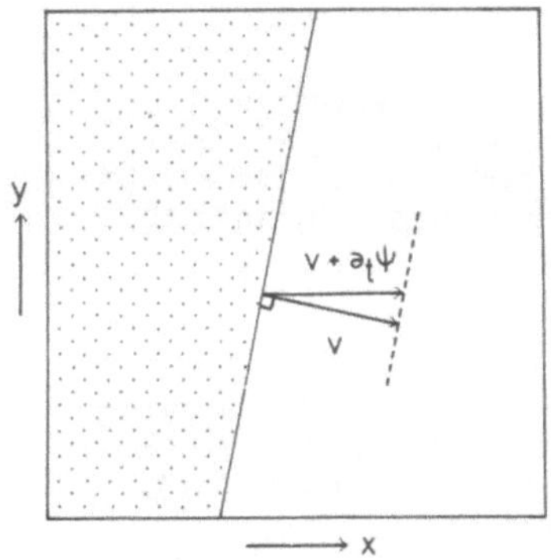

Fig.8

to the y axis (see Fig.8). Then we put

$$\partial_y\psi = c , \tag{22}$$

where c is a sufficiently small number. All the other derivatives $\partial_y^n\psi$ $(n\geq 2)$ are vanishing by assumption. Since the wavefront here is an essentially undistorted one, the propagation velocity normal to the front is v, and if observed along the x direction, it is $v+\partial_t\psi$ by definition. From Fig.8 it is seen that these quantities, v and $v+\partial_t\psi$, are related to each other through the quantity c, and we readily find

$$\partial_t\psi = vc^2/2 . \tag{23}$$

But the quantity $\partial_t\psi$ may independently be calculated from (21) via (22) and the property $\partial_y^n\psi=0$ $(n\geq 2)$, and we get

$$\partial_t\psi = \beta(\partial_y\psi)^2 = \beta c^2 \tag{24}$$

for sufficiently small c. By comparison of (23) and (24), we get

$$\beta = v/2 . \tag{25}$$

Thus we could determine the coefficients of the first three terms in (21). Fortunately enough, no other terms than the above need be taken into consideration when $|\alpha|$ is sufficiently small, namely, when the stability of the front is near its criticality. The reason is seen as follows. The solution ψ of the truncated equation

$$\partial_t\psi = \alpha\partial_y^2\psi + \beta(\partial_y\psi)^2 - \gamma\partial_y^4\psi \tag{26}$$

should be of the scaling form

$$\psi = |\alpha|\tilde{\psi}(|\alpha|^{1/2}y, \alpha^2 t) \tag{27}$$

with respect to the small parameter α. The above scaling form enables us to estimate the order of magnitude of each term in (21). It is easy to see that the terms $\partial_t\psi$, $\alpha\partial_y^2\psi$, $(\partial_y\psi)^2$ and $\partial_y^4\psi$ are $\sim\alpha^3$, while all the neglected terms are $\sim\alpha^4$ or higher. Thus, our truncation has proved to be consistent. In what follows, we shall work with the truncated equation (26).

6. Shock Solutions

In order to get some feeling for (26), let us first consider the normal case that α is positive. Then the term $\alpha\partial_y^2\psi$ represents ordinary diffusion, so that the additional dissipation term $\partial_y^4\psi$ will not play a very important role. Let us therefore neglect it for the moment. The resulting equation is

$$\partial_t\psi = \alpha\partial_y^2\psi + \beta(\partial_y\psi)^2 . \tag{28}$$

Calculation shows that (28) admits a family of shock solutions of the form

$$\psi = \beta(a^2+b^2)t + ay + \alpha\beta^{-1}\ln[\cosh\{b\beta\alpha^{-1}(y+2a\beta t)\}] . \tag{29}$$

Qualitatively, the solution above describes a pair of nearly uniform wavefronts colliding obliquely to form a shock, as illustrated in Fig.9. The parameters a and b in (29) are related to the slope of the wavefront at infinity:

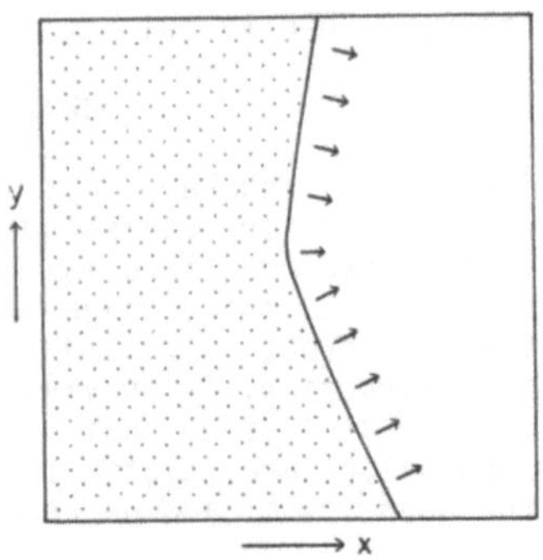

Fig.9 A shock pattern corresponding to (29)

$$a \pm b = \lim_{y\to\pm\infty} \partial_y\psi \, . \tag{30}$$

People often speak of shock waves in connection with the Burgers equation [32], a model for one-dimensional turbulence. Indeed, the replacement $\partial_y\psi=U$ transforms (28) into the Burgers equation

$$\partial_t U = \alpha\partial_y^2 U + 2\beta U\partial_y U \, . \tag{31}$$

Therefore, the shock solution (29) is mathematically equivalent to the Burgers shock. Eq.(31) may further be transformed to the simple diffusion equation

$$\partial_t f = \alpha\partial_y^2 f \tag{32}$$

through the Hopf-Cole transformation

$$\exp[\alpha^{-1}\beta\int U dy] = f \, . \tag{33}$$

If α were negative, (33) would be meaningless, and at the same time its equivalence (28) would lose meaning. Then, the term $\partial_y^4\psi$, which becomes the only term contributing to dissipation, comes to play an essential role.

7. Scaling and Symmetry Properties

In this section and in the next, we concentrate on the behavior of ψ obeying (26) when α is negative (and vanishingly small). The study has been done numerically. Before showing the results, however, some scaling and symmetry properties of the system have to be explained.

The interesting scale of the system length L is considered to be

$$L = |\alpha|^{-1/2} L_0 \, , \tag{34}$$

where L_0 is an α-independent quantity. In fact, by the above choice, the number of the linearly unstable modes obtained from the dispersion equation (11) is made finite in the limit $\alpha\to -0$; For any other scaling choice, this number will become zero or infinite. Let us make the rescaling

$$\psi \to \beta^{-1}\gamma(L/L_0)^{-2}\psi \, , \quad y \to (L/L_0)y, \quad t \to \gamma^{-1}(L/L_0)^4 t \, . \tag{35}$$

Then (26) becomes

$$\partial_t\psi = -\sigma\partial_y^2\psi + (\partial_y\psi)^2 - \partial_y^4\psi \, , \tag{36}$$

where $\sigma=\gamma^{-1}>0$, and the system length has now been reduced to L_0. The choice of the value L_0 is at our disposal, and it may be chosen according to the computational easiness. Thus the only parameter is σ which has a normal magnitude.

The numerical calculation was carried out directly for the partial differential equation (36) with suitable discretization of space and time, assuming the condition $\partial_y\psi=0$ at the boundaries y=0 and L_0. The numerical data may conveniently be analyzed by means of a mode picture. Indeed, ψ may be decomposed into many cosine functions as

$$\psi(y, t) = \sum_{n=0}^{\infty} A_n(t)\cos(n\pi y/L_0) \, , \tag{37}$$

and one may imagine the phase space spanned by the mode amplitudes A_n. Practically, it will be appropriate to work with a subspace spanned by the amplitudes of a few important modes. How to choose a meaningful subspace of this kind may partly be implied by the symmetry property of the system, which is now explained briefly. Our system clearly has such a symmetry that if $\psi(y,t)$ is a solution of (36) satisfying the aforementioned boundary condition, then the function $\psi^*(y,t)$ obtained by its spatial inversion should also be a solution satisfying the same boundary condition. If ψ is decomposed into the spatially symmetric and antisymmetric parts as

$$\psi(y,t) = \psi_s(y,t) + \psi_a(y,t) \ , \tag{38}$$

then the spatial inversion means the inversion of the sign before the antisymmetric part ψ_a. From this fact together with the fact that ψ_a is the part corresponding to odd n in (37) (and ψ_s to even n), it is obvious that the spatial inversion is equivalent to the transformation

$$(A_{even}, A_{odd}) \rightarrow (A_{even}, -A_{odd}) \tag{39}$$

The last statement is now illustrated, by taking a two-dimensional odd-mode space, e.g. A_1-A_3 (see Fig. 10). A solution $\psi(y,t)$ in the limit $t\to\infty$ will form a limiting trajectory (fixed point, limit cycle or more complicated attractor) in the infinite-dimensional phase space, and its projection onto the A_1-A_3 subspace is now represented by C. If C is point-symmetric (solid curve in Fig. 10), it is implied that the corresponding state still preserves the above mentioned inversion symmetry, although the full odd-mode space has to be examined to make it completely sure; If not point symmetric (broken curve in Fig.10), the state realized is definitely symmetry-breaking. In the latter case, another limiting trajectory C^* must exist, which is the thing obtained from the rotation of C by the angle π. The pair (C,C^*) of course recovers the system's symmetry.

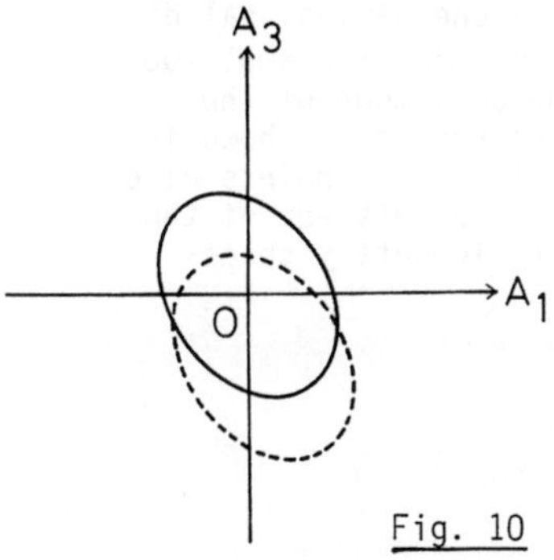

Fig. 10

In the following section, we will work mainly with the subspace A_1-A_3, and if necessary, we add the third axis corresponding to the even mode A_2 or A_4.

8. Wavefront Chaos

In our numerical calculation, we have put L_0=10.1 which is the value of no special significance. As we vary σ from small to large, the front ψ shows a number of bifurcations starting with a uniform stationary pattern. The first several bifurcations generate either oscillating or non-oscillating states, and in each of these both spatially symmetric and asymmetric states are possible. The sequence of these bifurcations looks rather non-systematic, and does not seem to have a direct relevance to the chaotic behavior coming afterwards. Thus we start with the value σ= 2.600, where the system has already experienced seven times of bifurcations, and is now in an oscillating state. The corresponding trajectory on the A_1-A_3 plane is shown in Fig.11. The orbit is point-symmetric, hence the spatial-inversion symmetry is considered to be preserved here. By increasing σ, however, this limit cycle splits into a pair of limit cycles, each of which is no longer point symmetric. Figure 12 shows one of these limit cycles for σ=2.720. What happens after this is a cascade of bifurcations in each (symmetry-broken) limit cycle. At first it is transformed into a double limit cycle (Fig.13), then quadruple and (probably) octaple and so forth. Thus the system soon comes to realize a chaotic state as shown in Fig.14a, where σ=2.745. Figure 14b shows the side-view of the same trajectory on the A_1-A_2 plane.

The process leading to the chaotic behavior above is strongly reminiscent of a spiral chaos, a simplest type chaos discovered by Rössler for three-variable first-

Fig.11

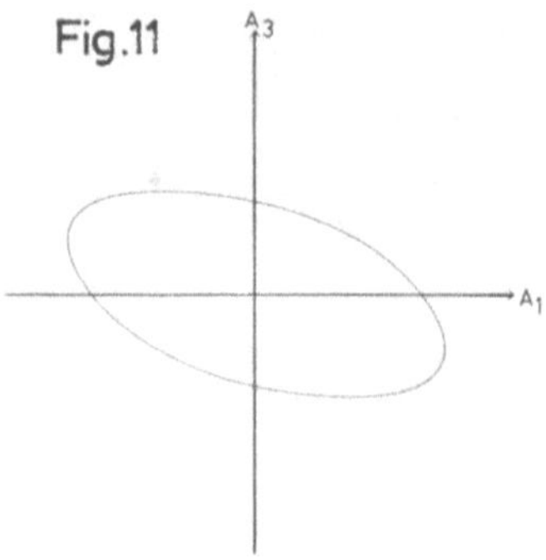

Fig.12

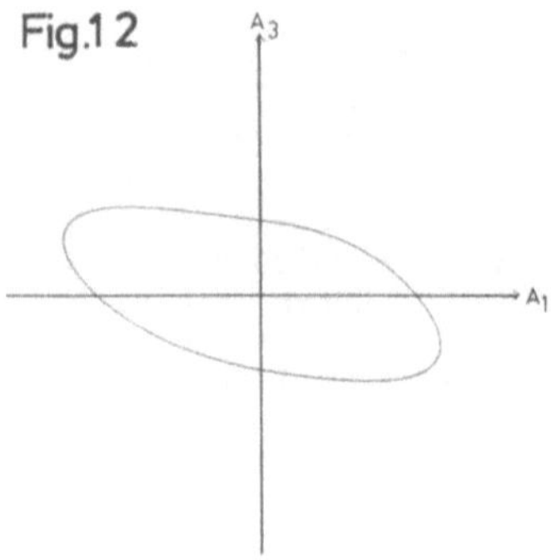

order ordinary differential equations [15,33-36]. The analysis by means of the Poincaré map actually makes it sure that the present chaos is essentially a spiral chaos. To see this, we introduce the third axis A_4 vertical to the plane in Fig.14a, and take the Poincaré plots at the sections P_1 and P_2. The results are shown in Fig.15, where X denotes the radial coordinate, i.e. $X=\sqrt{A_1^2+A_3^2}$, of the points on P_1 or P_2. As was anticipated, the map P_1 represents a typical walking-stick or horseshoe map, while the folding structure of the map is hardly visible for P_2.

From the fact that $A_4(X)$ is almost a single valued function as to the map P_2, it is reasonable to expect that the present chaos may further be reduced to a one-dimensional discrete chaos. In fact, we plot the amplitude X as a function of the amplitude of the preceding step, and the result is shown in Fig.16. It is seen that all the points nicely lie on a curve which is reminiscent of the logistic model, a most elementary chaos-producing map.

Fig.13

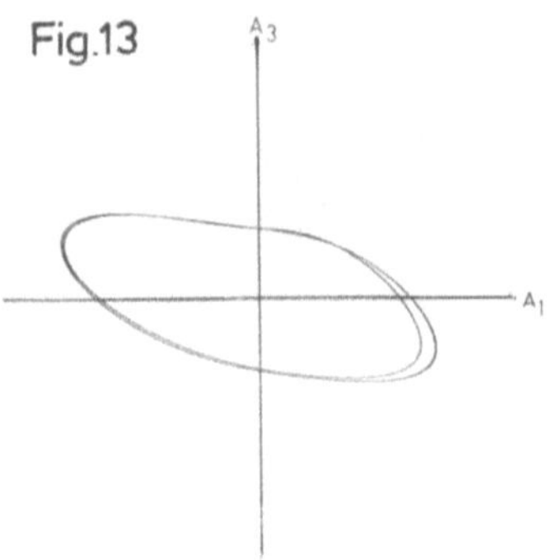

Fig.14(a) (b)

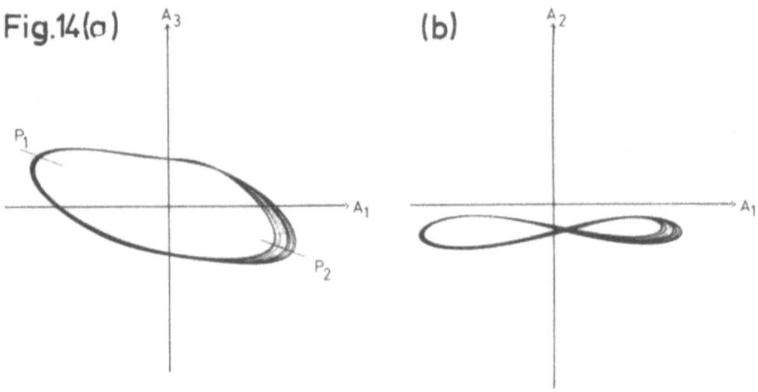

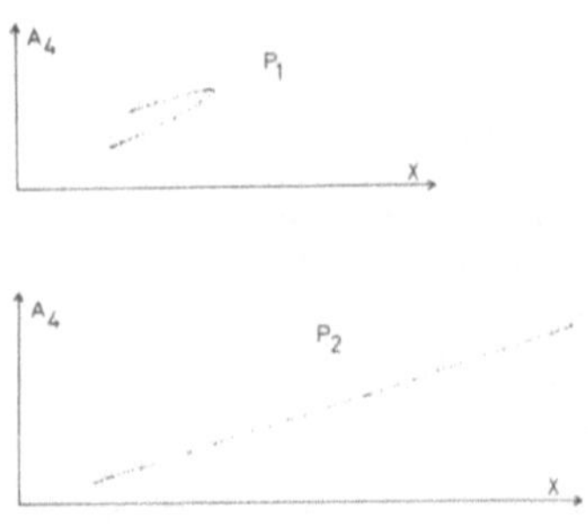

Fig.15

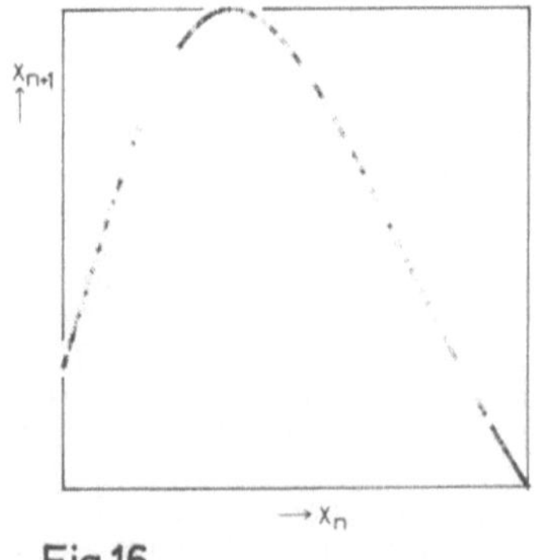

Fig.16

Fig.17 (a) (b)

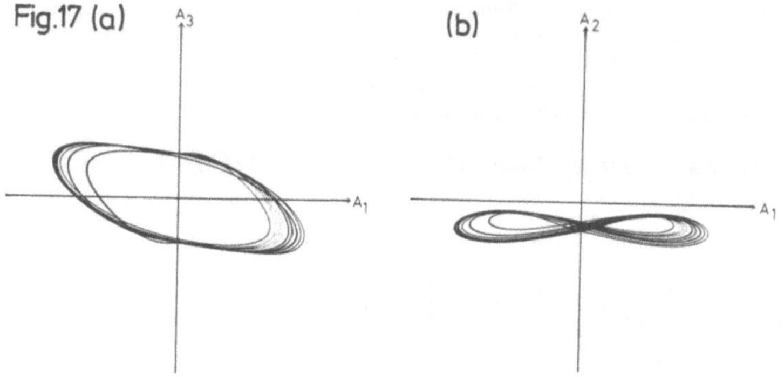

Further bifurcations beyond the above have not been analyzed yet, except the preliminary result of the one coming first. Figure 17a and b show the chaotic trajectories at σ=2.750. Here we have only one chaotic basin, while previously (the case of Fig.14) we had a pair of basins each symmetry-breaking. They are now united to one. It is expected, therefore, that the spatial-inversion symmetry has been recovered in statistical sense.

9. Concluding Remarks

We have started with the stability analysis of a laterally uniform wavefront of certain chemical waves. Then we have given an analytical demonstration that the front instability is possible, together with an intuitive argument. When the wavefront is only weakly unstable, we have succeeded in deriving the nonlinear evolution equation obeyed by the wavefront. This equation has turned out to show successive bifurcations and chaos. The chaos obtained belongs to a simplest type, and we could reduce it to a one-dimensional discrete chaos.

It seems that the present theory has to be extended in some directions. Among several possible extensions, we would like to point out one. An important problem yet untouched is how to understand further complications generated by a partial differential equation. Indeed a partial differential equation contains potentially an infinite number of degrees of freedom, and as we go deeper and deeper into the post-critical regime, there appear more and more active modes contributing to further complexity. Rössler proposed the notion *hyperchaos* [37-39], namely, the chaos with the orbital instability in multiple directions. In this respect, it will be an interesting future problem to study whether the increasing complexity of a chaotic solution from a partial differential equation is related to the increasing number of the Liapounov characteristic exponents with positive real part.

References

1. H.L. Swinny, P.R. Fenstermacher, J.P. Gollub: in *Synergetics, A Workshop* (ed. H. Haken; Springer-Verlag, Berlin-Heidelberg-New York, 1977) 60
2. G. Ahlers, R.P. Behringer: Prog. Theor. Phys. Suppl. 64, 186 (1978)
3. Y. Kuramoto, T. Yamada: Prog. Thoer. Phys. 56, 679 (1976)
4. Y. Kuramoto: in *Synergetics, A Workshop* (ed. H. Haken; Springer-Verlag, Berlin-Heidelberg-New York, 1977) 164
5. Y. Kuramoto: Prog. Theor. Phys. Suppl. 64, 346 (1978)
6. O.E. Rössler: Z. Naturforsch. 31a, 1168 (1976)
7. O.E. Rössler: in *Synergetics, A Workshop* (ed. H. Haken; Springer-Verlag, Berlin-Heidelberg-New York, 1977) 174
8. T. Yamada, Y. Kuramoto: Prog. Theor. Phys. 56, 681 (1976)
9. T. Yamada, H. Fujisaka: Z. Phys. B28, 239 (1977)

10. T. Yamada, H. Fujisaka: Prog. Theor. Phys. Suppl. 64, 269 (1978)
11. A.T. Winfree: in *Theoretical Chemistry* 4 (ed. H. Eyring, D. Henderson; Academic Press, New York, 1978) 1, and the references therein
12. J. Tyson: *The Belousov-Zhabotinsky Reaction*, Lecture Notes in Biomathematics 10 (ed. S. Levin; Springer-Verlag, Berlin-Heidelberg-New York, 1976) and the references therein
13. A. Nitzan, P. Ortoleva, J. Ross: Faraday Symp. Chem. Soc. 8, 241 (1974)
14. P. Fife: J. Chem. Phys. 64, 554 (1976)
15. O.E. Rössler: in *Bifurcation Theory and Applications in Scientific Disciplines* (ed. O. Gurel, O.E. Rössler; New York Academy of Science, 1978) 376
16. R.M. May: Nature 261, 459 (1976)
17. T.Y. Li, J.A. Yorke: Am. Math. Mon. 82, 985 (1975)
18. P. Ortoleva, J. Ross: J. Chem. Phys. 63, 3398 (1975)
19. Y. Kuramoto: Submitted to Prog. Thoer. Phys.
20. H.P. McKean: Adv. Math. 4, 209 (1970)
21. J. Rinzel, J. Keller: Biophys. J. 13, 1313 (1973)
22. H. FitzHugh: in *Biological Engineering* (ed. H.P. Schwan; McGraw Hill, New York, 1969) 1
23. J. Nagumo, S. Arimoto, S. Yoshizawa: Proc. IRE. 50, 2061 (1962)
24. E.C. Zeeman: in *Toward a Theoretical Biology* 4 (ed. C.D. Waddington; Edinburgh University Press, Edinburgh, 1972) 8
25. N. Rashevsky: Bull. Math. Biophys. 2, 15, 65, 109 (1940)
26. A.M. Turing: Phil. Trans. Roy. Soc. London, B237, 37 (1952)
27. G. Nicolis, I. Prigogine: *Self Organization in Nonequilibrium Systems* (Wiley-Interscience, New York, 1977)
28. A. Gierer, H. Meinhardt: Kybernetika 12, 30 (1972)
29. H. Meinhardt, A. Gierer: J. Cell Sci. 15, 321 (1974)
30. H. Meinhardt: in *Synergetics, a Workshop* (ed. H. Haken; Springer-Verlag, Berlin-Heidelberg-New York, 1977) 214
31. H. Haken: *Synergetics, An Introduction* (Springer-Verlag, Berlin-Heidelberg- New York 1977) Chap.7
32. J.M. Burgers: *The Nonlinear Diffusion Equation* (D. Reidel Publ. Co. Dordrech-Boston, 1974)
33. O.E. Rössler: Phys. Letters 57A, 397 (1976); 60A, 92 (1977)
34. O.E. Rössler: Bull. Math. Biol. 39, 275 (1977)
35. O.E. Rössler: Z. Naturforsch. 31a, 1664 (1976)
36. O.E. Rössler: in *Synergetics, Far From Equilibrium* (ed. A. Pacault, C. Vidal; Springer-Verlag, Berlin-Heidelberg-New York, 1978) 107
37. O.E. Rössler: in *Lectures in Applied Mathematics* 17 (ed. F.C. Hoppensteadt, Amer. Math. Soc. 1979)
38. O.E. Rössler: in *Structural Stability in Physics* (ed. W. Güttinger, Synergetic Series, Springer-Verlag, Berlin-Heidelberg-New York 1979)
39. O.E. Rössler : this issue

Chaos and Turbulence

O.E. Rössler

Institute for Physical and Theoretical Chemistry, University of Tübingen
D-7400 Tübingen, Fed. Rep. of Germany and
Institute for Theoretical Physics, University of Stuttgart
D-7000 Stuttgart, Fed. Rep. of Germany

Abstract

A special type of turbulence, called boiling type turbulence, can be obtained by coupling a set of identical single-threshold relaxation oscillators in such a manner that the slow variables are locally cross-inhibitory. If the cross-inhibition is overcritical, leading to 'morphogenesis' between the slow variables, a sudden slackening occurs wherever the the slowly moving-up 'skyline' hits the threshold (with the consequence of a rearrangement of the skyline); and so forth. A simple equation is considered on a ring in a cellular approximation. One cell is not chaotic; two are capable of chaos; three apparently produce higher chaos of the first order; and so forth. In the 3-cellular case, geometrical arguments backed by simulation suggest the presence of a cross-section which is folded over (between one passage and the next) in several independent directions. By implication, the formation of singular sets of Alexandrov type can be expected.

1. Introduction

In 1963, Lorenz showed that turbulent convective motion in a certain rotation-symmetric set-up governed by the Navier-Stokes equation can be approximately described by an ordinary 3-variable differential equation. The Lorenz equation produces 'typically 3-variable behavior,' namely, chaos (cf., for example, Rössler, 1977). Later, Ruelle and Takens (1971) postulated that fluid turbulence can be understood as being governed by an axiom A attractor in the sense of Smale (1967), because whenever four or more nonlinear oscillators are coupled arbitrarily weakly, a near-quasiperiodic strange (axiom A) attractor can be expected; and quasiperiodicity of an uncoupled type was thought to occur easily. Herewith, once more a qualitative type of dynamical behavior not possible in two-dimensional dynamical systems but arising in principle with the third dimension (Plykin, 1974) was invoked as a qualitative explanation for turbulence in general.

More recently, 'chaotic' motions were also observed in 2-variable reaction diffusion systems without convection, both in a continuous (Kuramoto and Yamada, 1976) and in a discrete (Rössler, 1976) setting. Again, typically 3-variable behavior of Lorenz type (Kuramoto, 1979) as well as typically 3-variable behavior of another type (Kuramoto, 1980) was found for certain parameter values. For a more restricted class of reaction-diffusion systems (with concentrations held fixed at the boundaries), the formation of a spatial rather than temporal chaotic regime was shown for the neighborhood of a certain singular (codimension two) bifurcation point (Guckenheimer, 1979).

It is therefore straightforward to speculate that if turbulence can at all be put into a larger equivalence class of behaviors, applying to many kinds of systems, it might perhaps be characterized best as presenting, neither typically 3-variable behavior (simple chaos) nor typically 4-variable behavior (hyper chaos), but typically n-variable behavior, that is, higher chaos of the (n-3)rd order.

In the following, one special class of turbulent systems is considered for which the presence of higher order chaos can be demonstrated.

2. Boiling-Type Turbulence

If spinach sauce is being heated in a narrow saucepan, after a while the whole content can be seen being lifted by a huge vapor bubble forming underneath. The process would be exactly repetitive if there were no spilling over when the lift-up spinach sauce reaches the top of the saucepan. That is to say, the whole process can, ideally, be described by a relaxation oscillation. The simplest equation yielding such an oscillation is

$$\dot{x} = 1, \qquad x \bmod 1. \tag{1}$$

If the saucepan is replaced by one of larger diameter (a frying pan), the effects of heating spinach sauce are slightly different. The reason is that the formerly laterally homogeneous situation during the buildup of the vapor layer now has undergone a morphogenetic (spatial symmetry breaking) bifurcation: there is the typical picture of a landscape with more or less sinusoidal hills and valleys as described by Turing (1952) for reaction diffusion systems. If one were able to halt the heat flow through the saucepan at the right moment, this profile would be stable. With the heat transfer going on, however, it is clear what happens: the biggest 'hill' that has formed is suddenly being punctured by a hole through which much steam escapes while the hill collapses. After the hole is 'healed,' the landscape rearranges itself, with the consequence that a different spot now takes the lead in the overall growth; and so forth. The resulting nonrepetitive pattern of waxing and waning may be termed boiling-type turbulence.

There are, of course, many natural phenomena obeying the same pattern. A horizontal handrail standing in the rain, with fine droplets falling on its top, develops a chain of budding large drops on its bottom side. Again, a very short piece of handrail shows a periodic relaxation oscillation, due to a single drop forming and falling off at a time; and a somewhat longer piece shows a symmetry breaking (there is never a horizontal bar of water forming); and so forth.

3. An 'Ideal' Example

The simplest equation likely to produce the 'boiling-type' phenomenon is

$$\begin{aligned} \dot{x} &= 2.1 - y - z, \qquad x \bmod 1 \\ \dot{y} &= 2.1 - x - z, \qquad y \bmod 1 \\ \dot{z} &= 2.1 - x - y, \qquad z \bmod 1\ . \end{aligned} \tag{2}$$

Here three relaxation oscillators of the type of Eq.(1) are coupled pairwise in a cross-inhibitory fashion. If only the linear parts are considered (omitting the mod 1 constraints), the cross-inhibition present between the three components is responsible for the formation of an unstable symmetrical steady state of saddle type in three-space, located at $x = y = z = 1.05$. All symmetrical initial conditions tend towards this point. The saddle has three unstable eigendirections, each leading to an increase in one variable and a simultaneous decrease in the two others. The flow in the lower portion of the positive orthant of state space therefore looks like the 'dust threads' of certain flowers, being grouped around the flower's axis and veering away from the latter the more strongly the farther out they start.

Adding the 'modulo' constraints in Eq.(2) means that there is a maximum height which can be reached in any of the three directions of space. Thus, the flower has been imbedded into a cube, stretching within it from one corner (the origin) across the center toward the opposite corner (1,1,1). If one of the three upper sides of the cube is being hit by a motion that originally came from one of the three lower sides, the system's state suddenly is displaced toward the opposite lower side. From there, it then starts moving up again, and so forth.

Nonetheless, there is a geometric argument showing that the flow described thus far cannot produce a higher form of chaos. Due to the fact that the sides of the cube are pair-wise identified, and the flow inside also invertibly connects those sides, at the specific parameters assumed, the whole flow necessarily takes place on a 3-torus. Its cross-section therefore is a 2-dimensional invertible map (homeomorphism or diffeomorphism) rather than a noninvertible map (endomorphism). Two-dimensional maps must be nonuniquely invertible, however, if they are to generate higher chaos.

This drawback is a consequence of the fact that Eq.(2), as written, is too much idealized. There are several ways to make it more realistic while leaving it simple. A straightforward change consists in reducing the constant 2.1 to 1.1. Since the saddle has now been pulled into the cube, trajectories no longer only enter the three lower sides of the cube, but may also leave the cube again through the upper parts of the lower sides. They nonetheless hit their mod 1 threshold, only now outside the cube. If they have not been veering away too strongly, they land still close enough to the cube in order not to leave it (or at least not to leave it equally far) next time, if the parameters are right. As a consequence, the lids of the cube now become endomorphisms rather than diffeomorphisms. The resulting flow now indeed can show hyperchaotic behavior - but, unfortunately, only as a transient. At the parameter values tested so far (around 1.1), the at first hyperchaotic appearing flow either escapes after a while toward an external limit cycle (because some trajectories are veering away too strongly after having left the cube), or finally settles for an asymptotic toroidal attractor inside. Thus, purely quantitative (from a qualitative point of view accidental) reasons prevent Eq.(2) from showing the desired behavior.

There are several possibilities to amend this situation. For example, one might slant the upper sides of the cube. Or, very realistically, one might round off the sharp edges that apply between the 3 upper sides of the cube. This can be done by inserting a slanted flat 'rim' at the edges. As a consequence, there is no longer complete independence between the thresholds of the three suboscillators throughout the whole range. The second solution has been tested and functions nicely. Both solutions nonetheless completely destroy the appealingly simple appearance of Eq.(2). Remains the possibility to take into account the fact that each relaxation oscillator in reality always involves two variables, a fact which implies both 'rounding off' and 'delay' effects. Such an explicit example follows below.

An alternative way of proceeding would be to stick to Eq.(2) as written above and simply increase the number of cells. Possibly, again a hierarchy of more and more complicated motions, beginning at a somewhat higher cell number, will be found when studying such (piecewise linear, C^0 type) flows on n-tori.

4. A Smooth Example

The following 6-variable equation is a 'smooth' realization of Eq.(2):

$$
\begin{aligned}
\dot{x} &= 20 - y - z - w \\
\dot{y} &= 20 - x - z - v \\
\dot{z} &= 20 - x - y - u \\
\dot{w} &= 0.01 + w(x - 10) \\
\dot{v} &= 0.01 + v(y - 10) \\
\dot{u} &= 0.01 + u(z - 10) \; .
\end{aligned}
\qquad (3)
$$

There are once more three 'morphogenetically' coupled linear variables (x, y, z) and three 'switches' (w, v, u), one for each linear variable. That is, Eq.(3) can again be said to consist of three single-threshold relaxation oscillators coupled in a cross-inhibitory manner. The fact that each suboscillator has a stable focus (which is a consequence of the extreme simplicity of the subsystems; a second quadratic term would be needed to make the individual relaxation oscillators undamped) has no consequences farther away from the unstable symmetrical state. The behavior of the system's trajectories in the three-dimensional 'slow' subspace (x, y, z) is displayed in Fig.1.

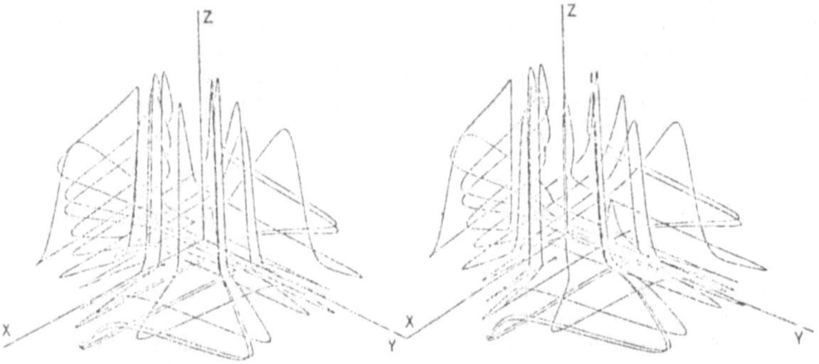

Fig.1 Trajectorial behavior of three single-threshold relaxation oscillators whose slow variables are coupled in a cross-inhibitory fashion, projected into the slow subspace. Numerical simulation of Eq.(3) on a HP 9845A desk computer, using a standard Runge-Kutta-Merson integration routine. The left-hand picture gives a symmetric view along the symmetry axis of the 'cube' (see text); the right-hand picture is slightly rotated (stereoplot). (To obtain a stereoscopic impression, one may first look at a pencil held about 15 cm before one's eyes halfway in front of the picture. Of the aligned, 4 blurred pictures behind the pencil, the two innermost can be merged by slightly adapting the position of the pencil. Thereafter, it is only a matter of waiting until the merged picture spontaneously ceases being blurred.) Initial conditions: $x(0) = y(0) = w(0) = 0.05$, $z(0) = 0.07$, $v(0) = u(0) = 0$. $t_{end} = 49.6$. Axes: -25...45 for x, y, z.

The behavior in this sub-state space is very similar to that displayed by Eq.(2). The main difference is that the upper portion of the 'cube' no longer consists of three orthogonal pieces separated by sharp edges. The 'cap' now has a rounded-off, soft shape.

There is a bunch of trajectories diverging from the middle of the left-hand picture (corresponding to the origin of the 'cube'). And there is a reinjection 'back downstairs' once the three upper sides of the cube are hit. By tracing back trajectories from one of the three upper sides, one verifies that there is non-uniqueness of trajectorial origin at some places (implying overlap of the underlying map), and so along different directions. See, for example, the first bundle to the right of the z-axis.

Fig.2 shows the time behavior of the six variables. Focusing only on the 'switches' (displayed in the 3 lower rows), one is tempted to interpret the spikes as falling drops (in terms of the above handrail example). One sees that nearly always two suboscillators are engaged in a mutual oscillation of 'push pull' type, with the third being idle. These sub-regimes last for differing periods and alternate irregularly.

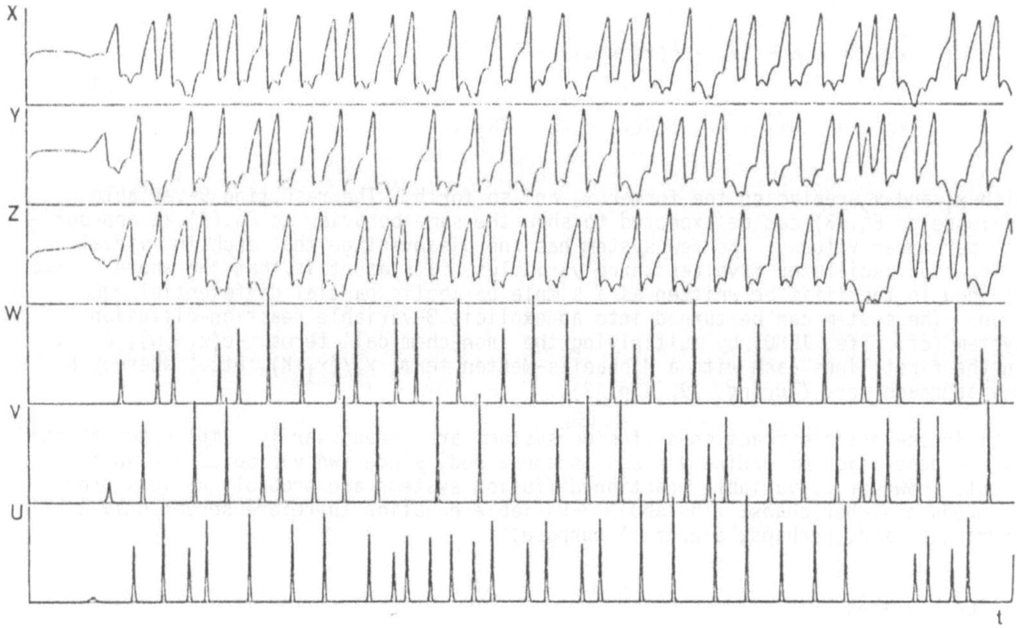

Fig.2 Time behavior of Eq.(3). Numerical simulation as in Fig.1. Axes: -25...55 for x, y, z; 0...600 for w, v, u; 0...100 for t.

Whereas Eq.(3) is too complicated to allow an analytical proof that its cross-section has the form of a towel that is folded in a more or less orderly way, the argument which led to Eq.(3), via Eq.(2) as a vehicle, suggests that there exists an ideal version of Eq.(3), with infinitely fast relaxation (that is, with a structure similar to Eq.(2)), for which the presence of an endomorphism of folded towel shape can be proven analytically.

Assuming this to be the case with the present ring (or plane) of three cells, it is natural to expect that the principle will carry through (with shape-dependent modifications) to the case of 4 and more cells. The trajectories, hitting now the sides of hyper cubes in order to be reinjected toward the opposite side, should again do so in a not uniquely invertible manner as far as the slow subspace is concerned, and with more and more directions opening up at the same time in the underlying, higher-dimensional maps. ('Folded-in cloud map,' etc.)

4. A Reaction Diffusion Analogue

Turing (1952) showed that cross-inhibition with formation of a 'morphogenetic' saddle point in state space, occurs also in linear systems of symmetric type in which the variables do not cross-inhibit each other directly (as in the above linear subsystem), but indirectly. In such cases at least one more linear variable is required in each compartment. The additional variable can then be coupled to its neighbors via linear diffusion (which is the opposite of cross-inhibitory coupling). Due to the fact that the second variable acts as an inhibitor of the first variable in each compartment (cf. Othmer, 1980), the diffusion coupling then indirectly gives rise to a cross-inhibition between the first variables.

Making use of this principle allows one to replace each linear variable in Eq.(3) by a set of two variables of the following form

$$\begin{aligned} \dot{x}_1 &= a + bx_1 - c(x_2 + w) \\ \dot{x}_2 &= dx_1 - ex_2 + D(y_2 + z_2 - 2x_2), \end{aligned} \tag{3a}$$

with x_1 and x_2 replacing the former x, and so forth. The resulting 9-variable analogue to Eq.(3) can be expected to show the same behavior as Eq.(3) at appropriate parameter values. The new system has the disadvantage that each individual relaxation oscillator involves three variables. Its asset is that the whole system can now in the limit be written as a simple parabolic partial differential equation. The system can be turned into an explicit 3-variable reaction-diffusion system (cf. Fife, 1980) by multiplying the 'non-chemical' terms, $-c(x_2+ w)$, etc., in the first lines each with a Michaelis-Menten term, $x_1/(x_1+K)$, etc., whereby K may approach zero (Turing, 1952, p.42).

Three-variable reaction-diffusion systems are unnecessarily complicated as far as the generation of ordinary chaos is concerned, since two variables are sufficient. However, 2-variable reaction-diffusion systems are probably no less prone to produce higher chaos. The above 3-variable equation therefore serves only a heuristic (and, perhaps, didactic) purpose.

5. Conclusions

The behavior found in the above cellular system was proposed to belong to a 'higher' class of chaotic phenomena. While in ordinary chaos, Cantor sets of singular solutions are formed (Smale, 1967), higher-order chaos is characterized by the formation of singular sets which are higher-order analogues of Cantor sets. The simplest such set is the Alexandrov cross which is a metric product of two Cantor sets (Alexandrov, 1948). It looks like a Swiss flag, with 4 little Swiss flags inserted, and then 16, and so on. A 4-variable ordinary differential equation in which the formation of such a set can be proven (because the cross-section is known explicitly in the limit of a parameter going to zero) has been found recently (Rössler and Mira, in preparation). Nonetheless, many implications of folded-towel maps (and their higher analogues) are still unknown. For example, it is an open problem how to assimilate the fact that topologically, Alexandrov discontinua are homeomorphic to an ordinary Cantor set (Alexandrov, 1948). The feature most easy to test in applications will be the presence of hyperbolic trajectorial instability in more than one eigen-direction for most trajectories for most of the time (Rössler, 1979).

While the practical significance of 'levels of chaos' remains to be seen, the theoretical prediction that a series of transitions toward more and more complicated behavior occurs if more and more identical relaxation oscillators are coupled in one or two space dimensions, is of interest in its own right in the context of studying cooperative phenomena (Haken, 1980; see also Babloyantz, 1980).

References

1. Alexandrov, P.S. (1948): Introduction to Set Theory and the Theory of Real Functions (in German), VEB-Verlag der Wissenschaften, Berlin, 1956, p. 224. (First Russian ed.: Ogie, Moscow-Leningrad, 1948.)

2. Babloyantz, A. (1980): These proceedings.

3. Fife, P. (1980): These proceedings.

4. Guckenheimer, J. (1979): On a codimension two bifurcation, Preprint.

5. Haken, H. (1980): These proceedings.

6. Kuramoto, Y. (1979): Diffusion-induced chaos in reaction systems, Progr. Theor. Phys. Suppl. 64, 346.

7. Kuramoto, Y. (1980): These proceedings.

8. Kuramoto, Y. and T. Yamada (1976): Turbulent state in chemical reactions, Progr. Theor. Phys. 56, 679-683.

9. Lorenz, E.N. (1963): Deterministic nonperiodic flow, J. Atmos. Sci. 20, 130-141.

10. Othmer, H. (1980): These proceedings.

11. Plykin, R.V. (1974): Math. Sbornik 94, 243-246.

12. Rössler, O.E. (1976): Chemical turbulence: chaos in a simple reaction-diffusion system, Z. Naturforsch. 31a, 1168-1172.

13. Rössler, O.E. (1979): An equation for hyperchaos, Phys. Lett. 71A, 155-157.

14. Ruelle, D. and F. Takens (1971): On the nature of turbulence, Commun. Math. Phys. 20, 167-192.

15. Smale, S. (1967): Differentiable dynamical systems, Bull. Amer. Math. Soc. 73, 747-817.

16. Turing, A.M. (1952): The chemical basis of morphogenesis, Phil. Trans. Roy. Soc. London B 237, 37-72.

Part VI

Self-organization of Biological Macromolecules

Self-organization of Biological Macromolecules and Evolutionary Stable Strategies

P. Schuster

Institut für Theoretische Chemie und Strahlenchemie, Universität Wien
A-1090 Wien, Austria

K. Sigmund

Institut für Mathematik, Universität Wien, A-1090 Wien, Austria

1. Introduction

Autocatalysis is a rather exceptional phenomenon in chemical kinetics and, when it appears in reaction networks, various characteristic phenomena like oscillations, spatial pattern formation or eventually chaotic behaviour are commonly observed. Biochemistry and biology, in contrary, have to deal with a class of molecules for which selfreplication became obligatory. These molecules, autocatalysts in a sense, are the polynucleotides, the nucleic acids or, later on in evolution, the genes. Self-organization, the most interesting attribute of the biosphere, appears to be essentially based on the capability of self-replication.

Selfreplication may occur either directly as it is the case with DNA in bacteria or via an intermediate "negative-copy" as with single stranded viral RNA. The difference between both modes of replication does not play a central role on the general features of selfreplication and we shall not refer to complementary copying here, in particular since this question has already been discussed previously by EIGEN [1,pp.492-494].

In order to make the theory accessible to experimental tests we have to design a simple enough system which may serve to study molecular selfreplication. The cell, the universal selfreplicating unit in biology is an enormously complicated system and even in case we restrict ourselves to the processes involved in DNA replication and translation we hardly have a chance to set up the proper kinetic equations not to speak of analysing them by mathematical techniques. Thus we have to go one step further and consider only the process of RNA (or DNA) polymerization with a complete enzymatic machinery given. The first experiments of this kind have been performed by SPIEGELMAN [2]. Refining and idealizing the experiment we may end up at an evolution reactor of the type shown in Fig. 1 (see also KÜPPERS [3]).

Let us visualize what happens in an evolution reactor which contains a set of independently replicating species (I_i, $i = 1,\dots,n$) The mechanism of copying is assumed to be free of errors at first ($\Gamma_i = k_i x_i$). Then (natural) selection will lead to an enrichment of the species I_m which replicates fastest ($k_m = \max(k_i, i=1,\dots,n)$, the largest rate constant) and, finally only this species will be present in the reactor.

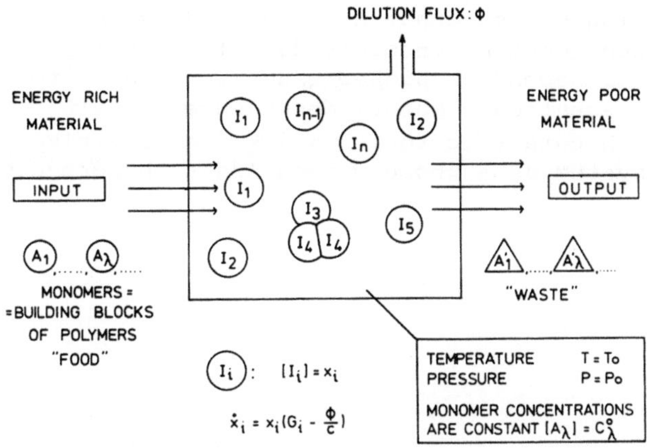

Fig.1 The evolution reactor. This kind of flow reactor consists of a reaction vessel which allows for temperature and pressure control. Its walls are impermeable to the selfreproductive units (biological macromolecules like polynucleotides - e.g. phage RNA - bacteria or, in principle, also higher organisms). Energy rich material ("food") is poured from the environment into the reactor. The degradation products ("waste") are removed steadily. Material transport is adjusted in such a way that food concentration is constant in the reactor. A dilution flux ϕ is installed in order to remove the excess of selfreplicative units produced by multiplication. Thus the sum of population numbers or concentrations,

$$[I_1] + [I_2] + \ldots\ldots + [I_n] = \sum_{i=1}^{n} x_i = c,$$

may be controlled by the flux ϕ. The differential equations describing the time development of the distribution of molecules present in the reactor can be formulated in general form:

$$\dot{x}_i = \Gamma_i - \frac{x_i}{c}\,\phi\;,\; i = 1,2,\ldots,n \tag{1}$$

Stationarity or, in other words, constant total concentration c is acchieved by proper adjustment of the flux, $\phi = \sum_i \Gamma_i$. According to the laws of mass action kinetics the growth function Γ_i can be represented by polynomial expansion:

$$\Gamma_i = k^{(i)} + \sum_{j=1}^{n} k_j^{(i)} x_j + \sum_{j=1}^{n} \sum_{l=1}^{n} k_{jl}^{(i)} x_j x_l + \ldots\ldots \tag{2}$$

In case of pure or error-free selfreplication several constants vanish and we obtain

$$\Gamma_i = k_i x_i + \sum_{j=1}^{n} k_{ij} x_i x_j = x_i \left(k_i + \sum_{j=1}^{n} k_{ij} x_j\right) \tag{3}$$

The experimental verification of evolution reactors has been discussed recently by KÜPPERS [3].

Mutations will complicate the situation somewhat: in favourable cases, as we shall see, a master sequence will be selected together with its frequent mutants. They form an ensemble called "quasispecies" [4]. A single quasispecies, thus, is the result of a long term competition in the evolution reactor provided the frequency of copying errors lies below a precisely defined threshold.

Coexistence of selfreplication units cannot be acchieved by the simple mechanism of template dependent polynucleotide polymerization. To enforce cooperative behaviour second or higher order terms have to be introduced into the kinetic equations. Mutual dependence of self-replicating units linked by such catalytic terms results from positive cyclic feedback (Fig.2). The system as a whole was called a "hypercycle" [1,4-6] .

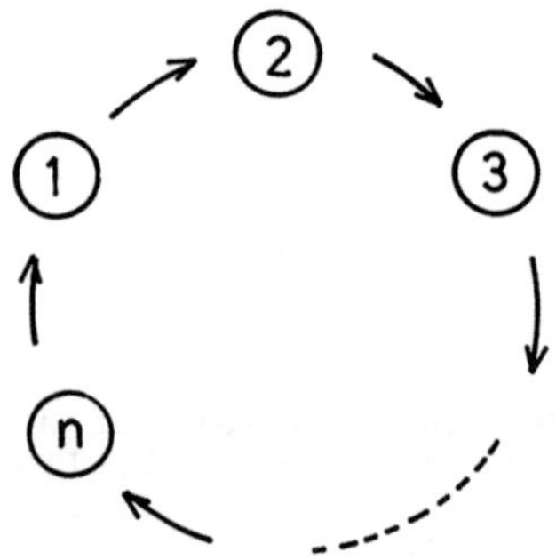

Fig.2 Sequence of catalytic interactions in a hypercycle. The simplest example is represented by
$\Gamma_i = k_i x_i x_j$, $i = 1,\dots,n$ and $j = i-1+n\delta_{i1}$
(see Fig.1)

In this contribution we shall be concerned first with the degree of organization in general. Two special problems of selfreplication, error propagation and cooperative behaviour will be discussed in some detail. Finally, we shall briefly mention two important fields of application: prebiotic evolution and social behaviour of animals.

2. Three features of organization

Self-organization commonly is considered as a process leading spontaneously from lower to higher levels of organization. Presumably, nobody will hesitate to call plants or animals more highly organized systems than bacteria. There are, however, more conflicting situations, in which the lack of any quantitative measure or ordering scale for the notion of organization will strongly hinder fruitful discussions. This point has been stressed already by VON NEUMANN [7] about thirty years ago. The organization of whole organisms is such an enormously complex dynamical phenomenon that the chance to pick out the essential and appropriate features for comparison is exceedingly low at present. Therefore, we report an attempt to search for quantities which may represent a useful measure for the degree of organization in a system as simple as an evolution reactor. Three features seem to serve well for this purpose.

(1) Dimension N: The number of cooperative selfreplicating macromolecules or species is called the dimension N. Cooperativity is characterized here as the absence of competition. Cooperative behaviour can be recognized by the existence of an attractor inside the simplex (or subsimplex) S_N (see Fig.3a).

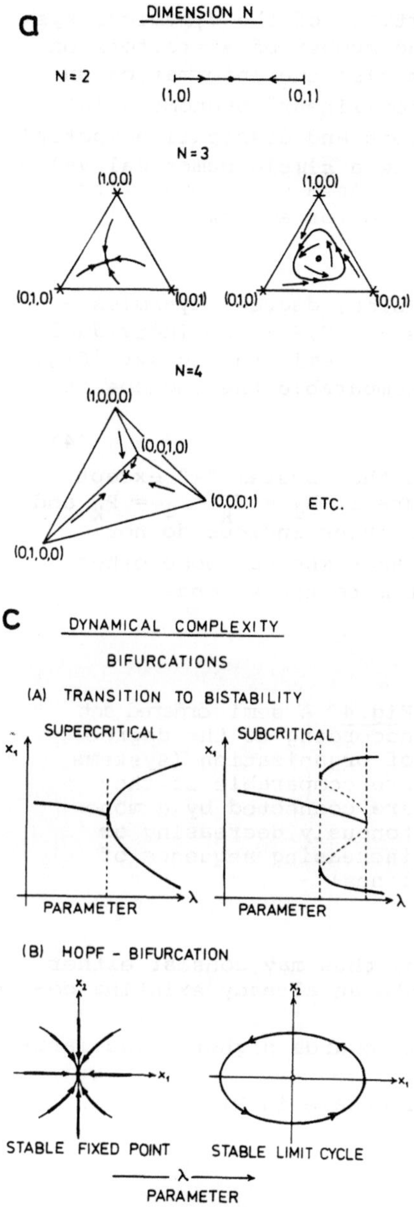

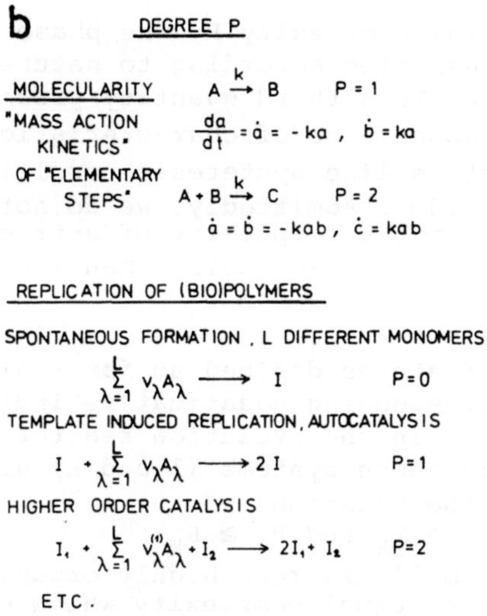

Fig.3 Three features of organization (*a*) dimension N; the system is mapped on the "concentration" simplex S_N [5] which is defined in terms of relative concentrations $\xi_i = x_i/c$, $S_N = \{\xi = (\xi_1,\dots,\xi_N): \Sigma\xi_i = 1,\ \xi_i \geq 0$ for $i = 1,\dots,N\}$, (*b*) degree P and (*c*) dynamical complexity)

(2) Degree P: The process of selfreplication follows a certain mechanism to which we assign a "molecularity" or degree P. This "molecularity" counts only the number of macromolecules which participate in the process of replication. It is identical with the power of the leading term in the functions Γ_i $(i=1,\dots,n)$. Thus, P = 0 represents spontaneous, template-free polynucleotide synthesis, P = 1 corresponds to template induced replication and finally, P = 2 involves catalysed, template induced replication. An example for the latter case is represented by protein catalysed DNA or RNA replication where the enzyme itself is a translation product of some gene (Fig.3b).

(3) Dynamical complexity R: The phase portrait of the dynamical system may be classified according to nature and number of attractors on the simplex S_N. This third quantity presents also the information on the possible appearance of characteristic "non-linear" phenomena in chemical kinetics like hysteresis, oscillations and dissipative spatial structure (Fig.3c). Admittedly, we do not know a simple numerical value corresponding to the complexity of attractors. As a very rough first approximation we may use their dimension (R = 0 for a sink, R = 1 for a limit cycle etc.).

The three features defined so far - dimension, degree, dynamics - and their corresponding quantitative indices - N,P,R - map individual systems studied in the evolution reactor onto a semi-ordered set (Fig.4) Therefore only those systems (l and k) are comparable the indices of which fulfil the relations

$$N_l \geq N_k, \; P_l \geq P_k \text{ and } R_l \geq R_k \tag{4}$$

Clearly, system "l" is more highly organized than system "k" except the situation of equal complexity which occurs if $N_l = N_k$, $P_l = P_k$ and $R_l = R_k$. Two systems are incomparable if the three indices do not change monotonously, i.e. if one index increases whereas some other index decreases on going from the first system to the second.

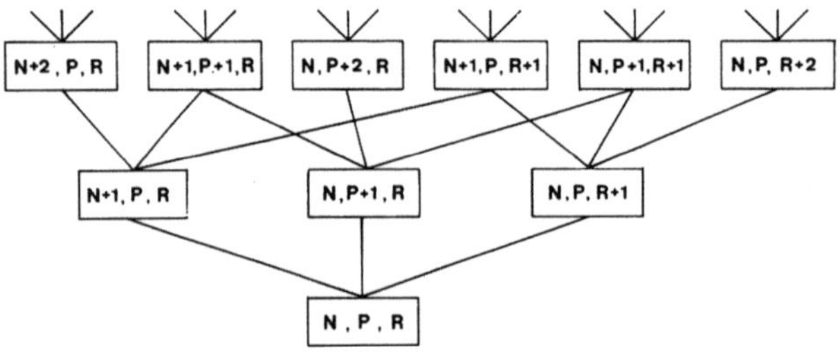

Fig.4 A semi ordered set according to the degree of organization (systems are comparable if they are connected by a monotonously decreasing or increasing sequence of lines)

An increase in the degree of organization, thus may consist either
(1) in the incorporation of a new species into an already existing cooperative system (increase in N) or
(2) in a change of the replication mechanism towards higher molecularity (increase in P), or
(3) in a change towards a richer dynamics (increase in R)

3. Selfreplication with Errors

No physical process can occur with absolut accuracy. In reality we have to deal always with a certain frequency of copying errors which give rise to mutant distributions. At first we consider the stochastic aspect of the problem by means of an illustrative simple model (Fig.5)

The errors are assumed to occur independently and with equal probability for every digit. Assigning a single digit accuracy of q - q is the probability of incorporation of the correct digit - to the replication process we obtain a probability

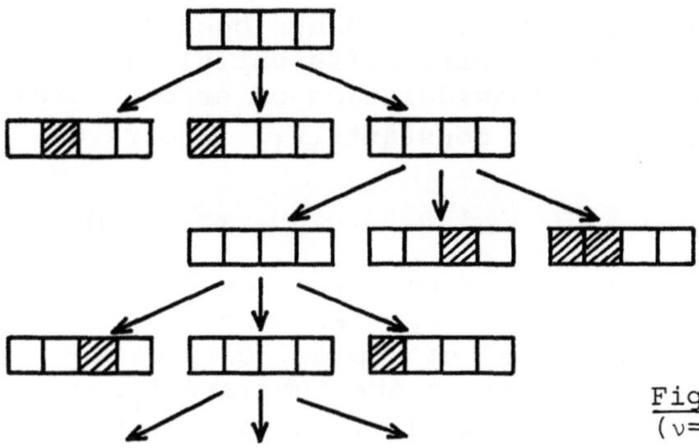

Fig.5 Selfreplication with errors (ν=4, Σ=3, q=0.83)

$$Q = q^{\nu} \tag{5}$$

for the correct replication of a polymer with ν segments.Every correct sequence is assumed to create a progeny of Σ copies per generation and to live for just one multiplication cycle. All error copies, i.e. mutants in this model are lethal.

Under these assumptions the problem can be treated as a multiplicative chain (see BARTLETT [8]). The necessary and sufficient condition for certain extinction of the master copy is $m \le 1$, where m represents the expectation value of reproductive success, i.e. the mean number of descendants per generation from a single correct copy. The number of offspring is a random variable X with binomial distribution $B(\Sigma,Q)$. This means that the probability

$$\text{Prob}\ \{ X = k \} = \binom{\Sigma}{k} Q^k (1 - Q)^{\Sigma-k} \tag{6}$$

For the expectation value we obtain

$$m = \Sigma.Q = \Sigma.q^{\nu} \tag{7}$$

The condition for a non-zero probability of survival simply reads

$$Q.\Sigma > 1 \quad \text{or} \quad Q > \Sigma^{-1} \tag{8}$$

respectively. There is a sharply defined minimum accuracy of replication, $Q_{min} = \Sigma^{-1}$, below which the system goes certainly extinct. For a given value of q this minimum accuracy corresponds to a maximum chain length of the polymer

$$Q_{min} = q^{\nu_{max}} \quad \text{or} \quad \nu_{max} = -\frac{\ln \Sigma}{\ln q} \sim \frac{\ln \Sigma}{1-q}\ , \text{ if } q \sim 1 \tag{9}$$

A given polymerization mechanism is characterized by a certain q value which in turn sets a critical limit to the information content that can be transmitted from generation to generation by the copying mechanism. An increase in the number of digits exceeding ν_{max} or, more generally speaking, an increase in the information content transferred via replication requires an improvement of the replication machinery.

The simple stochastic model discussed so far suffers mainly from two deficiencies: (1) an extension to more general situations is not easy and (2) the assumption that all mutations are lethal is highly unrealistic and inevitably leads to wrong conclusions in some questions. Therefore, we mention briefly also the deterministic approach to template induced replication with errors which has been studied extensively in the past [1,4,9,10]. We start with differential equation (1) and insert $\Gamma_i = \sum_j k_{ij}x_j$. The explicit consideration of errors makes it necessary to specify the "diagonal" rate constants k_{ii} in more detail [1] :

$$k_{ii} = A_i Q_i - D_i \tag{10}$$

Q_i is the quality factor for the replication of the sequence I_i. As before, it describes the probability for correct replication of the whole polymer. The rate constant A_i gives the number of copies, correct and erroneous, which are synthesized per time unit on the template I_i. D_i is the rate of hydrolysis of these templates. Finally, the constants k_{ij} represent the rates at which I_i is obtained as an error-copy of I_j.

It is useful to define an excess productivity $E_i = A_i - D_i$. From mass conservation we obtain the global relation

$$\sum_i A_i(1-Q_i)x_i = \sum_i \sum_{k \neq i} k_{ik}x_k \tag{11}$$

Accordingly, the mean excess productivity becomes identical to the dilution flux ϕ except a factor c^{-1}:

$$\bar{E} = \frac{1}{c}\sum_i E_i x_i = \frac{1}{c}\phi \tag{12}$$

The remaining differential equation

$$\dot{x}_i = [k_{ii} - \bar{E}(t)]x_i + \sum_{k \neq i} k_{ij}x_j \tag{13}$$

has been analysed in detail by EIGEN [1], THOMSON and MC BRIDE [9] and JONES et al. [10]. We briefly summarize the most important results which were derived by a transformation of variables and second order perturbation theory [10] .

Selection takes place in the evolution reactor during the whole transient period. By successive elimination of the slowly growing molecules the mean excess productivity, $\bar{E}(t)$, increases steadily until it reaches an optimal value:

$$\lim_{t\to\infty} \bar{E}(t) = \lambda_{max} = k_{mm} + \sum_{j \neq m} \frac{k_{mj}k_{jm}}{k_{mm}-k_{jj}} \tag{14}$$

By I_m we denote the species which has the largest value of k, $k_{mm} = \max(k_{ii},\ i = 1,\ldots,n)$. The results of second order perturbation theory are good approximations to the exact solutions provided $k_{im} \ll k_{mm}-k_{ii}$ for all $i \neq m$.

The stationary distribution of the "quasispecies", an ensemble consisting of a master copy and its mutants, is given by

$$\tilde{x}_m = c\,\frac{k_{mm}-\bar{E}_{-m}}{E_m-\bar{E}_{-m}} = \frac{c}{1-\sigma_m^{-1}}\,(Q_m-\sigma_m^{-1}) \tag{15}$$

and $$\tilde{x}_i = \tilde{x}_m \frac{k_{im}}{k_{mm}-k_{ii}}\ ;\ i = 1,\ldots,n\ ;\ i \neq m \tag{16}$$

Stationary concentrations are indicated by tilde (˜). For short notation we define the quantities

$$\bar{E}_{-m} = \frac{\sum_{i \neq m} E_i x_i}{c - x_m} \quad \text{and} \quad \sigma_m = \frac{A_m}{D_m + \bar{E}_{-m}} \quad . \qquad (17a,b)$$

From (15) and (16) we recognize immediately that a stationary population can only exist if $Q_m > \sigma_m^{-1}$. This condition reminds us of (8) which has been derived from the purely statistical aspects of replication with errors. Again we find a lower limit for the accuracy of the copying process below for which there is no stable transfer of information:

$$Q_m > Q_{min} = \sigma_m^{-1} \qquad (18)$$

The lower limit of accuracy, of course, is equivalent to an upper limit for the chain length of the polynucleotide

$$\nu_m < \nu_{max} = - \frac{\ln \sigma_m}{\ln \bar{q}_m} \sim \frac{\ln \sigma_m}{1 - \bar{q}_m} \qquad (19)$$

For this purpose we introduce an average single digit accuracy for polynucleotide replication:

$$Q_m = \bar{q}_m^{\nu_m} \qquad (20)$$

In this way the single digit accuracy becomes indirectly dependent on the nature of the digit and its neighbours in the sequence. In case this dependence is small we have $\bar{q}_m \sim q$.

It seems interesting to compare (8) and (18). At a first glance there is complete analogy. Looking more closely, however, we observe a substantial difference. Let us translate the assumptions of the stochastic model (Fig.5) into the superiority parameter σ_m. All sequences, correct and erroneous copies, live just for one reproductive cycle. All mutations are assumed to be lethal or, in other words, only the master sequence is allowed to reproduce: $A_i = 0$; $i = 1,\ldots,n$; $i \neq m$ and $A_m = \Sigma$. Accordingly we find $D_m - \bar{E}_{-m} = 0$ and $\sigma_m = \infty$. Thus, complete lethality of mutants implies replicability of infinitely long polynucleotide sequences in the deterministic approach whereas there exists still an error threshold within the stochastic treatment.

Finally we mention briefly some implications of (19) in realistic systems. The dependence of the maximum degree of polymerization (ν_{max}) on "superiority" σ_m and single digit accuracy $\bar{q}_m$ is summarized in Fig.6.

INFORMATION STORAGE IN DARWINIAN SYSTEM

ERROR RATE PER DIGIT $1 - \bar{q}_m$	SUPERIORITY σ_m	MAXIMUM DIGIT CONTENT ν_{max}	BIOLOGICAL EXAMPLES
$5 \cdot 10^{-2}$	2	14	ENZYME-FREE RNA REPLICATION
	20	60	
	200	106	t-RNA PRECURSOR, ν=80
$5 \cdot 10^{-4}$	2	1386	SINGLE-STRANDED RNA REPLICATION VIA SPECIFIC REPLICASES
	20	5991	
	200	10597	PHAGE Qβ, ν = 4500
$1 \cdot 10^{-6}$	2	$0{\cdot}7 \cdot 10^6$	DNA REPLICATION VIA POLYMERASES AND PROOFREADING
	20	$3{\cdot}0 \cdot 10^6$	
	200	$5{\cdot}3 \cdot 10^6$	E.COLI, $\nu = 4.10^6$
$1 \cdot 10^{-9}$	2	$0{\cdot}7 \cdot 10^9$	DNA REPLICATION AND RECOMBINATION IN EUCARIOTIC CELLS
	20	$3{\cdot}0 \cdot 10^9$	
	200	$5{\cdot}3 \cdot 10^9$	VERTEBRATES (MAN), $\nu = 3.10^9$

Fig.6 Examples of template induced replication of polynucleotides and nucleic acids

We realize that ν_{max} is particularly sensitive to the latter quantity. Three mechanisms of template induced replication are considered explicitly

(1) Due to its low accuracy enzyme-free replication sets a low limit of about 50 - 100 digits to information storage in prebiotic polynucleotides. The length of present day t-RNA's seems to be hardly exceedable (see [4], p.562)

(2) In case of RNA replication of Qβ phages in bacterial cells and in "in vitro" systems WEISSMANN and coworkers [11,12] succeeded to determine the mean single digit accuracy to $\bar{q} \sim 0.9997$. The corresponding value of ν_{max} lies close to the actual length of the viral genome. Apparently, in phage replication nature approaches the error limit as closely as possible (see [4], pp. 550, 557-559).

(3) The replication of DNA in the bacterial genome involves more enzymes than phage-RNA replication. In particular there are repair mechanisms which increase the accuracy and reduce the probability of "single-digit-errors" by about three orders of magnitude. Accordingly, a bacterial genome can be as long as a few million base pairs (see [4], pp.559-562).

4. Cooperation Between Selfreplicating Units

Selfreplication in systems with linear growth functions $\Gamma_i = \Sigma k_{ij} x_j$ leads to competition between the replicating units. The final outcome of such a selection process, as we have seen, is either a single quasi-species or, in case the accuracy was too low, a non-stationary steadily changing ensemble of sequences. In order to intoduce cooperativity we have to make use of second order terms at least, $\Gamma_i = \Sigma\Sigma\, k_{jl}^{(i)} x_j x_l$. We are primarily interested in general features and hence we restrict ourselves to error free replication ($\Gamma_i = \Sigma\, k_{ij} x_i x_j$). Then the differential equation to be analysed has the form

$$\dot{x}_i = (\sum_j k_{ij} x_j - \frac{1}{c}\,\phi)\, x_i \; , \quad i=1,\ldots,n \tag{21}$$

wherein $\phi = \sum\sum_{jl} k_{jl} x_j x_l$ (22)

This equation and special cases of it were studied in previous papers [13-15]. A sufficient criterion for cooperation of the n members was found to be the formation of a closed positive feedback loop. The system originating from loop closure was called a hypercycle:

$$k_{ij} = k_i \delta_{jl} \; ; \quad l = i-1+n.\delta_{i1} \tag{23}$$

In the subsequent contribution we present a proof for this conjecture.

5. Prebiotic Evolution

In this contribution we can present only a very brief outline of the application of the ideas put forward in the last two sections to the problems of prebiotic evolution. For more details we have to refer to the literature [1,4-6]. A schematic model of the logical sequence of processes is shown in Fig.7 . By "logical sequence" we want to emphasize that every step requires the products of the preceding ones. There is

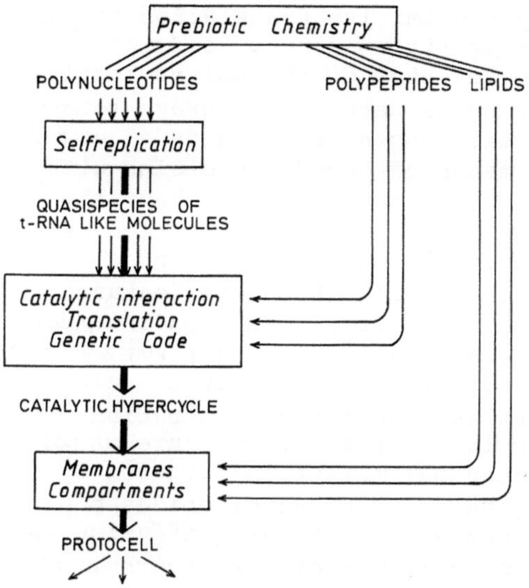

Fig.7 A schematic model for the logics of prebiotic evolution

no need, however, for the earlier process to stop precisely when the system enters a new phase of evolution. Historically, most of the processes described here may have taken place at the same time.

Prebiotic evolution my be subdivided into several periods:
(1) During the first phase of prebiotic chemistry the building blocks of biopolymers (aminoacids, purines, pyrimidines, sugars, constituents of lipids etc.) are formed spontaneously from small molecules in the prebiotic, reducing atmosphere [16].

(2) The building blocks form primitive polymers by means of condensation reactions [16,17]. The first oligopeptides and oligonucleotides appear in the primordial solutions. The latter represent the first templates for replication [18]. When they come into play a new dimension for evolution opens up.

(3) Selfreplication introduces competition into the primordial soup. Darwinian evolution in a wider sense starts at the level of molecules. Now, the results of section 3 become relevant. Polynucleotides may grow to give longer sequences but transfer of information is restricted to polynucleotides which are similar in lengths to present day t-RNA's. Under these conditions the fittest replicating templates will be selected according to their (secondary and) tertiary structures. "Fitness" in this context refers to rate and accuracy of replication as well as resistence against hydrolysis as it is expressed by the combination of rate constants and quality factors: $k_{mm} = A_mQ_m - D_m$. Master copies are accompanied by mutant distributions in the sense of "quasispecies".

(4) Replication has come to a "dead end" when the fittest sequence appeared in the solution. Further evolution requires cooperation of t-RNA like molecules which at first may be members of the same quasispecies. At this stage hypercyclic coupling becomes important.The coup-

ling is brought about most naturallv by a primitive translation mechanism which involves the development of a primitive genetic code at the same time [6]. The encoded polypeptides act back on the polynucleotides via their catalytic functions and present the catalytic linkage between the replicating units thereby. Further development results in a certain perfection of molecular properties: by means of encoded proteins with RNA polymerase activity the accuracy of replication is improved, the translation machinery is further developed and target recognition is optimized. Evolution in homogenous solution leads to a dead end at this stage. Still the system has not yet acchieved all the basic features found with Darwinian evolution at the cellular level. Perhaps the most serious deficiency has been pointed out in previous papers [1,6] and by MAYNARD-SMITH [19]: the system has no efficient possibility to select for favourable translation products. Let us assume that a mutation leading to a better polymerase crops up in an organized mixture of polynucleotides and polypeptides. The advantageous mutant is favoured by the more efficient enzyme no more than the other polynucleotides present. There is no preference for the better gene except in case diffusion is slow. Then a (slight) local enrichment of polynucleotides occurs in that volume element where the mutation appeared. Weak selection results from spatial separation due to incomplete mixing.

(5) The solution to the problem outlined above indeed is spatial separation in a more efficient way than hindered diffusion. The formation of compartments actually is the last step in our proposed logical sequence leading from molecules to the first protocells. The compartments may be formed by precursors of our present day cell membranes: hydrophobic material and lipid-like compounds presumably were around to some extent in prebiotic solutions. This step, however, requires more than just unorganized closure of compartments. Ultimately, this development culminates in the elaboration of a mechanism for organized cell division. At present, our knowledge concerning the properties and the biosynthesis of membranes is improving fast and steadily. Hence, there is some hope to be able to conceive physically relevant models also for this step of evolution in the near future.

6. Evolutionary Stable Strategies

In order to explain the evolution of genetically determined social behaviour within a single animal species MAYNARD-SMITH [20] introduced the concept of an evolutionary stable strategy (ESS). Strategies (I_i, $i=1,\ldots,n$) for animal contests are treated within the frame of game theory. If $A=\{a_{ij}\}$ is the payoff matrix then a_{ij} is the expectation value of the payoff for pure strategy I_i played against pure strategy I_j. Mixed strategies are defined by probability vectors, e.g.

$$I_q = \sum_{i=1}^{n} q_i I_i \qquad (24)$$

with $q = (q_1,\ldots,q_n)$, $\sum_i q_i = 1$ and $q_i \geq 0$ for $i = 1,\ldots,n$.

The payoff for I_i played against the mixed strategy I_q is simply given by $\sum_j a_{ij}q_j$. Finally, for playing I_p defined by $p = (p_1,\ldots,p_n)$ against I_q we obtain a payoff

$$p\,A\,q \;=\; \sum_i \sum_j p_i a_{ij} q_j \qquad (25)$$

All possible strategies I_x can be mapped on the simplex
$S_n = \{ x = (x_1,\dots,x_n);\ \sum_j x_j = 1,\ x_j \geq 0 \text{ for } j = 1,\dots,n\}$.

MAYNARD-SMITH [20] defines an ESS as a state $p \in S_n$ for which either

$$p\,A\,p \;>\; q\,A\,p \qquad \text{for all } q \neq p \qquad (26a)$$

or

$$p\,A\,p \;=\; q\,A\,p \quad \text{and} \quad p\,A\,q \;>\; q\,A\,q \qquad (26b).$$

Following TAYLOR and JONKER [21] we may identify "by definition" the payoff for a strategy in animal contests with a change in the rate of increase of the population using this (genetically programmed) strategy. If we assume further that this rate of increase for a strategy I_i in a population x is equal to the difference between the expected payoff for I_i and the average payoff for population x we obtain the differential equation

$$\dot{x}_i/x_i = \sum_j a_{ij}x_j - \sum_j \sum_k a_{jk}x_j x_k \qquad (27)$$

and hence we are back to our equations (21,22) with $\sum x_i = 1$.

It is easy to see that the condition for I_p being an ESS is equivalent to

$$p\,A\,x \;>\; x\,A\,x$$

for all $x \in S_n$, $x \neq p$ in a sufficiently small neighborhood of p. From this follows by simple argument [22,23] that every ESS is an asymptotically stable fixed point. The converse, however, is not true.

The dynamical representation of animal contests seems to offer some advantage when compared to the game theoretical approach:

(1) There are asymptotically stable points which are no ESS

(2) There are examples of equilibria which are neither ESS nor asymptotically stable fixed points but nevertheless play a role for dynamical self-organization of animal behaviour. An example for such a situation is presented by conflicts between two mutually dependent groups of populations like males and females in the battle of parental investment [24].

(3) There are more general situations without pointattractors like oscillations or aperiodic fluctuations to which game theory has no access.

(4) It seems that the equilibrium notions from game theory are often too static for a proper description of animal conflicts. In particular the notion of a PARETO equilibrium (a pair of strategies for two players such that no player has an advantage for changing his strategy as the other player sticks to his) appears to be irrelevant in conflicts where players cannot be assumed to behave rational and where changes in strategies are introduced by random events like mutations.

Finally, it is in line with the general ideas of synergetics to stress that the same equations crop up at the two very ends of evolution, namely self-organization of biological macromolecules and the development of strategies in animal societies.

References

1 M.Eigen, Selforganization of Matter and the Evolution of Biological Macromolecules, Naturwissenschaften 58 (1971), 465-526

2 S.Spiegelman, An Approach to the Experimental Analysis of Precellular Evolution, Quart.Rev. Biophysics 4(1971),213-253

3 B.O.Küppers, Towards an Experimental Analysis of Molecular Self-Organization and Precellular Darwinian Evolution; Naturwissenschaften 66 (1979), 228-243

4 M.Eigen and P.Schuster, The Hypercycle: A Principle of Natural Self-Organization Part A: Emergence of the Hypercycle, Naturwissenschaften 64 (1977), 541-565

5 M.Eigen and P. Schuster, The Hypercycle: A Principle of Natural Self-Organization Part B: The Abstract Hypercycle, Naturwissenschaften 65 (1978), 7-41

6 M.Eigen and P.Schuster, The Hypercycle: A Principle of Natural Self-Organization Part C: The Realistic Hypercycle, Naturwissenschaften 65 (1978), 341-369

7 J.v.Neumann, The General and Logical Theory of Automata, Collected Works, Pergamon Press, Oxford, Vol.5 (1951),312-318

8 M.S.Bartlett, An Introduction to Stochastic Processes, 3rd ed., Cambridge University Press (1978), 41-48

9 C.J.Thomson and J.L.Mc Bride, On Eigens Theory of Self-Organization of Matter and Evolution of Biological Macromolecules, Math.Bioscience 21 (1974), 127-142

10 B.L.Jones,R.H.Enns and S.S.Ragnekar, On the Theory of Selection of Coupled Macromolecular Systems, Bull.Math.Biol. 38 (1976), 15-28

11 E.Domingo, R.A.Flavell and Ch. Weissmann, In Vitro Site-directed Mutagenesis: Generation and Properties of an Infectious Extrecistronic Mutant of Bacteriophage Qß, Gene 1 (1976), 3-26

12 E.Batschelet, E.Domingo and Ch.Weissmann, Proportion of Revertant and Mutant Phage in a Growing Population as a Function of Mutation and Growth Rate, Gene 1 (1976), 27-32

13 P.Schuster,K.Sigmund and R.Wolff, Dynamical Systems Under Constant Organization,Part I: A Model for Catalytic Hypercycles, Bull.Math. Biol. 40 (1978), 743-769

14 J.Hofbauer,P.Schuster,K.Sigmund and R.Wolff, Dynamical Systems Under Constant Organization, Part II: Homogeneous Growth Functions of Degree P=2, SIAM J.Appl.Math.C (1979), in press

15 P.Schuster, K.Sigmund and R.Wolff, Dynamical Systems Under Constant Organization, Part III: Cooperative and Competitive Behaviour of Hypercycles, J.Diff.Equs. 32 (1979), 357-368

16 S.Miller and L.Orgel, Origins of Life on Earth, Prentice Hall, New Jersey (1974)

17 H.L.Sleeper and L.Orgel, The Catalysis of Nucleotide Polymerization by Compounds of Divalent Lead, J.Mol. Evol. 12 (1979), 357-364

18 R.Lohrmann and L.Orgel, Studies of Oligo Adenylate Formation on a Poly(U) Template, J.Mol.Evol. 12 (1979), 237-257

19 J.Maynard-Smith, Hypercycles and the Origin of Life, Nature 280 (1979),445-446

20 J.Maynard-Smith, The Theory of Games and the Evolution of Animal Conflicts, J.Theor.Biology 47 (1974), 209-221

21 P.Taylor and L.Jonker, Evolutionary Stable Strategies and Game Dynamics, Mathem.Bioscience 40 (1978), 145-156

22 E.C.Zeeman, Population Dynamics from Game Theory, Proc.Int.Conf. on Global Theory of Dynamical Systems, North Western University, Evanston (1979)

23 J.Hofbauer, P.Schuster and K.Sigmund, A Note on Evolutionary Stable Strategies and Game Dynamics, J.Theor.Biology 80 (1979), in press

24 P.Schuster and K.Sigmund, Coyness,Philandering and Stable Strategies, Animal Behaviour, submitted for publication

Acknowledgement: This work has been supported financially by the Austrian "Fonds zur Förderung der wissenschaftlichen Forschung", Project Nr.3502

A Mathematical Model of the Hypercycle

P. Schuster

Institut für Theoretische Chemie und Strahlenchemie, Universität Wien
A-1090 Wien, Austria

K. Sigmund

Institut für Mathematik, Universität Wien, A-1090 Wien, Austria

The emergence of life can be studied under two aspects, as a historical investigation or as an engineering problem.

The historical approach has to bridge a gap of some four billion years. No fossiles have been found in rocks older than 3.5 billion years; one may, however, also study variations in the structure of biological macromolecules, as caused by consecutive mutations. Many attempts have been made to reconstruct phylogenetic trees from characteristic proteins and nucleic acids (see e.g. [1] and [6]). Recently, M.EIGEN and R.WINKLER-OSWATITSCH tried to obtain "the first genetic information" as the common ancestor of present-day t-RNA's (to be published).

Wherever this program meets insuperable difficulties, it may be replaced by the more modest engineering approach. Instead of asking: How did it happen? one might ask: How could it conceivably have happened? An early example of this second line of thought is J.v.NEUMANN's famous essay [4] on selfreproducing automata, which deals in a very abstract way with the theoretical possibility of living organisms.

The recent theory of EIGEN and SCHUSTER [2] combines both approaches. Its central notion is that of the hypercycle. We won't describe this biochemical mechanism here, but mention only two of its aspects.

(a) The hypercycle solves an engineering problem, namely that of crossing a certain level of complexity. Indeed the minimal complexity of a biochemical system with enzymatic selfreproduction can be estimated. This complexity has to be reached through the selforganization of simpler systems whose copying mechanism is enzyme-free, and hence so fraught with errors that their information content is rather low. The hypercycle, then, is a closed loop of catalytic reactions between such macromolecular information carriers, each species catalysing the selfreproduction of the next one (Fig 1).

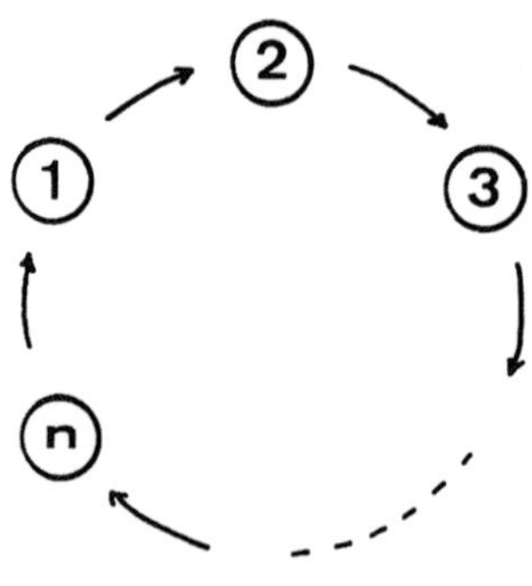

Fig 1

The coexistence of such polymers allows the integration of their information and leads therefore to a higher level of complexity.

(b) The historical aspect satisfied by hypercycles is their strongly competitive behaviour which is particular to prebiotic evolution. This evolution, indeed, must have included a sequence of decisions of the "once for ever" type - otherwise the essential uniqueness of the self-reproducing cellular mechanism would hardly be understandable. Instead of the familiar many branched descendency tree of Darwinian evolution (Fig 2) which led to the millions of species coexisting on earth today, the descendency tree of prebiotic evolution must have been of a one-branched type (Fig 3) leading to a unique genetic code and unique chiralities.

Fig 2

Fig 3

Briefly, hypercycles are biochemical devices satisfying two objectives

(a) they integrate the information of several species of selfreproducing polymers;

(b) they evolve through a one-branched descendency tree.

In this paper, we resume some of the results in [5] in order to show how a very simple mathematical model reflects these two aspects.

Thus let us consider, as a rough first approximation to the situation studied in [2], the macromolecular species 1,...,n in a chemical reactor representing some "primordial soup". In order to have evolution, we need some selection pressure, and we choose the most simple-minded one, namely the constraint that the total concentration of all the species is constant. This can be realized experimentally in a flow reactor, see [3].

If x_i denotes the concentration of species i, we obtain for its time evolution an equation of the form

$$\dot{x}_i = G_i - x_i\phi \qquad (1)$$

where $G_i = G_i(x_1,\ldots,x_n)$ describes the effect of the chemical interactions in the reactor and $\phi = \phi(x_1,\ldots,x_n)$ corresponds to a dilution flow which is regulated in such a way that the total concentration $x_1+\ldots+x_n$ is constant. It is easy to see that if we normalize it to 1, then we must have

$$\phi = \sum_{i=1}^{n} G_i \tag{2}$$

In this case

$$(\Sigma x_i)^{\cdot} = \sum_{i=1}^{n} G_i(1 - \sum_{i=1}^{n} x_i) = 0$$

if $x_1+\ldots+x_n = 1$. Hence we study equation (1) restricted to the invariant set

$$S_n = \{(x_1,\ldots,x_n) \in \mathbb{R}^n : x_i \geq 0,\ \Sigma x_i = 1\}$$

S_n is called the concentration simplex, and (1) is the general form of an equation with the constraint of constant concentration.

In a hypercycle, species 1 acts catalytically on the selfreproduction of 2,2 similarly on 3 etc. and finally n acts on 1, closing the loop. If we assume mass action kinetics for the catalytic reaction, then $G_i = k_i x_i x_{i-1}$, where $k_i > 0$ is a rate constant. (We count the indices cyclically, so that $x_o = x_n$). Hence (1) becomes:

$$\dot{x}_i = x_i(k_i x_{i-1} - \phi) \qquad i=1,\ldots,n \tag{3}$$

with

$$\phi = \sum_{j=1}^{n} k_j x_j x_{j-1} \tag{4}$$

Note first that if $x_i(0) = 0$ then $x_i(t) = 0$ for all t. Thus if some species is missing, it will not be created by the chemical reactions corresponding to (3): it is only by a supplementary mechanism, like a mutation, that it can be introduced into the chemical reactor. Mathematically, this means that the boundary of the concentration simplex S_n is invariant.

Of course we are more interested in what happens in the interior of S_n, where all species are present. Let us first ask whether we can find equilibrium states there. Since, in this case, all $k_i x_{i-1}$ must have the common value ϕ, we see that there exists a unique equilibrium in the interior, which is given by the solution of the linear equations

$$k_1 x_n = k_2 x_1 = \ldots = k_n x_{n-1} \tag{5}$$

$$\sum_{i=1}^{n} x_i = 1 \tag{6}$$

The next question is whether this equilibrium is stable or not. This is usually solved by linearization. If a differential equation

$$\dot{x}_i = F_i(x_1, \ldots, x_n) \qquad i=1,\ldots,n$$

has equilibrium P, one evaluates the eigenvalues of the Jacobian, i.e. the $n \times n$-matrix

$$\left(\left. \frac{\partial F_i}{\partial x_j} \right|_P \right)_{i,j}$$

P is called a sink if all real parts of these eigenvalues are strictly negative. In this case the equilibrium is asymptotically stable, i.e. all orbits in a sufficiently small neighborhood of P converge to P. If some eigenvalue has a strictly positive real part, the equilibrium is unstable.

A direct computation of the eigenvalues of the Jacobian of (3) is rather awkward. A simple trick, however, leads to the rather surprising result that the stability of the inner equilibrium of (3) depends only on n and not on the rate constants. Indeed, let us introduce the change in coordinates

$$y_i = \frac{k_{i+1} x_i}{\sum_{j=1}^{n} k_{j+1} x_j} \tag{7}$$

This is obviously an invertible differentiable transformation of S_n onto itself. Equation (3) transforms into

$$\dot{y}_i = y_i (y_{i-1} - \sum_{j=1}^{n} y_j y_{j-1}) M(y_1, \ldots, y_n) \tag{8}$$

where M is a function strictly positive on S_n. Omitting M means just to make a change in velocity. We are then reduced to (3) again, with all k_i equal to 1. Thus, up to a barycentric change in coordinates and a change in velocity, we may assume that the hypercycle is symmetric, i.e. that species i acts on j like i+1 on j+1, so that its equation is given by

$$\dot{x}_i = x_i (x_{i-1} - \sum_{j=1}^{n} x_j x_{j-1}) \; . \tag{9}$$

The inner equilibrium is now $C = (\frac{1}{n},\dots,\frac{1}{n})$, the central point of the simplex. The Jacobian is obviously cyclic, i.e. of the form

$$\begin{bmatrix} c_0 & c_1 & c_2 & \dots & c_{n-1} \\ c_{n-1} & c_0 & c_1 & & c_{n-2} \\ \vdots & \vdots & \vdots & & \vdots \\ c_1 & c_2 & c_3 & \dots & c_0 \end{bmatrix}$$

and the eigenvalues are given by

$$\lambda_j = \sum_{k=0}^{n-1} c_k \exp\left(\frac{2\pi ikj}{n}\right) \qquad j=0,1,\dots,n-1.$$

An easy computation leads to

Theorem 1 The eigenvalues of (3) at the inner equilibrium are

$$\lambda_j = K \exp\left(\frac{2\pi ij}{n}\right) \qquad j=1,\dots,n-1 \qquad (10)$$

where K is some positive constant.

Note that we have only n-1 (and not n) eigenvalues. Indeed, one of them is dropped because we restrict the equation to the (n-1)-dimensional invariant subset S_n.

Let us now consider the phase portrait of (9).

For n=2, the eigenvalue is - K, the equilibrium is a sink and indeed a global attractor.

For n=3, the eigenvalues are two complex conjugate numbers in the left half plane and the equilibrium is a sink. This tells us what happens in a (possibly very small) neighborhood of the fixed point. In order to know the global picture, we look at the function

$$P = x_1 x_2 x_3$$

which vanishes on the boundary of S_3, is strictly positive in the interior and attains its unique maximum at C. The function $t \to P(t) = x_1(t)x_2(t)x_3(t)$ has as time derivative

$$\begin{aligned} \dot{P} &= \dot{x}_1 x_2 x_3 + x_1 \dot{x}_2 x_3 + x_1 x_2 \dot{x}_3 = x_1 x_2 x_3 \left[(x_3-\phi) + (x_1-\phi) + (x_2-\phi)\right] \\ &= P(1-3\phi) = P\left[(x_1+x_2+x_3)^2 - 3(x_1x_2+x_2x_3+x_3x_1)\right] \\ &= \frac{P}{2}\left[(x_1-x_2)^2 + (x_2-x_3)^2 + (x_3-x_1)^2\right] \qquad (11) \end{aligned}$$

and is therefore strictly positive in the interior of S_3, except at C. By the theorem of LJAPUNOV it follows that every orbit in the interior of S_3 converges to C.

For n=4, one eigenvalue is real negative, and the other two are on the imaginary axis. An argument similar to the previous one shows that again, every orbit in the interior of S_3 converges to C.

For $n \geq 5$, there is at least one pair of complex conjugate eigenvalues in the positive half-plane, and C is unstable. Hence:

Theorem 2 The inner equilibrium of (3) is asymptotically (and even globally) stable if and only if $n \leq 4$.

For convenience, we call a hypercycle with $n \leq 4$ a short hypercycle. Thus all short hypercycles equilibrate. What happens with the long hypercycles? Numerical evidence suggest that we have a unique limit cycle i.e. a stable periodic orbit. We don't know how to prove this, however. We can only show a partial result. namely that no orbit in the interior of the concentration simplex converges to the boundary. This is important in view of aspect (a) of the hypercycle, its integration of information. If some of the species would drop extremely low in concentration, then a small fluctuation might wipe it out alltogether and hence erase its information. That this cannot happen follows from

Theorem 3: All hypercycles are cooperative. More precisely, there exists a $\delta > 0$ such that if $x_i(0) > 0$ for i=1,...,n, then $x_i(t) > \delta$ for i=1,...,n, provided it is sufficiently large.

We sketch the proof in Fig 5, for n=3 (where of course it makes no sense).

(A) One first shows that if the initial value $\underline{x}(0)$ is on the boundary of S_n, then $\underline{x}(t)$ converges to the set F of fixed points on the boundary, which is just the set of points where ϕ vanishes. (For n=3, F consists of the three vertices. For larger n, it contains also higher dimensional faces.)

(B) One next considers the relation

$$\dot{P} = P(1-n\phi) \qquad (12)$$

which is obtained as in (11). The set

$$A = \{\underline{x} \in S_n : 1-n\phi > \tfrac{1}{2}\}$$

is an open neighborhood of F. Let us now consider

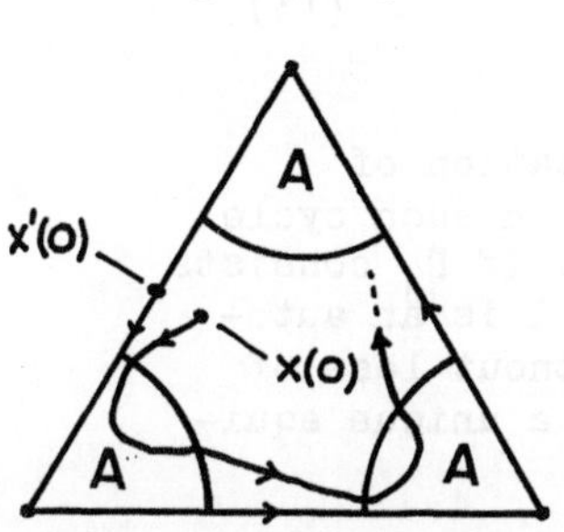

Fig.4: The corners correspond to F, the sections to A.

a thin layer in the interior of S_n, covering its boundary. Assume $\underline{x}(0)$ to be in the layer. It is near some point $\underline{x}'(0)$ on the boundary, which converges to F. Hence $\underline{x}(0)$ is whisked into A. As long as $\underline{x}(t)$ is not in A, $P(\underline{x}(t))$ may decrease: but as soon as $\underline{x}(t)$ is in A, $P(\underline{x}(t))$ increases exponentially with a factor at least $\frac{1}{2}$. Of course $\underline{x}(t)$ need not remain forever in A: but if we choose the layer thin enough, every amount of time that $\underline{x}(t)$ spends outside of A is followed by a much longer time in A, so that every eventual loss in P is soon compensated again. It takes some time to spell this out in detail, but is quite elementary and implies theorem 3.

Problem I Describe the attractors of (3) for $n \geq 5$.

The sketch of the proof of theorem 3 suggests some features pertinent to the formation of a hypercycle. Suppose that not all of the n species of the hypercycle are yet present in the chemical reactor - or the "primordial soup". Thus the state of the system will be on the boundary and hence, by (A), near the fixed point set F. The state moves very slowly and seems inert. If a mutation now introduces one of the missing species, without completing the hypercycle yet, the system remains on the boundary. After a period of transition and some possibly drastic changes in concentration, the system will be near F and hence inert again. But if a mutation introduces the slightest amount of the last missing species, the system will be in the interior of the concentration simplex and the picture changes completely: after some time, all species will be around in sizeable quantity, so that small perturbations will not be able to interrupt the cycle of reactions. If $n \geq 5$, instead of an apparently inert steady state, we obtain a pulsating form of dynamical cooperation - actually a biological clock if the attractor is indeed of the simplest possible kind, namely a limit cycle.

Let us now consider the competition of several hypercycles under the constraint of constant concentration (see Fig 5).

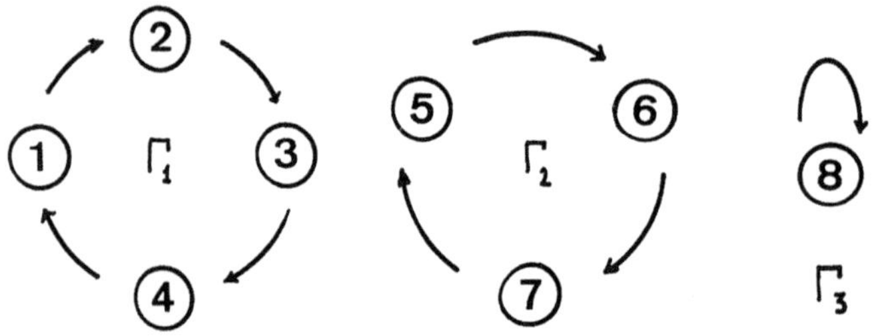

Fig 5

We describe it by

$$\dot{x}_i = x_i(k_i x_{\pi(i)} - \phi) \qquad i=1,\ldots,n \qquad (13)$$

where the $k_i > 0$ are rate constants and π is a permutation of $\{1,\ldots,n\}$. π consists of the cycles $\Gamma_1,\ldots,\Gamma_m$ and each such cycle corresponds to a hypercycle in the chemical reactor. If Γ_j consists of a unique element i, which is fixed under π, then i is an autocatalytic species. Just as before, we may assume without loss of generality that all k_i are equal to 1. There exists a unique equilibrium state which is easily seen to be unstable if π consists of more than one cycle.

If each cycle Γ_k has length $|\Gamma_k| \leq 4$, then (13) is easy to analyse, and one obtains

<u>Theorem 4</u> If several short hypercycles compete, then for almost all initial conditions, the concentrations of the species belonging to all but one of the hypercycles will vanish.

In such a competition, there is a unique "winner", therefore. The outcome depends on the initial conditions: each hypercycle has a certain "basin of attraction". Numerical evidence suggests that this is still valid for long hypercycles, but we don't know a proof of this. Hence we have

<u>Problem II:</u> Analyse the competition of hypercycles in general

Let us consider now several hypercycles (long or short) in a chemical reactor and assume again that not all species are yet present and that mutations (which are supposed to be neither large nor frequent) introduce one missing species after another. In this case, it is quite easy to see that the first hypercycle which is completed will win the competition: indeed, even if a later mutation completes another hypercycle, it doesn't stand a chance against the first one. The concentrations of the second hypercycle will be small initially and will further decrease to 0.

At first glance, this result may look quite disappointing. The first hypercycle will probably be a short one, and no further evolution seems possible. This is not so, however. We have assumed, up to now, that the competing hypercycles are disjoint, i.e. that they have no species in common. It is conceivable that hypercycles have some species in common (see Fig 6) and that species 1, say, catalyses both 2 and 3 belonging to different hypercycles.

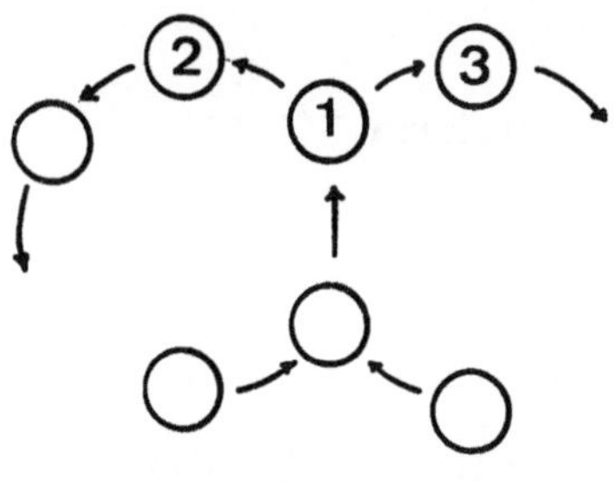

Fig 6

The equations are

$$\dot{x}_2 = x_2(k_2x_1 - \phi)$$
$$\dot{x}_3 = x_3(k_3x_1 - \phi) \qquad (14)$$

In this case, the outcome of the contest depends on the rate constants. If $k_2 > k_3$, then the hypercycle with species 2 will win, regardless of the initial concentration. Even if the hypercycle with species 3 is completed first, it will vanish as soon as a mutation has completed the hypercycle with species 2. This follows from

$$\left(\frac{x_2}{x_3}\right)^{\cdot} = \frac{x_3\dot{x}_2 - x_2\dot{x}_3}{x_3^2} = \frac{x_3x_2(k_2x_1-\phi) - x_2x_3(k_3x_1-\phi)}{x_3^2} = \frac{x_2}{x_3}\,x_1\,(k_2-k_3)$$

Thus we see how hypercycles evolve. There is no stable form of coexistence between several hypercycles, but one hypercycle may supersede another one, provided it has some species in common. In this case, it will kill its predecessor, but inherit the common species. The steps of this evolution are of the type "once for ever" and satisfy aspect (b) of the hypercycle. The descendency tree will be of the one-branched form. The path of the evolution will depend on the order of occurrence of mutations, i.e. on chance.

Thus we see that this simple model of the hypercycle behaves just as it should to fit the theory of Eigen and Schuster. Of course the justification of this theory has to be biochemical. Besides, the equations considered here are rather coarse descriptions of the hypercyclic mechanism. But it is reassuring to see how a very simple-minded mathematical model features already some of the main aspects for a theory of prebiotic evolution.

1 M.O.Dayhoff, Evolution of proteins, in R.Buvet and C.Ponnameruma, Chemical Evolution and the Origins of Life, Molecular Evolution vol 1, (1971) North Holland, p.392-419.

2 M.Eigen and P.Schuster, The Hypercycle, a Principle of Natural Selforganization, (1979) Springer Heidelberg-New York.

3 B.O.Küppers, Towards an experimental analysis of molecular selforganization and precellular Darwinian evolution, Naturwissenschaften 66 (1979), 228-243

4 J.v.Neumann, The general and logical theory of automata, (1951) Collected works vol 5, Pergamon Press 1963.

5 P.Schuster, K.Sigmund and R.Wolff, Dynamical Systems under Constant Organization.
Part 1: A model for catalytic hypercycles, Bull.Math.Biophysics 40 (1978) 743-769.
Part 2: (with J.Hofbauer) Homogeneous Growth Functions of Degree p=2, SIAM J.App..Math., to appear.
Part 3: Cooperative and Competitive Behaviour of Hypercycles, J.Diff.Equs. 32 (1979) 357-368.

6 R.M.Schwartz and M.O.Dayhoff, Origins of Prokaryotes, Eukaryotes, Mitochondria and Chloroplasts, Science 199 (1978) 395-403.

Acknowledgement: This work has been supported financially by the Austrian "Fonds zur Förderung der wissenschaftlichen Forschung", project Nr.3502

Part VII

Dynamics of Multi-unit Systems

Self-organization Phenomena in Multiple Unit Systems

A. Babloyantz

Université Libre de Bruxelles, Chimie-Physique II
B-1050 Bruxelles, Belgium

1. Introduction

Since the pioneering work of TURING [1] there has been great interest in the study of selforganizing systems. These systems present coherent behaviour that will not be possible when considering only the properties of the subunits of the ensemble. The relevance of these problems to the study of biological systems has been viedly recognized. These questions have been reviewed in several monographs [2,3].

Reaction diffusion processes had been the focus of these studies even when the system under consideration is composed from an ensemble of subunits separated from each other by membranes [4]. Inherent to this approach are the following assumptions.

(a) One completely neglects the existence of separating barriers between units. This approximation is based on the following observation. The substances enter and leave a cell by a permeation process, proportional to their concentration and following discontinuous Fick's law of diffusion. For a large number of cells and small cell volume one recovers a reaction diffusion system. In all these treatments it is implicitly assumed that at each instant the amount of substance entering a cell is compensated by an outflow.

(b) In order to transmit chemical information throughout the entire system the molecules actually have to travel a considerable distance between subunits.

(c) The molecules responsible for the selforganization processes must be relatively small in order to be able to permeate through the cell membrane. Moreover two active chemical species must react according to a specific non linear kinetics.

(d) A given unit cannot be directly in contact but with its first neighbour. In most biological systems of interest these assumptions are violated. A cell is surrounded by a complex structured membrane performing active transport; forming gates, pores and channels. Receptors on its surface can be activated or inactivated, influencing the concentration of the substances inside the cells actually without molecules leaving the cells where they are produced.

These mechanisms involve highly non linear kinetics. The Fick's law with constant coefficients is a poor approximation for these systems. There is an important class of systems where a given subunit is in direct contact with many other subunits not necessarily in their immediate neighbourhood.

One could ask the question, it is possible to develop a formalism that will extend the concept of selforganization to describe the more realistic situations. The aim of the present paper is to suggest such a formalism.

In Sect.2 we shall introduce a general formalism that shows selforganization is possible outside of reaction-diffusion processes. In Sect.3 we extend these concepts to the case of multiply connected systems. Section 4 is devoted to the study of epileptic seizure in the framework of these selforganizing concepts.

2. Selforganization Resulting from First Neighbor Interactions

Let us first consider the case where chemical information is transmitted into a given unit via first neighbor interactions only. The multicellular system is composed of N cells. In each cell there are m reacting chemical substances. For simplicity a one dimensional field is examined although the generalization to the three-dimensional system is straightforward.

Assume that a cell can influence another via contact interactions, i.e., molecules on the surface of one cell affect rates of reactions in its neighbours without any molecules actually passing out of the one cell into the other. In particular, let us assume that the concentration X_i of a certain non-diffusible substance in the ith cell of a line affects, through surface contact interactions, the concentration of X_{i-1} in the $(i-1)^{th}$ cell and that of X_{i+1} in the $(i+1)^{th}$ cell. Then the rate of change of concentration X_i in the ith cell will include a term $g(X_{i-1})$ and $g(X_{i+1})$ depending on the concentration of the $(i+1)^{th}$ and $(i-1)^{th}$ cells.

The kinetic equations describing the rate of change with time of the concentration of m reacting substances in the ith cell of a line of N cells are then:

$$\dot{X}_i^{\ell} = f^{\ell}(X_i^1 \ldots X_i^m) + g^{\ell}(X_{i+1}) + g^{\ell}(X_{i-1}) \quad , \quad \ell = 1,\ldots,m \quad ,$$
$$\ell = 1,\ldots,N \quad ; \qquad (1)$$

the function f refers to the non-linear chemical reactions proceeding inside a given cell.

In general g is a non linear function of variables. No general method of solution for the set of eq.(1) is available. Each specific case must be examined separately. And in most cases the equations can only be solved by numerical methods.

Recently a specific model with $g = D[X_{i+1} + X_{i-1}] - D'X_i[X_{i+1} + X_{i-1}]$ has been studied numerically by BABLOYANTZ [5]. It has been shown that the homogeneous steady state of the system bifurcates into a space dependent solution. In some cases analytical results can be shown for these systems. This is the case when the function $g(X_{i-1})$ and $g(X_{i+1})$ can be written as linear functions of the concentrations DX_{i-1} and DX_{i+1} where D is the intercellular interaction constant. Then eq.(1) reduces to:

$$\dot{X}_i^\ell = f^\ell(X_i^1,\ldots,X_i^m) + D^\ell X_{i+1} + D^\ell X_{i-1}$$

$$= h^\ell(X_i^1,\ldots,X_i^m) + D^\ell[X_{i+1} + X_{i-1}] - 2D^{\ell\prime}X_i \quad , \quad \ell = 1,\ldots,m \quad ,$$

$$i = 1,\ldots,N \quad , \qquad (2)$$

where

$$h = f + 2D'X_i \quad .$$

The cell contact operator $D[X_{i+1} + X_{i-1}] - DX_i$ forms a Jacobi matrix with known eigenvalues and eigenfunctions. For a two-variable system, a standard linear stability analysis can be performed and conditions can be given for the occurrence of bifurcating solutions out of an initially homogeneous state (LEMARCHAND and NICOLIS [7]).

The eigenfunctions and the eigenvalues are

$$\dot{X}_i^k = X_i^0 \sin\left[\frac{i\pi k}{(N+1)}\right] \quad ,$$

$$\lambda_i^k = D2\cos\left[\frac{\pi k}{(N+1)}\right] - D' \quad , \quad k = 1,\ldots,N \quad . \qquad (3)$$

Using these expressions, the linear stability analysis gives

$$\Lambda_k = \begin{vmatrix} \left(\frac{\partial h_1}{\partial X_1}\right)_{\substack{X_1=X_1^0\\X_2=X_2^0}} - \lambda_k D_1 & \left(\frac{\partial h_1}{\partial X_2}\right)_{\substack{X_1=X_1^0\\X_2=X_2^0}} \\ \left(\frac{\partial h_2}{\partial X_1}\right)_{\substack{X_1=X_1^0\\X_2=X_2^0}} & \left(\frac{\partial h_2}{\partial X_2}\right)_{\substack{X_1=X_1^0\\X_2=X_2^0}} - \lambda_k D_2 \end{vmatrix} \quad ,$$

where X_1^0 and X_2^0 are the bifurcating homogeneous steady state concentrations.

Let us take $\Delta = \mathrm{Det}.\Lambda_k$ and $T = T_r\Lambda_k$, then $\rho = T^2 - 4\Delta$. Now if $\rho > 0$, $\Delta < 0$ and $T < 0$ the steady state is a saddle point. For $\rho < 0$ and $T > 0$ the steady state is an unstable focus. In both cases the homogeneous steady state becomes unstable and bifurcates into a self-organized solution.

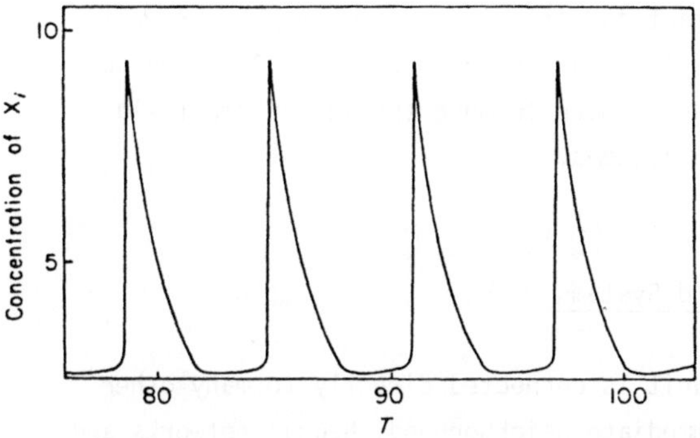

Fig.1. Oscillation in concentration of X_i in time inside a given cell at steady-state regime as the result of wave propagation in the system. The boundary cells were held at the homogeneous steady states $X_i = 2.454$, $Y_i = 3.563$ with $A = 2$, $B = 8.6$, $D_X = 0.04$, $D_X' = 0.02$, $D_Y = 0.1$, $D_Y' = 0.05$

The knowledge of the eigenfunctions and eigenvalues of the linear contact operator enables one to use the bifurcation scheme of AUCHMUTY and NICOLIS for the calculation of the steady states of the set of eq.(2). Under proper boundary conditions one finds that the homogeneous steady state bifurcates into a space dependent solution.

Figure 1 gives a numerical result showing time periodic solution in a system with linear contact operator. For the details of the model one must consult the original publication [5].

We can also convince ourselves of the existence of bifurcation in this system in the following manner.

If we take D = D' then eq.(2) becomes

$$\dot{X}_i^{\ell} = h^{\ell} + D^{\ell}[X_{i+1} + X_{i-1} - 2X_i] \quad . \tag{4}$$

Equation (4) is formally identical to a reaction diffusion mechanism where the diffusion is approximated by a discrete representation of Fick's law of diffusion and describes the permeation of X_i from cell to cell. A complete identity between eq.(4) and a reaction diffusion equation can be established if we take

$$D_i = (N/L)^2 d_i$$

where L is the length of the reaction diffusion systems, N the number of cells and d_i is the Fick coefficient. In the limit $N \to \infty$ the contact operator tends to the Laplacian operator describing diffusion phenomenon and the solution to eq. (3) is expected and does reduce to that of a reaction diffusion mechanism much studied in the literature [6] HERSCHKOWITZ-KAUFMAN and NICOLIS [8]. Thus all the self-organizing properties demonstrated for the reaction diffusion systems may be shown to exist for the ensemble of cells in contact and described by eq.(4). One need not assume that at least two chemical substances have to move throughout the field which is the case for reaction diffusion systems.

3. Selforganization in Multiply Connected Systems

It may happen that in a system a given unit is connected directly to many other units that are not necessarily in its immediate neighborhood. Neural networks and social systems are examples of such systems. Again the time evolution of the petinent variables describing such systems can be written in terms of a set of first order coupled non linear differential equations. However, the coupling between different variables is more intricate.

Starting from a given homogeneous steady state we want to investigate the possibilities of selforganization in such systems. Moreover we want to relate the different modes of selforganization to the degree of connectivity of the system. That is we choose the number of connectivity as a new bifurcating parameter while keeping all the other parameters of the system unchanged. Let us stress again that the change in connectivity does not change the homogeneous steady state.

The system under consideration is the same as in the previous section. However, in the present case cells are in direct interaction not only with their immediate neighbours but also with 2n-1 distant units. We shall assume that the interaction between two units is a linear function of concentrations.

The time evolution of the variables of the system can be described by the following set of differential equations.

$$\dot{x}_i = f(x_i,y_i) + c_x\left[\frac{1}{n}(x_{i-n} + \dots + x_{i-1} + x_{i+1} + \dots + x_{i+n}) - 2x_i\right]$$

$$y_i = h(x_i,y_i) + c_y\left[\frac{1}{n'}(y_{i-n'} + \dots + y_{i-1} + y_{i+1} + \dots + y_{i+n'}) - 2y_i\right] \qquad (5)$$

$$i = 1,\dots,N \quad .$$

The functions f and h are the same as in preceding section. c_x and c_y are the contact coefficients denoting the strength of the influence of one cell on the other. n and n' are number of connectivity. The contact operator is defined in such a way as when the number of connections are increased the homogeneous steady state itself remains unchanged. In other words, for a given cell when the number of connecting neighbors increases, the strength of each individual connection decreases. The total amount of input and output per cell, therefore, remains constant at the steady state.

One way of looking for the selforganization in these systems is to study the stability properties of the homogeneous steady states of the set of eq.(5). If the state is unstable for a given set of parameters their bifurcation into a new time periodic, or spatially inhomogeneous or spatio-temporal state becomes possible. The stability of set of eq.(5) around the uniform steady state can be studied using the technique of normal mode analysis [9]. In order to perform this analysis we need to examine the eigenvalues of the linear contact operator. Two sets of boundary conditions will be considered.

(a) The system interacts at each of its boundaries with n other cells [-(n-1), ...,-1,0] and (N + 1,...,N + n) within which the concentrations of x_i and y_i are kept fixed at their uniform steady state value.

(b) The ensemble of cells form a closed ring. This is the case of periodic boundary conditions.

For periodic boundary conditions it can be shown that the eigenvalues $(-\lambda^k)$ of the contact operator are given by

$$\lambda^k = 2\left[1 - \frac{1}{n}\,\frac{\cos[(n+1)w_k/2]\sin(nw_k/2)}{\sin(w_k/2)}\right], \quad w_k = 2\pi k/N \quad k = 1,\ldots,N \quad . \tag{6}$$

Figure 2 shows the first four values of λ^k as a function of n for a 40 unit ensemble. It is seen that all λ^k increase with increasing values of n. However, the Nth eigenvalue is always zero $(\lambda^N = 0)$ and is independent of n.

For fixed boundary conditions the eigenvalues must be found by numerical methods (BABLOYANTZ and KACZMAREK [10]).

Again one can see that for this case also λ^k increases with increasing n. But there is no more possibility for a zero eigenvalue.

It can be shown [6] that the analysis for the entire system can be reduced to the study of a Jacobian matrix

$$\Lambda_k = \begin{bmatrix} f_x - c_x\lambda_x^k & f_y \\ h_x & h_y - c_y\lambda_y^k \end{bmatrix}; \; f_x = \frac{\partial f}{\partial x} \;, \; f_y = \frac{\partial f}{\partial y} \;, \; h_x = \frac{\partial h}{\partial x} \;, \; h_y = \frac{\partial h}{\partial y} \quad .$$

In order to perform the linear stability analysis we need the following expressions. Det $\Lambda_k = \Delta_s + \Delta_k$, $T_r\Lambda_k = T_s + T_k$ and $\rho = T^2 - 4\Delta = \rho_s + \rho_k$. Here the subscript s and k refers to those contributions independent of and dependent of λ^k, respectively. The possibility and nature of unstabilities will depend crucially on the nature of

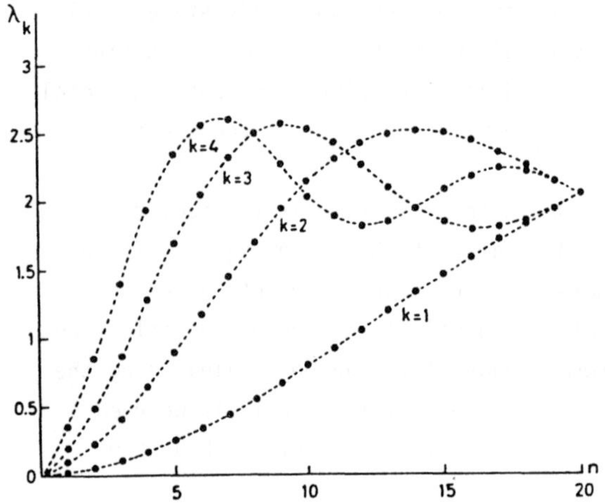

Fig.2. A plot of the first four values of λ^k from eq.(3) against n for a 40 unit ensemble with periodic boundary conditions. λ^1 corresponds to the lowest value of λ at all n except for λ^{40} which is always zero

functions f and h. Two special cases will be examined here. (I) The functions f and h are such that within a certain domain of parameters the steady state for an unconnected unit bifurcates to a limit cycle solution, i.e., the steady state is an unstable focus. (II) The uniform steady state for the unconnected ensemble is a stable focus.

Case (I). An unstable focus is characterized by $T_s > 0$ and $\rho_s < 0$ with $\Delta_s > 0$. The condition $T_s > 0$ can be satisfied in four ways. We shall only consider two situations that may arise more often

(a) $f_x > 0 \quad h_y < 0 \quad f_x > |h_y|$

(b) $f_x < 0 \quad h_y > 0 \quad |f_x| < h_y$.

From relation (6) and Fig.2 it can be seen that, since λ_x^k and λ_y^k are always positive, the formation of connections will always decrease the value of T, i.e., T_k will always be negative. For the smallest values of λ any further increase in connectivity will cause an even greater decrease in T as λ_x^k and λ_y^k increase monotonically with n. When this negative term T_k becomes dominant, T becomes negative for this and for all other λ^k. The orginal unstable focus then takes on the character of a stable focus, stable node or a saddle point depending on the values of ρ.

To investigate the influence of the number of connectivity on ρ we first consider the case where both variables have the same number of interconnections. Thus

$$\rho_k = (\lambda^k)^2(c_x - c_y)^2 + 2\lambda^k(c_x - c_y)(h_y - f_x) \quad . \tag{7}$$

Moreover if we choose $C_n = C_y$, ρ_k vanishes and ρ remains negative. The unstable focus is transformed into a stable one and the oscillatory property of the individual unconnected units is lost. If $C_x > C_y$, ρ_k is always positive for case (b) by eq.(7). If $|C_x - C_y|$ is sufficiently large, ρ may become positive and the steady state will be transformed to either a saddle point ($\Delta < 0$) or a stable node ($\Delta > 0$). An exactly analogous situation applies to case (a) with $C_x < C_y$. When $(C_x - C_y)$ and $(h_y - f_x)$ are of opposite sign i.e., case (a) with $C_x > C_y$ or case (b) with $C_x < C_y$, the sign of ρ_k will depend on the relative magnitude of these two terms.

The interested reader will find in the original paper [10] a similar discussion for the case where the degree of interconnection of the two variables is not identical.

Case (II). For any system in which the stability properties of an unconnected unit are those of a stable focus we have $T_s < 0$ and $\rho_s < 0$. Changes in the connectivity of either or both variables will always drive T more negative and, in this sense always make the system more stable. Periodic phenomena will be excluded. The influence of multiple connections on the sign of ρ_k is, however, exactly the same as for case (I) and the steady-state may remain a stable focus or be transformed to a stable node or saddle point. Therefore, although temporal oscillations can never emerge as a result of increased connectivity it is possible for spatial patterning to be produced when the number of interconnections is increased.

4. Neural Network Model for Epileptic Seizures

In this section we shall try to apply the ideas of the preceding section to a concrete biological problem. We want to model the emergence of spatio-temporal patterns during an epileptic seizure.

Figure 3 shows the electroencephalographic activity (EEG Waves) for a normal cortical state and during a seizure. In the normal conditions and in the absence of external stimuli it seems that there is no correlation between individual units in the same cortical domain [11]. However, during a seizure the neurons fire in phase. There appears a large increase in the amplitude of the EEG followed by an extremely regular sharp and biphasic wave.

We propose the following simple synaptic model to account for this phenomenon. The network is formed by a set of 2 N excitatory and inhibitory neurons in equal numbers. The variables are the set of mean membrane potentials across the electrically unexcitable dendritic membranes of excitatory cells and that of the inhibitory neurons.

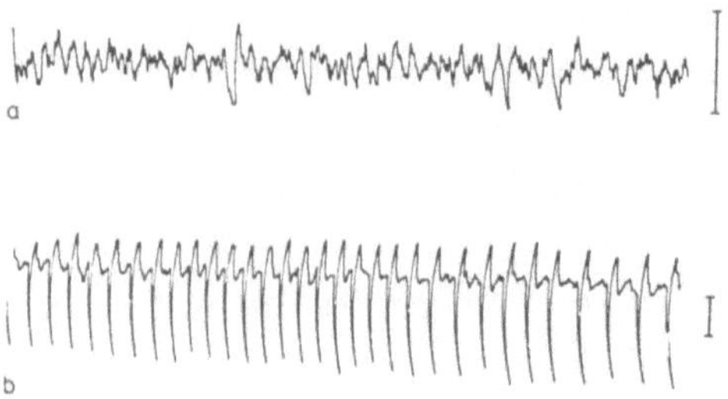

Fig.3a,b. Normal cortical electroencephalographic activity (EEG). (b) Activity during a seizure. (Data from KACZMAREK and ADEY, 1973)

The time evolution of these variables is given by the set of equations

$$\frac{dx_i(t)}{dt} = \kappa[V - x_i(t)] + [\varepsilon - x_i(t)] \sum_{j=1}^{N} a_{ij} f[x_j(t-\tau)] + [E_{Cl} - x_i(t)] \sum_{j=1}^{N} b_{ij} f[y_j(t-\tau)]$$

$$\frac{dy_i(t)}{dt} = \kappa[V - y_i(t)] + [\varepsilon - y_i(t)] \sum_{j=1}^{N} c_{ij} f[x_j(t-\tau)]\ ,\quad i = 1 \ldots N \quad . \tag{8}$$

For the details of the model the original paper of KACZMAREK and BABLOYANTZ must be consulted [12]. Note that the firing rate f plays here the role of multiunit contact operator and is responsible for the nonlinear character of the ensemble. Its functional form can be found from experimental data [13]. As in preceding section we shall assume that the total input into a given neuron is constant. This assumption introduces the following conservation relations.

$$\sum_{j=1}^{N} a_{ij} = A\ ,\quad \sum_{j=1}^{N} b_{ij} = B\ ,\quad \sum_{j=1}^{N} c_{ij} = C\ ,\quad i = 1 \ldots N$$

The set of eq.(8) together with these conditions admits a uniform steady state. For low excitatory input A there is only one stationary state with low values of x and y. As excitation increases for a critical value of A multiple steady states appear with high values of x and y. However, they still correspond to a homogeneous steady state situation. The linear stability analysis shows that for critical values of $A = A_c$ the upper branch may become unstable and the system will bifurcate into an oscillatory solution. The direct integration of eq.(8) by computer confirmed these predictions. The computer results are shown in Fig.4 for $A > A_c$.

It is seen that the potential of excitatory cells shows regular oscillation in time. The oscillation now also has a markedly biphasic appearance with each cycle having a sharp negative going peak, during which the cells are firing, followed very abruptly by a peak in the direction of depolarization. The black bars in the upper part of Fig.4 show the firing time of cells. It is seen that both types of cells fire in unison. After each firing period, there follows a period of spike

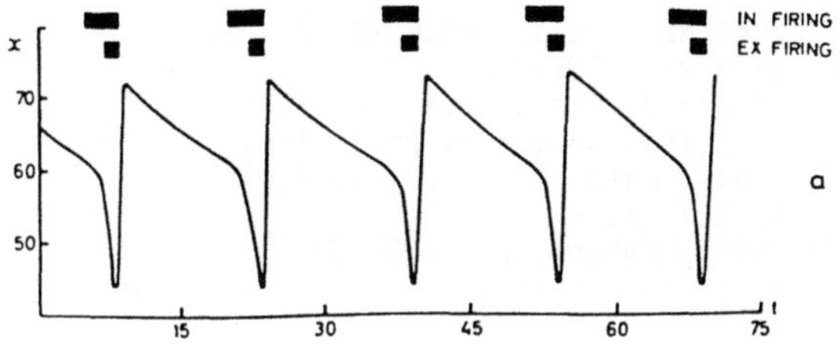

Fig.4. Upper Diagram. A plot of the sharp biphasic oscillatory solution to eq.(8) for $\kappa=0.02$, $V = 20.0$, $\varepsilon=72.0$, $E_{Cl} = -20.0$, $A = 0.03$, $B = 0.002$, $C = 0.003$, $k = 10.0$, $\phi = 30.0$

inactivation until the potentials fall sufficiently to start repetitive firing and recommence the cycle. It can be shown that the oscillatory instability arises because of the delay terms in eq.(8). A critical value can be calculated above which oscillatory solutions may be found.

5. Discussion

We have shown that selforganizational properties can be seen in multiunit systems if the latter are described by a set of first order coupled non linear differential equations. All the properties of reaction diffusion systems such as time periodicity, spatial inhomogeneity and time periodic solutions can be shown for these systems. These formalisms enable one to extend the treatment successfully to the systems with multiple connections.

For these systems it is seen that increasing the connectivity of an ensemble of units with multiple contacts may transform the spatiotemporal patterns generated by the ensemble. These transformations are generally in the direction of greater stability. Temporal periodicity is nearly always sacrified by sufficiently increasing the multiplicity of contacts. Under certain conditions, however, purely spatial patterns may arise through an increased number of interconnections. With the help of this formalism we could reproduce the main features of the patterns that arise during an epileptic seizure, namely a biphasic periodicity. Also, in general agreement with the present analysis, complex spatiotemporal patterns were destroyed by increasing the spatial extent of excitatory interconnections to produce spatially homogeneous solutions.

References

1. Turing, A.M. (1962): Phil. Trans. R. Soc. Lond. B *237*, 37
2. Haken, H. (1976): *Synergetics* (Springer, Berlin, Heidelberg, New York)
3. Nicolis, G., Prigogine, I. (1977): *Self-Organization in Nonequilibrium Systems* (Wiley, London)
4. Babloyantz, A., Hiernaux, J. (1975): Bull. Math. Biol. *37*, 637-657
5. Babloyantz, A. (1977): J. Theor. Biol. *68*, 551-561

6. Nicolis, G., Auchmuty, J.F.G. (1974): Proc. Nat. Acad. Sci. USA *71*, 2748
7. Lemarchand, H., Nicolis, G. (1976): Physica *82*A, 521-542
8. Herschkowitz-Kaufman, M., Nicolis, G. (1972): J. Chem. Phys. *56*, 1890
9. Minorski, N. (1962): *Nonlinear Oscillations* (Van Nostrand, Princeton)
10. Babloyantz, A., Kaczmarek, L.K. (1979): Bull. Math. Biol. *41*, 193-201
11. Elul, R. (1972): Int. Rev. Neurobiol. *15*, 227-272
12. Kaczmarek, L.K., Babloyantz, A. (1977): Biol. Cybern. *26*, 199-208
13. Granit, R., Kernell, D., Smith, R.S. (1963): J. Physiol. (Lond.) *168*, 890-910

Synchronized and Differentiated Modes of Cellular Dynamics

H.G. Othmer

Department of Mathematics, Rutgers University
New Brunswick, NJ, USA

1. Introduction

Intercellular communication serves a variety of purposes in metazoan systems, including, (i) the recruitment of individual cells into multicellular aggregates, as in the transition from vegetative growth to the slug stage in the slime mold *Dictyostelium discoideum* (Dd hereafter), (ii) the initiation and/or synchronization of collective activities such as the repetitive contraction in the myocardium, (iii) the initiation and coordination of spatial differentiation in developing systems, and (iv) the guidance of cell movement during morphogenesis. It is convenient to classify the different modes of intercellular communication that have evolved for these purposes as either long-range or short-range, according as the distances involved are greater or less than about 0.25 mm, this being the distance for which the relaxation time for diffusion is of the order of one minute. Long-range signal transmission can occur via convective transport, as in hormonal interactions via a circulatory system, or it may involve diffusion coupled with a spatially-distributed mechanism for regenerating and relaying the signal, such as occurs in nerve impulse conduction and in the aggregation phase of Dd.

At least three distinct modes of short-range communication between cells have been identified: (i) direct exchange of diffusible substances such as ions, cyclic nucleotides, small metabolites, and neurotransmitters via gap junctions and synapses, (ii) indirect interactions by uptake of nutrients or other essential substances from a common pool or by release of substances that activate or inhibit cellular functions into the pool, and (iii) surface interactions that result from mechanical stresses or that occur via receptor molecules that are embedded in the cell membrane. Direct exchange occurs frequently in developing systems and plays a central role in reaction -diffusion models of pattern formation [1]. For instance, it has been found that the ionic permeability of gap junctions in developing insect epidermis is hormonally controlled and varies with the developmental stage [2]. Furthermore, it has been demonstrated that cell-to-cell communication via gap junctions is responsible for the synchronization of activities in the myocardium and in smooth muscle [3]. Indirect interaction occurs in suspensions of Dd, which releases CAMP into the external medium via vesicles [4], and in suspensions of the yeast cell *S. carlsbergensis* [5]. It is also used as an alternative to the conventional mode of synaptic transmission in certain types of neurons in the marine mollusk *Aplysia* [6].

In many other cases the mode of interaction between cells is not known because several modes may be possible and the molecular species involved has not yet been identified. Since it is to be expected that different modes lead to different dynamical behavior in a cellular aggregate, it may be helpful in identifying the mode used to contrast the dynamical behavior for the various modes, and this is done here for directly- and indirectly-coupled systems. The following section deals with the question of global synchronization and gives conditions under which all cells relax to the same instantaneous state, irrespective of whether that state is time-invariant, time-periodic, or shows some more complicated temporal behavior. These conditions prove to be very stringent and one can ask whether weaker conditions will ensure that a synchronized system remains synchronized in the presence of small disturbances. This question is dealt with in the third section. The fourth section is devoted to a detailed analysis of the simplest example of direct and indirect coupling, namely, a system of two cells with two active species in each cell. The range of dynamical behavior exhibited in this simple system illustrates the difficulty inherent in determining the complete structure of the solution set for more complicated systems.

2. Global Synchronization

For many purposes an aggregate of cells coupled by gap junctions can be regarded as a reacting continuum and this viewpoint is adopted here. Let Ω be a bounded region of R_p ($p = 1,2$ or 3) with outward normal $\underset{\sim}{n}$. We assume that the flux of species i is given by $j_i = -\bar{\mathcal{D}}_i \nabla C_i$ where the $\bar{\mathcal{D}}_i$ are positive constants, and that the boundary is impermeable. The governing equations are

$$
\begin{aligned}
\frac{\partial C}{\partial t} &= \bar{\mathcal{D}}\bar{\nabla}^2 C + \bar{R}(C) \quad \text{in} \quad \Omega \\
\underset{\sim}{n}\cdot\nabla C &= 0 \quad \text{on} \quad \partial\Omega \\
C(\underset{\sim}{r},0) &= C_0(\underset{\sim}{r}) \quad \text{in} \quad \Omega
\end{aligned}
\tag{1}
$$

where $C = (C_1,\ldots,C_n)^T$, $R(C)$ is the reaction rate vector, r is the space coordinate in Ω, and $\bar{\mathcal{D}}$ is the diagonal diffusion matrix. For the purpose of casting these equations into a dimensionless form, let L be a measure of Ω, let $\bar{C}_i$ be a reference concentration for species i, let κ^{-1} be a time scale characteristic of the kinetics, and let $\delta = \max \bar{\mathcal{D}}_i$. Set

$$\underset{\sim}{\zeta} = \underset{\sim}{r}/L, \quad \tau = \kappa t, \quad c_i = C_i/\bar{C}_i, \quad \mathcal{D}_i = \bar{\mathcal{D}}_i/\delta \quad \text{and} \quad \nabla = L\bar{\nabla}.$$

Then (1) becomes

$$
\begin{aligned}
\frac{\partial c}{\partial \tau} &= \Delta_1 \mathcal{D}\nabla^2 c + R(c) \quad \text{in} \quad \Omega \\
\underset{\sim}{n}\cdot\nabla c &= 0 \quad \text{on} \quad \partial\Omega \\
c(\underset{\sim}{\zeta},0) &= c_0(\underset{\sim}{\zeta}) \quad \text{in} \quad \Omega
\end{aligned}
\tag{2}
$$

where $\Delta_1 \equiv \delta/\kappa L^2$ and $R_i(c) = \bar{R}_i(\bar{C}_1 c_1,\ldots,\bar{C}_n c_n)/\kappa\bar{C}_i$.

When $R(c) \equiv 0$, the initial distribution relaxes to a spatially uniform one exponentially in time and the Fourier series solution of (2) in that case shows that the rate is controlled by the smallest (in magnitude) nonzero eigenvalue of the Laplacian on Ω. The same should be true when $R(c) \neq 0$, provided that the appropriate relaxation time for diffusion is short compared to that for reaction. To make this precise, we must assume that solutions of (2) are bounded for all τ, pointwise in ζ. Let $\|\cdot\|_E$ and $\|\cdot\|_{L_2}$ be the Euclidean and L_2 norms, respectively, set $\tilde{m}(\Omega) = \int_\Omega d\underset{\sim}{\xi}$, and let

$$\bar{c}(\tau) = \frac{1}{m(\Omega)} \int_\Omega c(\underset{\sim}{\zeta},\tau)d\underset{\sim}{\zeta}.$$

Further, let

$$\hat{k} \equiv \sup_c \left\| \frac{\partial R}{\partial c} \right\|_E$$

and let $-\alpha_1^2$ be the smallest nonzero eigenvalue of the problem

$$
\begin{aligned}
\nabla^2\phi &= -\alpha^2\phi \quad \text{in} \quad \Omega \\
\underset{\sim}{n}\cdot\nabla\phi &= 0 \quad \text{on} \quad \partial\Omega.
\end{aligned}
\tag{3}
$$

Then it can be shown [1,7] that if

$$\Delta_1 \alpha_1^2 (\min_i \mathcal{D}_i) > \hat{k} \tag{4}$$

then

$$\| c(\underset{\sim}{\xi},\tau) - \bar{c}(\tau) \|_{L_2} \to 0$$

exponentially in τ. Thus the appropriate relaxation times for reaction and diffusion are $\tau_R = (k\kappa)^{-1}$ and $\tau_D = L^2/\alpha_1^2(\min \bar{\mathcal{D}}_i)$, respectively, and (4) is equivalent to the condition $\tau_D < \tau_R$. Typically $\tau_R \sim 10$ seconds but for some metabolic processes it may be up to 9 hours [8]. In a one dimensional system of length L, $\alpha_1^2 = \pi^2$ and if $\min \bar{\mathcal{D}}_i \sim 10^{-7}$ cm²/sec, then the lengths for which (4) is satisfied for the foregoing τ_R are ~0.03 mm and 1.8 mm, respectively. A typical cell diameter is ~10μ and so one can expect synchrony over at least 3ϕ's *if all species in question pass through the gap junctions*.

A different mathematical description is required when diffusible substances can pass between cells only via the extracellular medium. While it is possible to formulate the governing partial differential equations for a random suspension of cells, it is difficult to extract information from them and a different approach is taken here. Imagine that the cells form a monolayer at the bottom of a petri dish and that the cells are covered by a thin layer of culture medium. Suppose that reactions occur only in the cells, that the composition within a cell is uniform, and that there is no direct communication between cells. Further, suppose that the overlying fluid layer is very thin compared to the dish diameter so that vertical nonuniformity decays rapidly compared to horizontal nonuniformity. If the flux between fluid and cells is linear in the concentration difference, then the governing equations, written in dimensionless form, are

$$\begin{aligned} \frac{\partial u}{\partial \tau} &= \Delta_1 \mathcal{D} \nabla^2 u + \varepsilon \Delta_2 H(v-u) \\ \frac{\partial v}{\partial \tau} &= -\Delta_2 H(v-u) + R(v). \end{aligned} \tag{5}$$

Here u is the vertical average concentration in the fluid, v is the composition in the cells, ε is the ratio of total cell volume to fluid volume, and H is a constant diagonal matrix whose entries are $h_i = \bar{h}_i/(\max \bar{h}_i)$. The $\bar{h}_i$ are the dimensional transfer coefficients, $\Delta_2 \equiv (\max \bar{h}_i)a/$, and a is the interfacial area per unit cell volume. These equations provide an exact description in the limit as the cell diameter tends to zero and otherwise are adequate whenever the scale of concentration variation in the fluid is greater than a cell diameter.

Since there is no direct exchange between cells, the conditions under which both phases relax to a uniform state will necessarily involve the h_i's as well as the diffusion coefficients. To formulate these, let V^c and V^f denote the cell and fluid volume, respectively, let $V = V^c + V^f$, set $\underline{h} = \min_i h_i$, and let

$$\bar{u}(\tau) = \frac{1}{V}\int_V u(\underset{\sim}{\xi},\tau)d\underset{\sim}{\xi} \qquad \bar{v}(\tau) = \frac{1}{V}\int_V v(\underset{\sim}{\xi},\tau)d\underset{\sim}{\xi}. \tag{6}$$

Here u and v are defined to be zero in V^c and V^f, respectively. Then it can be shown [9] that if

$$\underline{h} > 2\hat{k}/\Delta_2 \tag{7}$$

and

$$\Delta_1 \alpha_1^2 (\min \mathcal{D}_i) > \frac{\varepsilon \Delta_2^2 (1-\underline{h}^2) + 2\varepsilon \Delta_2 \hat{k}}{\Delta_2 \underline{h} - 2\hat{k}} \tag{8}$$

then

$$\| u(\underset{\sim}{\xi},\tau) - \bar{u}(\tau) \|_{L_2} \quad \text{and} \quad \| v(\underset{\sim}{\xi},\tau) - \bar{v}(\tau) \|_{L_2}$$

tend to zero exponentially in τ.

If we define a relaxation time for interphase transport as $\tau_T = [(\min \bar{h}_i)a]^{-1}$ and use the previous definitions of τ_R and τ_D, then (7) and (8) can be written as

$$\tau_T < \frac{1}{2}\tau_R \tag{9}$$

and

$$\tau_D < \frac{\underline{h}^2\tau_T}{\varepsilon}\left[\frac{\tau_R-2\tau_T}{\tau_R(1-\underline{h}^2)+2\underline{h}\tau_T}\right]. \tag{10}$$

As $\underline{h} \to 1$, in which case all h_i's are equal, (10) reduces to

$$\tau_D < \frac{\tau_R-2\tau_T}{2\varepsilon}. \tag{11}$$

Furthermore, if $\underline{h} = 1$ and $\bar{h}_i \to \infty$, $\tau_T \to 0$ and so (9) is always satisfied, while (11) reduces to $\tau_D < \tau_R/2\varepsilon$. This is to be compared with the inequality $\tau_D < \tau_R$ that ensures synchronization for the directly-coupled case. It should be noted that if $\underline{h} = 0$ or if $\underline{h} \in (0,1)$ and $\min \bar{h}_i \to \infty$, the right-hand side of (10) reduces to zero. Thus the synchronization conditions cannot be satisfied unless all $\bar{h}_i > 0$ and if $\min \bar{h}_i \to \infty$, all $\bar{h}_i$ must be equal as well. A later example will show that the uniform state can in fact be unstable when $\underline{h} = 0$.

3. Destabilization of a Synchronized State

The conditions that guarantee global asymptotic stability of the set of all uniform solutions provide estimates of the region in parameter space in which no spatial differentiation between cells can persist in time. Outside that region it may happen that a spatially-uniform solution is unstable to certain small-amplitude nonuniform disturbances and the problem is to determine conditions on the kinetic mechanism and the transport coefficients under which this is possible. We first consider the destabilization of uniform steady states in directly-coupled systems.

Suppose that $\bar{c}$ is such that $R(\bar{c}) = 0$ and set $x = c-\bar{c}$; then x satisfies

$$\frac{\partial x}{\partial \tau} = \Delta_1 \mathcal{D}\nabla^2 x + Kx + F(x) \quad \text{in} \quad \Omega$$
$$\frac{\partial x}{\partial n} = 0 \quad \text{on} \quad \partial\Omega \tag{12}$$

where K is the Jacobian of R at $\bar{c}$ and $\|F(x)\|_E \sim \mathcal{O}(\|x\|_E)$ as $\|x\|_E \to 0$. If asymptotic stability of $\bar{c}$ is defined in terms of either the L_2 or L_∞ norm, it can be shown that in noncritical cases stability is governed by the linear terms in (12). The linear equation gotten by dropping $F(x)$ has the solution

$$x(\underset{\sim}{\zeta},\tau) = \sum_{n=0}^{\infty} e^{(K-\mu_n\mathcal{D})\tau} y_n\phi_n(\underset{\sim}{\zeta}), \tag{13}$$

where $\mu_n \equiv \alpha_n^2\Delta_1$, and therefore asymptotic stability is governed by the spectrum of the set $\{K-\mu_n\mathcal{D}\}$. For any given smooth domain Ω there are only a countable number of matrices to test, but to arrive at results valid for any domain, we replace μ_n with a continuous parameter $\mu \in [0,\infty)$. Furthermore, we assume that the steady state is asymptotically stable as a solution of the kinetic equations $dc/d\tau = R(c)$, in which case the spectrum of K lies in the left-half plane. A number of special cases that are asymptotically stable at all wavelengths μ are known, including the case of symmetric K and the case where the diffusion coefficients do not differ too much from each other. Since we are interested in determining when instabilities can arise, we are primarily concerned with establishing necessary conditions for asymptotic stability at all wavelengths.

Let $\sigma(K)$ denote the spectrum of K, let LHP ($\overline{LHP}$) denote the open (closed) left-half complex plane, and let $K[i_1,\ldots,i_p]$ denote a $p \times p$ principal submatrix of K formed from rows and columns $i_1,\ldots,i_p$ for $1 \le p \le n-1$.

Theorem 1 Let $\mathcal{D}$ be diagonal with $\mathcal{D}_i \ge 0$. In order that $\sigma(K-\mu\mathcal{D}) \subset$ LHP for all such $\mathcal{D}$ and all $\mu \in [0,\infty)$, it is necessary that

(i) $\sigma(K) \subset$ LHP

(ii) $\sigma(K[i_1,\ldots,i_p]) \subset \overline{LHP}$ for all p^{th}-order submatrices of K, where $1 \le p \le n-1$.

We shall not give the proof here but will merely indicate how an instability arises when one of the conditions is violated. Suppose that there is a $p \times p$ principal submatrix of K whose spectrum intersects the open right-half plane. Without loss of generality we may assume that it lies in the first p rows and columns, and we partition K as follows

$$K = \begin{bmatrix} K_1 & K_2 \\ K_3 & K_4 \end{bmatrix} \tag{14}$$

Here K_1 is $p \times p$, K_4 is $(n-p) \times (n-p)$, etc. Since K_1 has at least one eigenvalue with a positive real part, choose $\mathcal{D}$ so that the first p $\mathcal{D}_i$'s are zero and the remaining n-p $\mathcal{D}_i$'s are one. Then it can be shown that for $\mu \to \infty$ the asymptotic expansions of the eigenvalues have the form

$$\lambda_j = \lambda_j^{K_1} + \mathcal{O}(\mu^{-\ell_j/p}) \qquad j = 1,\ldots,p$$

$$\lambda_j = -\mu + \mathcal{O}(1) \qquad j = p+1,\ldots,n$$

where $\ell_j \le p$. Thus $K-\mu\mathcal{D}$ will have at least one eigenvalue with a positive real part if μ is large enough. We have not required that K_1 be the smallest submatrix that has an eigenvalue with a positive real part and as a result, it can happen that for $n > 2$ either stationary or oscillatory instabilities arise, depending on the choice of p and $\mathcal{D}$.

The physical interpretation of these instabilities is as follows. The amplitude $y_k = (y_{1k}, y_{2k})^T$ of a small disturbance evolves according to the equation

$$\frac{d}{d\tau}\begin{bmatrix} y_{1k} \\ y_{2k} \end{bmatrix} = \begin{bmatrix} K_1-\mu_k\mathcal{D}_1 & K_2 \\ K_3 & K_4-\mu_k\mathcal{D}_2 \end{bmatrix}\begin{bmatrix} y_{1k} \\ y_{2k} \end{bmatrix},$$

and for $k = 0$, $y_{10}(\tau)$ grows exponentially if $y_{20}(\tau) \equiv 0$. Of course this is not possible in an isolated uniform system and the instability in the kinetic subsystem whose matrix is K_1 is suppressed through stabilizing interactions with the remainder of the network. For the earlier choice of $\mathcal{D}$, namely, $\mathcal{D}_1 = 0$ and $\mathcal{D}_2 = I$, diffusion has no direct effect on the evolution of the amplitudes of the components y_{1k}. However there is an indirect effect, because for large μ_k, an asymptotic analysis shows that y_{2k} decays rapidly and remains small and thus the stabilizing effect of the remainder of the network is lost. This analysis predicts that the largest wave numbers are fastest growing, which is due to the fact that $\mathcal{D}_1 = 0$, but if $\mathcal{D}_1 = \varepsilon I$ where $\varepsilon \ll 1$, an upper cut-off in μ exists and the preceding argument still goes through. CROSS [10] has given a proof of Theorem 1 based on Rouche's Theorem, but his proof does not show that the unstable modes arise from K_1 for large μ and so the preceding interpretation cannot be made. SEGEL and JACKSON [20] have discussed the origin of such instabilities in two-component systems.

Let us single out a particular parameter in (12) and label it p, and suppose that when p increases through p_0 a real eigenvalue of $K-\mu_k\mathcal{D}$ crosses from the left-half to the right-half plane for some k. Write (12) as

$$\frac{\partial x}{\partial \tau} = L(p)x + F(x,p) \tag{15}$$

and for $|p-p_0| \sim \mathcal{O}(\varepsilon)$, $\varepsilon << 1$, write

$$\sigma(L) = \sigma_1(L) \cup \sigma_2(L)$$

where

$$\sigma_1(L) = \{\lambda \in \sigma(L) \mid |\mathrm{Re}\,\lambda| \leq \gamma\}$$

$$\sigma_2(L) = \{\lambda \in \sigma(L) \mid \mathrm{Re}\,\lambda < -\gamma\}$$

and γ is a small positive constant. Such a separation is possible because L(p) has a compact inverse on $L_2(\Omega)$. Let P_1 and $P_2 = I-P_1$ be the projections associated with this decomposition of the spectrum and write $u = P_1u + P_2u \equiv \varepsilon(v+\bar{w})$; then v and $\bar{w}$ satisfy the equations

$$\begin{aligned}\frac{\partial v}{\partial \tau} &= L_1 v + \varepsilon F_1(v,\bar{w},p,\varepsilon)\\ \frac{\partial \bar{w}}{\partial \tau} &= L_2\bar{w} + \varepsilon F_2(v,\bar{w},p,\varepsilon),\end{aligned} \tag{16}$$

which are gotten by applying the projections to (15). Under the standing assumption that $\|u\|_{L_2} \sim \mathcal{O}(\varepsilon)$ for all $t \geq 0$, it can be shown that $\|\bar{w}\|_{L_2} \sim \mathcal{O}(\varepsilon)$ for $t > 0$ if $\|\bar{w}(\underline{r},0)\|_{L_2} \sim \mathcal{O}(\varepsilon)$ and γ is large enough. We assume that this is true and write $\bar{w} = \varepsilon w$; then

$$\begin{aligned}\frac{\partial v}{\partial \tau} &= L_1 v + \varepsilon F_1(v,\varepsilon w,p,\varepsilon)\\ \frac{\partial w}{\partial \tau} &= L_2 w + F_2(v,\varepsilon w,p,\varepsilon)\end{aligned} \tag{17}$$

If F(x,p) is C^k, $k \geq 3$, for (x,p) near $(0,p_0)$, F_1 and F_2 have the expansions

$$\varepsilon F_1(v,\varepsilon w,p,\varepsilon) = \varepsilon Q^1(v,v,p) + \varepsilon^2\{2Q^1(v,w,p) + C^1(v,v,v,p)\} + \mathcal{O}(\varepsilon^3)$$

$$F_2(v,\varepsilon w,p,\varepsilon) = Q^2(v,v,p) + \varepsilon\{2Q^2(v,w,p) + C^2(v,v,v,p)\} + \mathcal{O}(\varepsilon^2)$$

where Q^i and C^i are homogeneous of degree two and three respectively. Therefore, to $\mathcal{O}(\varepsilon^2)$ v evolves independently of w while to $\mathcal{O}(\varepsilon)$, w is 'forced' by v through the term $Q^2(v,v,p)$. This partial separation of the unstable or nearly unstable modes from the rapidly decaying forced modes is in effect a first step toward the complete separation that can be obtained formally by invoking the Center Manifold Theorem for flows generated by partial differential equations [11].

Since the eigenvalue that crosses zero at $p = p_0$ is simple, bifurcation is known to occur and the bifurcating solution can be constructed as an expansion in the amplitude parameter $\varepsilon \equiv \langle\psi^*,u\rangle_{L_2}$ [12]. If $p_1(\varepsilon)$ is defined by setting $p-p_0 = \varepsilon p_1(\varepsilon)$, then one finds that

$$p_1(0) = \frac{\langle y^*,Q(y,y,p_0)\rangle}{\langle y^*,\frac{\partial K}{\partial p}y\rangle} \int_\Omega \phi^3 d\Omega. \tag{18}$$

If this is nonvanishing then a nontrivial solution exists on both sides of $p = p_0$ and an exchange of stability occurs at p_0. Evidently $p_1(0)$ will vanish for purely geometric reasons if $\int \phi^3 d\Omega = 0$, which happens in one dimension because $\phi = \cos n\pi x/L$. In such cases higher-order terms must be examined. When the other factor in the numerator vanishes the kinetics have a degeneracy in the quadratic terms in the direction of y^*, and again the properties of the bifurcating solution are governed by higher order terms. AUCHMUTY and NICOLIS [13] and others have constructed the bifurcating solutions for the trimolecular reaction scheme.

Bifurcation is certain to occur at an eigenvalue of odd multiplicity but if the multiplicity is greater than one, the number of bifurcating branches cannot be determined a priori. In this case and the case of even multiplicity one has to examine the bifurcation equations to determine the number of solutions. One source of eigenvalue degeneracy is symmetry of the domain Ω and an example of this is given in [12]. For other results on bifurcation in the presence of symmetry see [14] and the references therein.

Next let us suppose that the kinetic equations have an orbitally asymptotically stable (OAS hereafter) periodic solution $\Phi(\tau)$ of least period T. Under Neumann boundary conditions $\Phi(\tau)$ is also a solution of (2), and we can write

$$c(\underset{\sim}{\zeta},\tau) = \Phi(\tau) + \sum_k y_k(\tau)\phi_k(\underset{\sim}{\zeta}). \tag{19}$$

For small disturbances the amplitudes satisfy the linear equation

$$\frac{dy_k}{d\tau} = (K(\tau)-\mu_k\mathcal{D})y_k \tag{20}$$

where $K(\tau+T) = K(\tau)$. It can be shown that if the $\mathcal{D}_i$'s are not too different or if all of them are large enough, then $\Phi(\tau)$ is an OAS solution of the partial differential equation as well [1], but we are again more interested in necessary conditions for stability, and a result that parallels Theorem 1 can be obtained as follows. Partition $K(\tau)$ as in (14) and choose $\mathcal{D}$ as before, replace μ_k by a continuous variable $\mu \in [0,\infty)$, and let $\varepsilon = \mu^{-1}$. Then (20) becomes

$$\begin{aligned} \frac{dy_1}{d\tau} &= K_1(\tau)y_1 + K_2(\tau)y_2 \\ \varepsilon\frac{dy_2}{d\tau} &= \varepsilon K_3(\tau)y_1 + (\varepsilon K_4(\tau)-I)y_2 \end{aligned} \tag{21}$$

and associated with this is the degenerate system

$$\begin{aligned} \frac{d\bar{y}_1}{d\tau} &= K_1(\tau)\bar{y}_1 \\ \bar{y}_2(\tau) &\equiv 0 \end{aligned} \tag{22}$$

obtained by setting $\varepsilon \equiv 0$. For initial conditions such that $y_1(0) = \bar{y}_1(0)$ and $y_2(0) = \bar{y}_2(0) = 0$, it can be shown, using results in [15], that for any τ in a closed subinterval of $\mathbb{R}^+$,

$$\begin{aligned} ||y_1(\tau)-\bar{y}_1(\tau)|| &\leq \Lambda(\varepsilon) \\ ||y_2(\tau)|| &\leq \Lambda(\varepsilon) \end{aligned} \tag{23}$$

Here $\Lambda(\varepsilon)$ is a continuous nonnegative function that vanishes at zero. It follows that if $K_1(\tau)$ has one or more characteristic exponents with a positive real part, the uniform periodic solution is unstable to disturbances of sufficiently short wavelength. This conclusion leads to the following set of necessary conditions for stability.

Theorem 2. Suppose that $\Phi(\tau)$ is an OAS periodic solution of the kinetic equations and that $\mathcal{D}$ is diagonal with $\mathcal{D}_i \geq 0$. Then $\Phi(\tau)$ is an OAS periodic solution of (2) (in the L_2 norm) only if all the characteristic exponents of every pth-order submatrix $K[i_1,\ldots,i_p](\tau)$ of $K(\tau)$ have nonpositive real parts for $1 \leq p \leq n-1$.

The similarity between the foregoing conditions and those in Theorem 1 is apparent; in each case the presence of a subsystem that would be unstable if it were isolated from the remainder of the kinetic network is sufficient for instability, given the proper choice of diffusion coefficients. Of course the criterion for stability of the subsystems is different in the two cases because the underlying basic solutions are different.

Theorem 2 enables us to delineate a large class of kinetic systems for which the uniform periodic solution can be destabilized by diffusion. Suppose that the OAS periodic solution $\Phi(\tau)$ of the kinetic equations bifurcates from the steady state $c^s(p)$ by a Hopf bifurcation as p increases through p_0, and that there is at least one species for which $k_{jj} > 0$ in a neighborhood of $c^s(p_0)$. Then the uniform periodic solution of (2) can be destabilized by diffusion if $p-p_0$ is sufficiently small and positive. Said otherwise, if there is autocatalysis present in the linearized kinetics at $c^s(p_0)$, then all small amplitude solutions can be destabilized. If there are only two species present, a Hopf bifurcation can *only* occur if there is autocatalysis in the linearized kinetics and $d(\text{trace } K)/dp > 0$, and therefore *a uniform OAS periodic solution of a two-species system can always be destabilized when the kinetic parameters are sufficiently close to the values that yield a Hopf bifurcation*. An example of such an instability is given in the following section.

The type of solution that exists beyond the critical parameter value p_c at which $\Phi(\tau)$ loses stability to nonhomogeneous disturbances depends on how the multipliers π_j cross the unit circle. In two-species systems they can only cross at ± 1, and the bifurcating solution is necessarily periodic [1]. In general, if $\pi_j^n(p_c) \neq 1$, $n = 1,2,3,4$, an invariant torus bifurcates from the periodic solution [16,17], and the bifurcating solution is quasi-periodic. There may still be a periodic solution in this case, or in some cases where the above non-resonance conditions aren't met, but in any event one must generally determine the parametric behavior of the multipliers numerically.

Finally, we consider the stability properties of uniform steady states of (5). We assume that $h_i > 0$ for all i and therefore these steady states are given by $u^s = v^s = R^{-1}(0)$. The differences $x \equiv u-u^s$ and $y \equiv v-v^s$ satisfy

$$\begin{aligned} \frac{\partial x}{\partial \tau} &= \Delta_1 \mathcal{D} \nabla^2 x + \varepsilon \Delta_2 H(y-x) \\ \frac{\partial y}{\partial \tau} &= -\Delta_2 H(y-x) + Ky + F(y), \end{aligned} \tag{24}$$

and the matrix of the linear system is

$$L = \begin{bmatrix} -\mu_n \Delta_1 \mathcal{D} - \varepsilon \Delta_2 H & \varepsilon \Delta_2 H \\ \Delta_2 H & K - \Delta_2 H \end{bmatrix} .$$

Now suppose that $\mathcal{D} = H = I$ and $\sigma(K) \subset \text{LHP}$. By Lyapunov's theorem there exists a positive definite W such that $WK + K^T W$ is negative definite, and if we set $A = \text{diag}\,\{I, \sqrt{\varepsilon} I\}$ and $B = \text{diag}\,\{W, W\}$, then it follows that

$$B(ALA^{-1}) + (ALA^{-1})^T B$$

is negative definite. Thus a kinetically-stable steady state can be destabilized in the present case only if the diffusion coefficients and the mass transfer coefficients are sufficiently different. This is analogous to the requirement for directly-coupled systems and there are other parallels. The characteristic equation of L is

$$\det\{K(\mu_n\Delta_1\mathcal{D}+\varepsilon\Delta_2 H)-\mu_n\Delta_1\Delta_2 H\mathcal{D}-\lambda((1+\varepsilon)\Delta_2 H+\mu_n\Delta_1\mathcal{D}-K)-\lambda^2 I\} = 0 \tag{25}$$

and the potential bifurcation loci for steady states are subsets of the loci along which

$$\det\{K(\mu_n\Delta_1\mathcal{D}+\varepsilon\Delta_2 H)-\mu_n\Delta_1\Delta_2 H\mathcal{D}\} = 0. \tag{26}$$

We have assumed that $h_i > 0$ for all i and therefore the multiplier of K is invertible for $\varepsilon > 0$ and $\Delta_2 > 0$. Thus (26) can be written

$$\det\{K-T^{(n)}\}\det\{\mu_n\Delta_1\mathcal{D}+\varepsilon\Delta_2 H\} = 0 \tag{27}$$

where $T^{(n)}$ is a diagonal matrix whose elements are

$$\frac{1}{T_i} = \frac{1}{\Delta_2 h_i} + \frac{\varepsilon}{\mu_n\Delta_1\mathcal{D}_i}. \tag{28}$$

This form stems from the fact that the transport steps are in series and either step can control the rate. Indeed, as $\Delta_2 \to \infty$ or $\mu_n \to 0$, $T_i \to \mu_n\Delta_1\mathcal{D}_i/\varepsilon$ and diffusion controls, while interphase transport controls for large n (short wavelengths) or as $\Delta_1 \to \infty$.

When it is expanded, $\det\{K-T^{(n)}\}$ is a polynomial in the T_i and hence a rational function in the h_i and $\mathcal{D}_i$, and one can prove [9] that it is nonvanishing under the following conditions.

Theorem 3. Let ε, Δ_1, and Δ_2 be positive and let μ_n be nonnegative. Then $\det\{K-T^{(n)}\} \neq 0$ for all $\mathcal{D}_i \geq 0$ and $h_i > 0$ if and only if

(i) $(-1)^n \det K > 0$

(ii) $(-1)^p \det K[i_1,\ldots i_p] \geq 0$ for all pth-order submatrices of K, where $1 \leq p \leq n-1$.

Under these conditions there can be no bifurcation of steady states from the uniform steady state (u^s,v^s). An analogous result, in which the conditions on the kinetics are identical, can be stated for directly-coupled systems.

Despite the foregoing similarities, there can be substantial differences between the dynamical behavior of directly- and indirectly-coupled systems. Some of these differences are illustrated in the example that follows.

4. A Comparison for a Model Reaction

The simplest system that will illustrate the differences is one with only two active chemical species and two cells. A number of back-activation mechanisms of the type used to model glycolytic reactions can produce oscillations for suitable parameter values, and the simplest of these gives rise to the following kinetic equations [18, 19]:

$$\begin{aligned}\frac{dx}{d\tau} &= \delta-\kappa x-xy^2 \equiv F(x,y)\\ \frac{dy}{d\tau} &= \kappa x + xy^2-y \equiv G(x,y).\end{aligned} \tag{29}$$

This system has the unique steady state $(x^s,y^s) = (\delta/\kappa+\delta^2,\delta)$ and when $\kappa < 1/8$ it has a periodic solution for $\delta \in (\delta_-, \delta_+)$, where

$$\delta_\pm = \frac{1-2\kappa \pm \sqrt{1-8\kappa}}{2} \quad . \tag{30}$$

Now suppose that these reactions occur in each of two identical cells that are separated by extracellular medium in which no reaction occurs. Let the volume of the cells and the medium be V_c and V_0, respectively, let $\varepsilon = 2V_c/V_0$, and let '0' denote concentrations in the extracellular medium. The governing equations are

$$\begin{aligned}
\frac{dx_i}{d\tau} &= F(x_i, y_i) + \mathcal{D}_x(x^0 - x_i) \\
& \qquad\qquad\qquad\qquad\qquad\qquad i = 1,2 \\
\frac{dy_i}{d\tau} &= G(x_i, y_i) + \mathcal{D}_y(y^0 - y_i) \\
\frac{dx^0}{d\tau} &= 2\,\varepsilon\,\mathcal{D}_x\left(\frac{x_1+x_2}{2} - x^0\right) \\
\frac{dy^0}{d\tau} &= 2\,\varepsilon\,\mathcal{D}_y\left(\frac{y_1+y_2}{2} - y^0\right)
\end{aligned} \tag{31}$$

Here $\mathcal{D}_x$ and $\mathcal{D}_y$ represent the diffusion coefficients across a double membrane, which accounts for the factor of two. When $V_0 = 0$ ($\varepsilon = \infty$) the cells are in direct contact and (31) reduces to

$$\begin{aligned}
\frac{dx_i}{d\tau} &= F(x_i, y_i) + \mathcal{D}_x(x_j - x_i) \\
& \qquad\qquad\qquad\qquad\qquad\qquad i \neq j \\
\frac{dy_i}{d\tau} &= G(x_i, y_i) + \mathcal{D}_y(y_j - y_i).
\end{aligned} \tag{32}$$

If gap junctions or other low-resistance pathways form when the cells come into contact, the diffusion coefficients may increase substantially.

It follows from (31) that the steady state solution $S \equiv (\bar{x}_1, \bar{y}_1, \bar{x}_2, \bar{y}_2, \bar{x}^0, \bar{y}^0)$ is independent of ε and is given by

$$\begin{aligned}
\bar{x}_{1,2} &= \frac{\delta \pm (1+2\mathcal{D}_y)}{\kappa + (\delta \pm \zeta)^2} \\
\bar{y}_{1,2} &= \delta \pm \zeta \\
\bar{x}^0 &= (\bar{x}_1 + \bar{x}_2)/2 \qquad\qquad \bar{y}^0 = (\bar{y}_1 + \bar{y}_2)/2,
\end{aligned} \tag{33}$$

where ζ is a solution of

$$\zeta[\zeta^4 + 2(\kappa - \delta^2 + \mathcal{D}_x)\zeta^2 + \frac{\kappa+\delta^2}{1+2\mathcal{D}_y}\det(K - 2\mathcal{D})] = \zeta[\zeta^4 + 2b\zeta + c] = 0$$

The first four components of S give the steady state solutions of (32). When $\zeta = 0$ the uniform steady state (USS) is recovered while $\zeta \neq 0$ gives a nonuniform steady state (NUSS). The number of NUSSes is given by the following proposition [19], and the regions in (ii) and (iii) are shown in Fig. 1.

Proposition Let

$$\mathcal{D}_y^* \equiv \frac{1}{2}\,\frac{1-\kappa/\mathcal{D}_x-\sqrt{(\kappa/\mathcal{D}_x)^2+2\kappa/\mathcal{D}_x}}{1+\kappa/\mathcal{D}_x+\sqrt{(\kappa/\mathcal{D}_x)^2+2\kappa/\mathcal{D}_x}} \quad .$$

Then

(i) if $\mathcal{D}_y > \max(\frac{1}{6},\mathcal{D}_y^*)$ there are no NUSSes for any $\delta > 0$.

(ii) if $\mathcal{D}_y^* > 1/6$ and $\mathcal{D}_y \in (1/6,\mathcal{D}_y^*)$ there exists a region in the $\mathcal{D}_x$-δ^2 plane in which there is a pair of NUSSes. There is never more than one pair.

(iii) if $\mathcal{D}_y \in [0,1/6)$ there is one region in which there are two pairs of NUSSes and a contiguous region in which there is one pair.

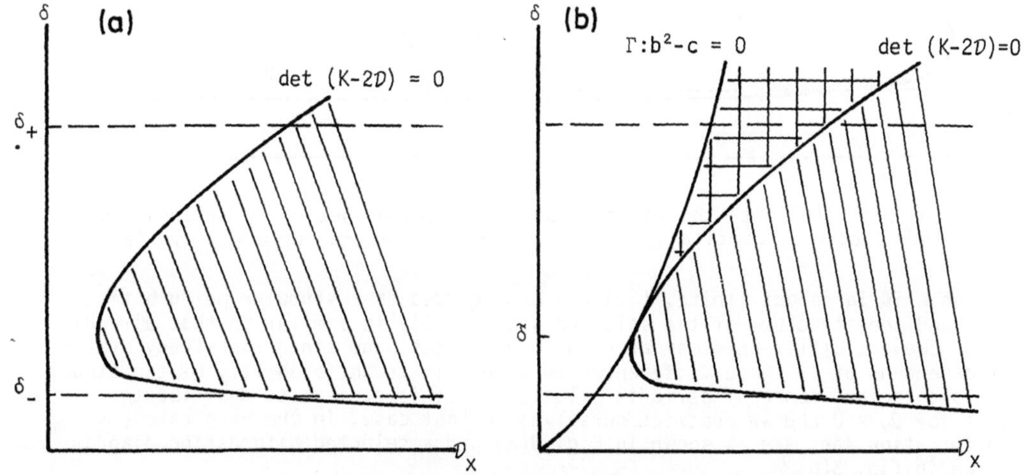

Fig. 1 The regions of multiple steady states: (a) case (ii); (b) case (iii). \\\\\\ one pair of NUSSes, xxxx two pairs of NUSSes, elsewhere; no NUSSes.

Suppose that δ is fixed and $\mathcal{D}_x$ is regarded as the bifurcation parameter. In (a) a pair of NUSSes bifurcates supercritically as $\mathcal{D}_x$ increases across det $(K-2\mathcal{D}) = 0$, and whether they or not they are stable depends on δ: if $\delta \in [\delta_-,\delta_+]$ they are unstable near the bifurcation point. In (b) the bifurcation along det $(K-2\mathcal{D}) = 0$ is subcritical and unstable for $\delta > \tilde{\delta}$ and these branches connect to a pair of stable branches along Γ. At the intersection of $\delta = \delta_+$ and det $(K-2\mathcal{D}) = 0$, the linear system derived from (32) has a zero real eigenvalue and a pair of complex conjugate eigenvalues with zero real part, and such systems exist only on a submanifold of codimension 2 in the space of all 4×4 real matrices. By perturbing such a system one may find solutions other than those that bifurcate along $\delta = \delta_+$ or along det $(K-2\mathcal{D}) = 0$, and an analysis of the case $\varepsilon = \infty$ has been done [19]. One finds, using a multiple-time-scale analysis, that there are two other bifurcation curves that emanate from the crossover point and on which a nonuniform periodic solution (NUPS) bifurcates. These curves, whose behavior near the crossover point is found analytically, can be continued numerically and the results are shown in Fig. 2(a). The curve $H_3 = 0$ corresponds to bifurcation of a NUPS from the NUSS. On the broken curve, whose location is schematic because only the solid points were computed, a NUPS bifurcates from the UPS. At the computed points the bifurcation is supercritical and the NUPS is stable, and we conjecture that this is true all along the broken curve. The results illustrate the earlier assertion that a UPS of a two-species system can be destabilized by diffusion sufficiently near the bifurcation point of such solutions. Fig. 2(b) shows the global structure of the solution set for a fixed value of δ. The amplitudes shown are only schematic. For further details and other bifurcation diagrams see [19].

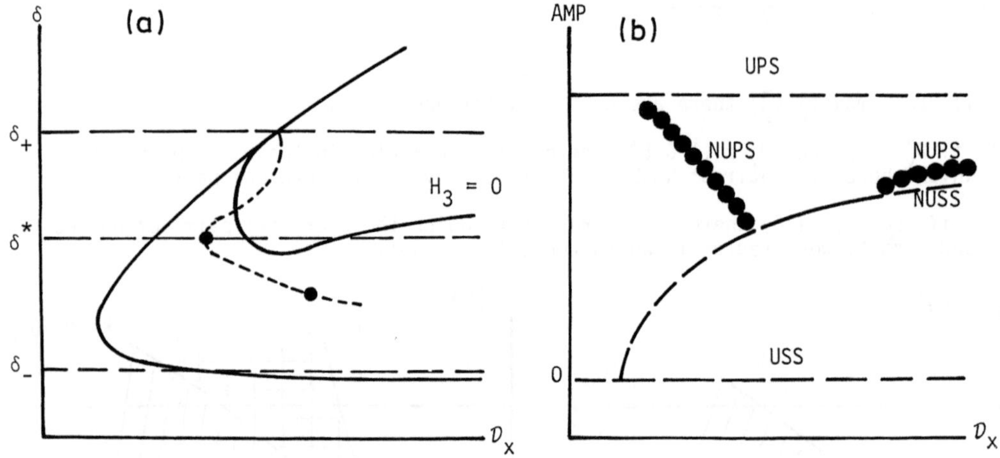

Fig. 2 (a) Fig. 1(a) with the addition of the curves of secondary bifurcation. (b) Amplitude (shown schematically) of the various solutions for $\delta = \delta^*$ in (a).

When ε is large but finite, (32) can be regarded as a singular perturbation of (31) and the structure of the solution set for (31) is similar to that already given. However, differences may arise as ε decreases, and since the steady states are independent of ε, these differences must pertain to the stability of the steady states and the existence of periodic solutions. It is easiest to describe the results for $\mathcal{D}_y = 0$ and we restrict ourselves to that case. In the base case $\varepsilon = \infty$ the bifurcation loci are as shown in Fig. 3(a) and a selected bifurcation diagram is shown in Fig. 3(b).

As before, the linearization of the four-dimensional system around a NUSS has a pair of complex conjugate eigenvalues with zero real part on the curve labelled $H_3 = 0$. On the broken portion of the curve the NUPS bifurcates from the intermediate

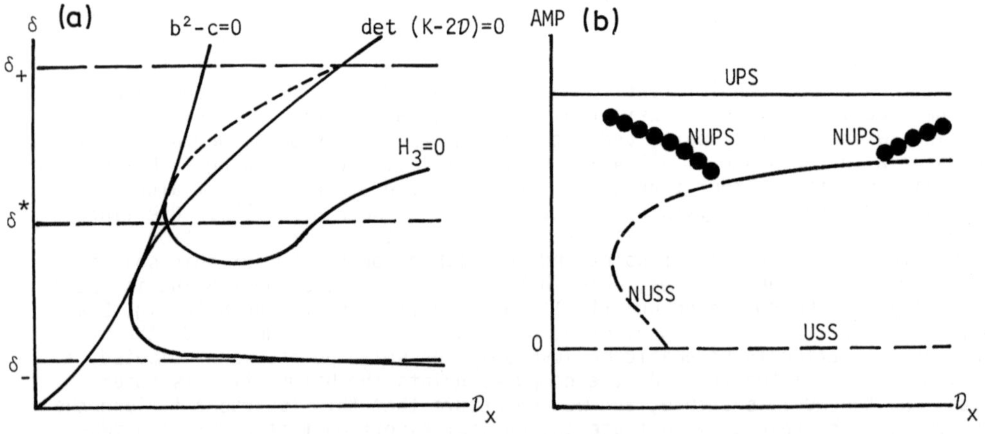

Fig. 3 (a) The bifurcation loci for $\mathcal{D}_y = 0$, $\varepsilon = \infty$. (b) A schematic of the amplitude of the various solutions at $\delta = \delta^*$.

(unstable) branch of NUSSes, while the solid portion corresponds to bifurcation from the upper branch of NUSSes.

Now let ε be finite and consider what happens to the synchronized periodic solution. This solution persists for large ε (apply a perturbation argument) but if ε is sufficiently small the steady state can be stabilized for a range of $\mathcal{D}_X$ and the periodic solution appears to be quenched. To be precise, the uniform steady state is asymptotically stable for $(\mathcal{D}_X,\varepsilon) \in (\frac{1}{2}T^K,\bar{\mathcal{D}}_X) \times (0,\bar{\varepsilon})$ where $\bar{\mathcal{D}}_X$ is the positive solution of det $(K-2\mathcal{D}) = 0$ and $\bar{\varepsilon}$ is the positive solution of

$$-4\varepsilon^2\mathcal{D}_X^2T^K-2\,\varepsilon\,\mathcal{D}_X[(2\mathcal{D}_X-T^K)T^K + 2\mathcal{D}_Xk_{22}]-\det\,(K-2\mathcal{D})(T^K-2\mathcal{D}_X) = 0. \tag{34}$$

Here $T^K \equiv$ trace K and δ is fixed in the interval (δ_-,δ_+). Numerical computations show that the trajectories beginning in a finite neighborhood of the USS converge to this steady state, but we have not proven that the periodic solution disappears. Nonetheless, it is certain that there are no small amplitude oscillations, and a degree of quenching exists.

Another effect of finite ε arises when $\delta > \delta_+$, in which case the kinetics have no periodic solution and the directly-coupled system has no uniform periodic solution. Equation (34) gives the locus on which the linearization of (31) has a pair of complex conjugate eigenvalues with zero real part and an analysis of this equation shows that when

$$\delta_+^2 < \delta^2 < 1-\kappa + \sqrt{1-4\kappa} \tag{35}$$

the bifurcation locus is as shown in Fig. 4. Numerical computations show that the periodic solution exists between the two branches of this locus and is stable when $\mathcal{D}_X < \bar{\mathcal{D}}_X$. Thus the cells can oscillate in synchrony if the volume of the intervening dead space is adjusted properly, and these oscillations are stable to both uniform and nonuniform disturbances. It has been suggested [18] that this model can account for the observed effect of cell density in yeast cell suspensions [5] and a somewhat more complicated model can be used for describing Dd suspensions.

In addition to the synchronized oscillations, there are several branches of asynchronous periodic solutions shown in Fig. 4. The bifurcating solutions are

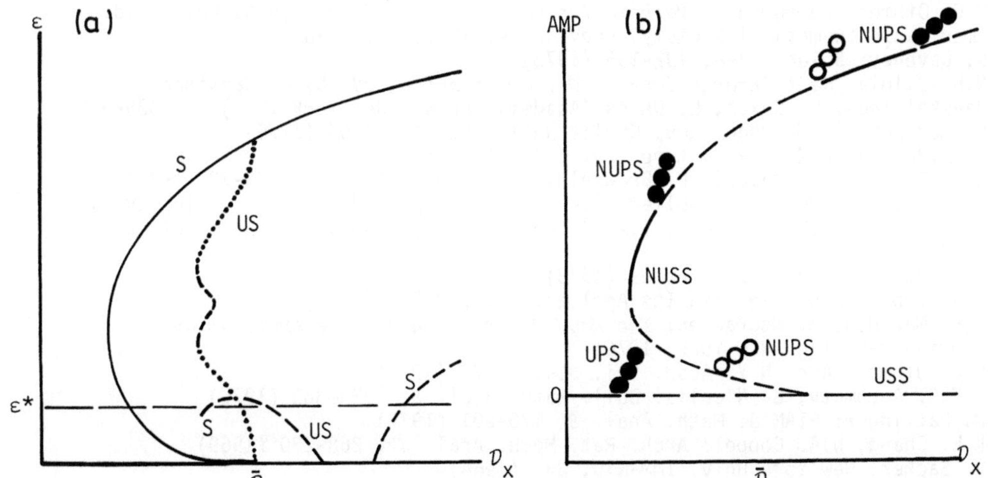

Fig. 4 (a) Bifurcation loci for fixed δ satisfying (35): —: bifurcation locus for UPS; ----(.....) locus for bifurcation of NUPS from upper (intermediate) NUSS. (b) A schematic of the amplitudes for $\varepsilon = \varepsilon^*$.

unstable along curves marked 'US' and stable on those marked 'S'. When its amplitude is small, the solution that bifurcates from the intermediate steady state is stable in a four-dimensional submanifold of the five-dimensional space, but it is not known whether it becomes stable at larger amplitudes. Similarly, it is not known whether any of the other unstable solutions become stable by a tertiary bifurcation. While some of the local features of the bifurcation diagrams are known (cf. Fig. 4(b)), the global structure has not been determined on any cross-section with $\delta > \delta_+$, and there are several singularities of codimension greater than one that must be analyzed. Moreover, it remains to patch together the bifurcation loci in the three-dimensional parameter space. This is easily done for those that correspond to the synchronous oscillations, using results in [18] and those given here.

5. Conclusions

The results given here illustrate how the dynamical behavior of subsystems at one level of organization can influence the dynamical behavior at a higher level of organization. Given a network of intracellular reactions that contains an unstable subnetwork, the diffusive coupling between cells can be chosen so as to destabilize the synchronized steady or time-periodic state of an aggregate of cells. Thus a variety of spatial and spatio-temporal patterns can be generated simply by varying the degree of communication between cells, and TURING first suggested that such patterns could serve as prepatterns for controlling cellular differentiation. The variety of behavior possible with only two cells and two active chemical species points up the fact that relatively tight control of the trajectory in *parameter space* will be needed to achieve a desired sequence of patterns in more complicated systems, but such close control can probably be achieved by turning communication on and off at the appropriate developmental stages. By readjusting internal kinetic parameters during periods in which the cells are uncoupled, quantum steps in the dynamical behavior can be achieved without passing through undesirable types of behavior, and the final destination in parameter space can be reached by a series of zig-zag steps.

References

1. H.G. Othmer: *Lectures on Mathematics in the Life Sciences* (S.A. Levin, ed. American Mathematical Society, Providence, 1977) pp. 57-85
2. S. Caveney: Science *199*, 192-195 (1978)
3. N.B. Gilula: *Cell Interactions in Differentiation*, ed. by M. Karkinen-Jaaskelainen, L. Saxen, L. Weiss (Academic Press, New York 1977) pp. 325-338
4. G. Gerisch, D. Malchow: Adv. Cyclic Nucl. Res. *7*, 49-65 (1976)
5. J. Aldridge, E.K. Pye: Nature *259*, 670-671 (1976)
6. W.D. Branton, E. Mayeri, P. Brownell, S. Simon: Nature *274*, 70-72 (1978)
7. E. Conway, D. Hoff, J. Smoller: SIAM Jour. on Applied Math. *35*, 1-16 (1978)
8. R. Heinrich, S.M. Rapoport, T.A. Rapoport: Prog. Biophys. Molec. Biol. *32*, 1-82 (1977)
9. H.G. Othmer: unpublished notes (1978)
10. G.W. Cross: Lin. Alg. and Its Applies. *20*, 253-263 (1978)
11. J.E. Marsden, M. McCracken: *The Hopf Bifurcation and Its Applications* (Springer-Verlag, New York, 1976)
12. H.G. Othmer: Ann. N.Y. Acad. Sci. *316*, 64-77 (1979)
13. J.F.G. Auchmuty, G. Nicolis: Bull. Math. Biol. *37*, 323–365 (1975)
14. D. Sattinger: SIAM J. Math. Anal. *8*, 179-201 (1977)
15. K.W. Chang, W.A. Coppel: Arch. Rat. Mech. Anal. *32*, 268-280 (1969)
16. R. Sacker: New York Univ. IMM-NYU, *333* (1964)
17. N. Fenichel: J. Diff. Eqns. *17*, 308-328 (1975)
18. H.G. Othmer, J. Aldridge: J. Math. Biol. *5*, 169-200 (1978)
19. M. Ashkenazi, H.G. Othmer: J. Math. Biol. *5*, 305-350 (1978)
20. L.A. Segel, J.L. Jackson: J. Theor. Biol. *37*, 545-559 (1972)

Dynamics of Cell-Mediated Immune Response

R. Lefèver

Chimie Physique II, C.P. 231, Université Libre de Bruxelles
B-1050 Brussels, Belgium

1. Introduction

The immune system regulates in a remarkably concerted way the production and properties of an extraordinary diverse collection of cells and molecules. The purpose of this activity is to protect the organism against various forms of aggression. This requires the circulation of cells and molecules between different organs, the discrimination between self and non self antigens, the capability to deliver highly specific responses, the capability of learning and of memorisation (for review articles, see [1]).

The interdependence of all the above mentioned processes, the fact that they involve organs disseminated over the entire body (bone marrow, spleen, lymph nodes, thymus), the fact that they may be influenced by a multitude of systemic and environmental factors, make the description of the immune system dynamics a quite difficult problem. Frequently, the approach which is adopted is global. It consists in describing the overall behavior of lymphocytes, target cells, antigens, antibodies,... as if they were interacting in a spatially homogeneous medium. This "well stirred tank reactor" approximation is useful from a general qualitative point of view. It is however inappropriate for the quantitative analysis of the in vivo in situ behavior and of the relation of this behavior with the mechanisms and properties discovered by in vitro experiments. In order to deal with these more precise questions, an approach unfolding the in situ behavior on the basis of a local description of events (cell proliferation, diffusion, cytotoxicity, ...) is necessary. The results presented here and which concern that function of the immune system named cell-mediated immune surveillance, have been obtained with this objective in mind.

The next section briefly introduces the phenomenon of immune cell-mediated surveillance and explains the role generally attributed to this phenomenon in the defense of the organism against cancer growth. To assess this role quantitatively as well as the contribution of other cells not derived from the immune system, the kinetics governing the interactions between cytotoxic (effector) cells and cancer (target) cells must be elucidated. This question is dealt with in the third section. It is reported that with a great degree of generality, cytotoxic reactions obey a michaelian kinetics. The cytotoxic kinetic parameters of various effector cells have been determined from the experimental data of the literature. A quantitative classification of effector cells according to their cytotoxicity can be made. In section 4, a kinetic analysis of the immunological rejection of MSV-M tumors confirms the assumptions on which the approach is based and predicts the minimum ratio of effector to target cell that must be reached within the tumor mass to obtain rejection. The contribution of other, non T-cells is discussed. On the other hand, if one takes inside a tumor a sufficiently small volume element of tissue, one expects to find that over periods of time which are long compared with the cell replication time the average cellular populations and properties of this volume element remain constant; i.e. even in tumors which as a whole are in a dynamic state of progression (or regression), the internal local properties are in a quasi-stationary state. An original aspect of our approach, is that it permits one to determine the stability of these local stationary state properties and to study the influence of these properties on the mechanism and rate of growth of the tumor. This question is treated in section 5.

2. Cell-mediated immune surveillance

The immune stem cells produced in the bone marrow differentiate into either B or T lymphocytes (see Fig. 1). We shall deal exclusively with immune responses which involve this latter category of cells. The problem is best introduced by briefly recalling, together with the Thomas-Burnet theory which has dominated immunology, some major experimental findings obtained by attempts to prove or disprove this theory [3].

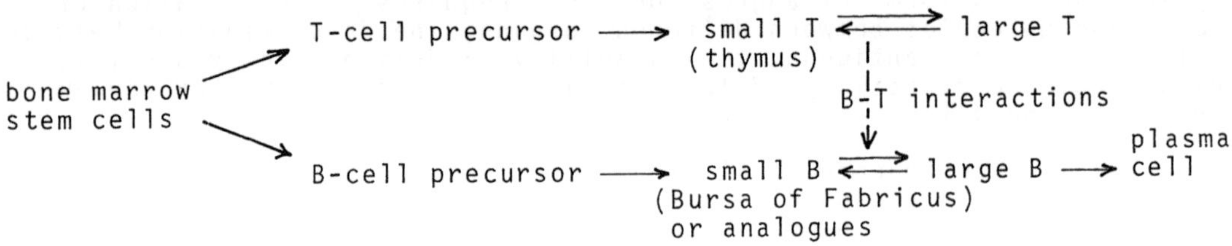

Fig. 1 Pathways of immune cells differentiation. The differentiation into small B-cells takes place into the bursa of Fabricus (birds) or in its mammals analogues. On encounter with an appropriate antigen, the small non replicating B cells transform into large B cells which proliferate and secrete antibodies. Subsequently they may revert back to the small B state in the form of memory cells or transform futher into plasma cells which have the highest rate of antibody secretion but do not replicate [2]. T-cells differentiate in the thymus, they may regulate, either enhance or suppress, the humoral response mediated by the B-cells. Besides that, they are involved in cell-mediated immune responses against grafts and cancer cells.

The initial discoveries in organ transplantation phenomena which begun to accumulate some thirty years ago, led Thomas to postulate [4] that "the phenomena of homograft rejection will turn out to represent a primary mechanism of defense against neoplasia". This idea was futher developed by Burnet [5-7] who elaborated a theory which in brief can be stated as follows:(i) spontaneous somatic mutations represent a continuous source of cellular transformation; (ii) most of the transformed cells have antigenic qualities different from those of the original cells from which they are derived; (iii) these antigenic differences are recognized as foreign and an immune response is developed. This immune response is mediated by T lymphocytes and is equivalent to homograft rejection.

As a consequence Burnet predicts that cancer arises:(i) under all conditions that correspond to a depression of the immune system (in the ante and perinatal period, in old age, upon the action of some chemicals, radiations,...); (ii) when transformed cells lack antgenicity; (iii) when the growth potentiality of the transformed cells is capable of overriding any immunological control.

Although in the beginning the experimental and clinical evidence seemed to confirm Burnet's theory, in the last few years several results have appeared contradicting some of its hypotheses. At present, the situation in respect to the above mentioned points summarizes as follows.

Source of transformed cells. In the last few years conclusive evidence has accumulated against the existence of a relation between spontaneous somatic mutations and carcinogenesis, in particular from the very precise and long experiments made with the nude mouse [8]. The nude mouse is a mutant which lacks the thymus-dependent immune system. Under germ free conditions, no spontaneous malignant tumors have been observed on 15700 nude mice studied for a total of approximately 5600 years of mouse life. On the contrary, the nude mice have been found to be much more sensitive to some oncogenic viruses. Thus, although it is still admitted that the primary event in the transformation of normal cells into cancerous ones is some sort of genetic damage, the point of view generally prevailing now is that this transformation needs the presence of environmental physical, chemical or biological agents [9,10].

Antigenicity of transformed cells. It was clearly seen many years ago that induced tumors, either by chemicals [11,12] or by viruses [13] were capable of manifesting antigenicity in the strain of origin. Antigenic changes consist in the loss of some normal antigen and / or in the appearance of new antigens on the tumor cell surface [14-17]. It was found later that the tumor cell surface is also immunogenic to the primary host [18]. As the Thomas-Burnet theory requires, this response is based in the thymus-dependent system. However only in certain rare cases does it lead to the destruction of the tumor.

Nature of the effector cell. As required by Burnet's formulation, a cytotoxic activity of T lymphocytes was demonstrated against a great variety of cancer cells either in vitro or in vivo [19-23]. The properties of this reaction can be summarized as follows: it is specific against the antigens to which the T lymphocytes were sensitized; it has an extremely high lytic efficiency; and it does not need the presence of helper cells [24,25]. However, in spite of the high cytotoxic activity commonly observed in vitro, the immune response elicited against tumor cells in vivo is frequently inappropriate for tumor rejection in syngeneic animals [26] and in the primary host [18]. This failure was assigned to some inhibitory mechanism operating in vivo. It seems that at least in some cases this failure is due to the presence of blocking factors which interact with the antigens of the tumor cell surface, the sensitized lymphocytes [18] or to some other kind of immunosuppressive agent probably operating in vivo [26,27].

Besides the questioned efficiency of the immune surveillance exerted by the T-cells, it was found that other cellular species have a cytotoxic activity. The role of macrophages in tumor immunity has been firmly established. Macrophages infiltrate developing tumors and their population may be up to more than 50% of the total number of cells present. The percentage of infiltration is proportional to the tumor resistance in syngeneically transplanted rats [26]. Their cytotoxicity depends on activation factors, some of which, like BCG and phytohemagglutinin are non specific while others are specific and released by the T-cells in contact with cancerous cells.

Other cellular species also may have a cytotoxic activity against cancerous cells. The involvment of natural killer cells (NK cells) and of antibody dependent effector cells (K cells) has been considered.

In conclusion of this experimental background, it seems thus possible to affirm, first that the idea of an endogeneous origin of cancer, related to an accumulation of somatic mutations, must be abandonned; second that the thymus-dependent immune system indeed is a component of the defense against cancer. T lymphocytes however are not the only cellular species responsible for it. Macrophages, natural killer cells also have a significant cytotoxic activity or participate in the regulation of the cytotoxicity of other cells.

In order to progress towards a quantitative understanding of the contributions of these various cells on the "tumor battle field", the following questions must be dealt with and are the object of the following sections: (i) what is the kinetics of cytotoxic reactions; (ii) how does the cytotoxicity of the different effector cells compare; (iii) under which conditions does cytotoxicity override the growth potentiality of the cancer cells; (iv) what is the rate of progression of a tumor submitted to the cytotoxic effect of a particular type of effector cells.

3. Kinetics of cytotoxic reactions

In vivo cytotoxic reactions take place in a multicompartmental system which can be described as follows. Far from the site of proliferation of the tumoral cells, effector cells are produced which are cytotoxic against these target cells. Effector cells such as immune T lymphocytes and natural killer cells are produced in the lymphatic ganglia and in the spleen. They reach the tumor by the vascular system. Other cells such as macrophages which normally exist in the connective tissue surrounding many solid tumors, may also exert a cytotoxic activity or modify either the replication constant of the target cells or the cytotoxicity of other effector cells. In the case of macrophages they are removed from the connective tissue surrounding the tumor. They infiltrate it by a non linear diffusion process the rate of which depends on the presence of agents such as lymphokines. This multicompartmental system can be visualized as in Fig. 2. The immunological response may then be viewed as a succession of steps in time: 1) Hidden cancer: a transformed cell has given rise to a clone of cells which proliferate freely without immune recognition. 2) Recognition: the antigenic neoplastic cells are detected by T lymphocytes which come back to lymphoid organs where they induce the clone proliferation of a given amount of immune T lymphocytes. 3) Immune response: the immune T lymphocytes reach the tumor by the vascular system. Besides their cytotoxic effect they may then trigger a complex series of phenomena such as for example the activation and infiltration of the tumor by other effector cells, e.g. macrophages. 4) Suppression: The production of suppressor T cells or other blocking agents inhibits the immune response.

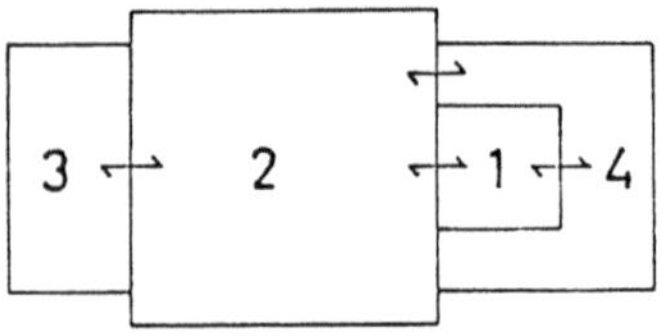

Fig. 2 Compartment 1 is the tumoral mass, 2 is the vascular compartment, 3 is the lymphoid organs where immune T lymphocytes and natural killer cells are produced and 4 is the tumor surrounding connective tissue from which activated macrophages are removed.

In the following I shall discuss processes taking place in compartment 1, during step 3 in time. During this step, the interplay of feedback mechanisms between cells inside the tumor and between the different compartments, produces considerable time variations in the nature of the effector cells present in the tumor. These variations affect both the relative proportions of effector cells and their immunological state (non sensitized, sensitized). However for a vast majority of situations it can be assumed that locally, over extended time intervals, i.e. much longer than the average time between successive tumoral cell replications, the properties of effector cells are in a quasi-stationary state (for a more complete justification see [28]). Given this assumption, a local formulation of the balance between the cytotoxic activity of effector cells and the replication of cancer cells may be proposed. For simplicity, the medium is taken unidimensional. We consider a small box (labelled i) of size Δr. At time t this box contains X_i cancer cells of diameter a and the maximum number of cells that it could contain is N. At time $t+\Delta t$, $X_i(t+\Delta t)$ is in first approximation given by (λ: replication constant of X):

$$X_i(t+\Delta t) = X_i(t) + \lambda\left[X_i\left(\frac{N-X_i}{N}\right) - \frac{X_i}{N}\left(\frac{2N-X_{i-1}-X_{i+1}}{N}\right) + \left(\frac{X_{i-1}+X_{i+1}}{N}\right)\left(\frac{N-X_i}{N}\right)\right]\Delta t \qquad (1)$$

In the limit $t \to 0$, $\Delta r = aN \to 0$, this yields the Fisher equation

$$\frac{\partial x}{\partial t} = \lambda x(1-x) + a^2\lambda\frac{\partial^2 x}{\partial r^2} \qquad \text{with} \quad x = X/N \qquad (2)$$

which approximates the behavior of the tumor during step 1 in time mentioned above.

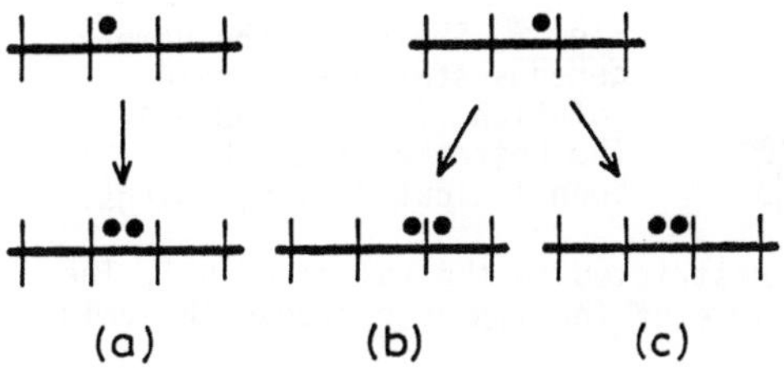

Fig. 3 Three events may occur when a cell of box i replicates: (a) the original cell is located inside the box and the new cell also belongs to i. The original cell is located at the boundary of i and the new cell belongs to $i\pm1$ (b) or to i (c).

On the other hand the cytotoxic activity of the effector cells corresponds to the two step process represented in Fig. 4. Under the assumption that the immune response has reached its quasi-stationary state $E_t = Y+Z$, k_1, k_2 are constants, the evolution is amenable to the set of equations [29]

$$\frac{\partial x}{\partial t} = (1-\theta x)x - \beta xy + a^2\frac{\partial^2 x}{\partial r^2} \qquad (3a)$$

$$\frac{\partial y}{\partial t} = \rho(z-xy) + \mu\frac{\partial^2 y}{\partial r^2} \qquad (3b)$$

$$\frac{\partial z}{\partial t} = \rho(xy-z) \qquad (3c)$$

with $t = \lambda t$, $x = k_1X/k_2$, $y = Y/E_t$, $z = Z/E_t$, $\beta = k_1E_t/\lambda$, $\theta = k_2/k_1N$, $\rho = k_2/\lambda$

$\mu = D/\lambda$

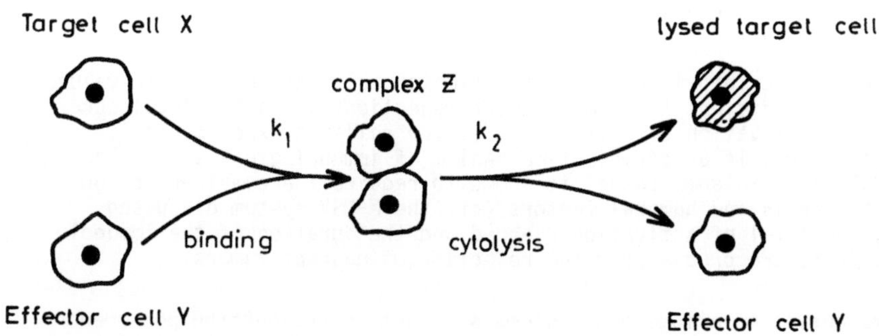

Fig. 4 Effector cell cytolysis cycle

D is the diffusion coefficient of the non-bound effector cells which are considered as the only cells which may diffuse.

It is obvious that equations (3) always admit the steady state solution for which $x = 0$. For $\beta<1$ this solution is unstable while for $\beta>1$ it is stable. The existence of a non nul steady state solution for x corresponds to a tissue in a tumoral state. Two situations arise according to the value of θ and are represented in Fig. 5.

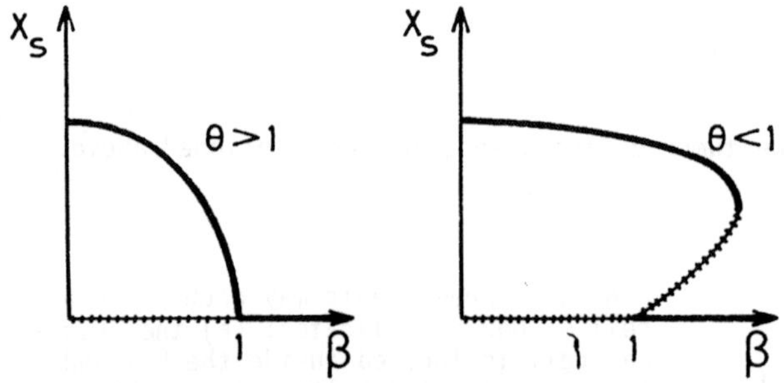

Fig. 5 Sketch of the homogeneous stationary state solutions for $\theta>1$ and $\theta<1$. The unstable states have been indicated by hatchings.

1. For $\theta>1$, the existence of a tumoral state is restricted to the values of $\theta>1$. The condition for rejection is simply that in the course of the immune response the value of β must become larger than one.
2. For $\theta<1$, the transition between tumoral and non tumoral states involves a bistability phenomenon. In the domain $1<\beta<\beta^+ = (1+\theta)^2/4\theta$, the nul state and the tumoral state are stable simultaneously.

Investigating the literature on cytotoxic tests we have come to the conclusion that both types of steady state properties may arise in reality. From this literature it has indeed been possible to calculate, very precisely for T lymphocytes and natural killer cells and somewhat less precisely for macrophages, the values of the constants k_1 and k_2 for a large number of effector target reaction systems (see for example Fig. 6). In table 1 a summary is given of the orders of magnitude which came out of this analysis. Taking into account futhermore that the representative values of the replication constant   are always in the interval $0.2<\lambda<1.5$ and that E_t ranges from a few percents in normal tissues up to 50% (sometimes even more) of the total cellular populations in certain tumors, the principal conclusions can be summarized as follows:

1) when the in vitro cytotoxic kinetic constants of different effector cells are compared the following scale of activity is obtained: activated macrophages < immune T lymphocytes and natural killer cells < allosensitized T lymphocytes.

2) in allogeneic systems the cytotoxicity of lymphocytes is fully sufficient to fully account of the observed rejection of tumors, even with very low intratumoral effector/target cells ratios.

3) in normal tissues, the cytotoxic activity of effector cells corresponds to values of $\beta<<1$. Thus highly insufficient to permit cancer rejection. In fact a local accumulation and cytotoxic activation of different effector cells is experimentally observed in nascent tumors. If an efficient mechanism of immunological surveillance operates in vivo, at least in some cases, it seems to require the combined action of different effector cells and humoral factors (cf. the M-MSV system discussed below). The delay, the level of activation reached and the duration of the immunological response are therefore crucial for the rejection of nascent tumors.

4) the steady state curves of allogeneic systems seem not to present the phenomenon of bistability. On the contrary, in syngeneic systems, at least for natural killer cells such a behavior appears possible.

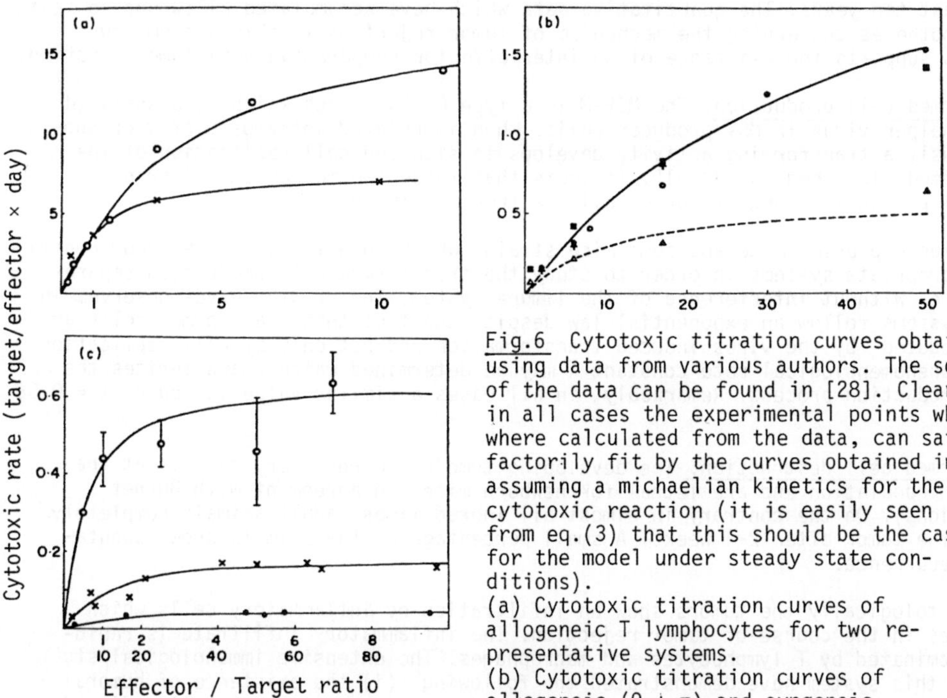

Fig.6 Cytotoxic titration curves obtained using data from various authors. The sources of the data can be found in [28]. Clearly in all cases the experimental points which where calculated from the data, can satisfactorily fit by the curves obtained in assuming a michaelian kinetics for the cytotoxic reaction (it is easily seen from eq.(3) that this should be the case for the model under steady state conditions).
(a) Cytotoxic titration curves of allogeneic T lymphocytes for two representative systems.
(b) Cytotoxic titration curves of allogeneic (0,+,■) and syngeneic natural killer cells.
(c) Cytotoxic titration curves of macrophages. MAF-activated macrophages(O), lower scale of effector/target ratio. BCG-stimulated (x) and non-stimulated (●) macrophages, upper scale of effector/target ratio

Table 1: Approximated values of cytotoxic and rejection parameters for each kind of effector cell

effector cell	k_1N (day^{-1})	k_2 (day^{-1})	θ	E_c
allosensitized T lymphocytes	18	18	1	0.01-0.1
immune T lymphocytes	0.425	0.85	1-2	0.5-3
syngeneic natural killer cells	>1-6	0.6-3	<0.1-3	<0.3-1.3
syngeneic activated macrophages	0.1-0.4	0.2-0.7	0.5-5	0.6-10

E_c is the minimum effector / target ratio for obtaining tumor rejection

4. The immunological rejection of MSV-M induced tumors

When inoculated in rodents, the Moloney murine sarcoma virus induces rapidly growing tumors with a high frequency of complete spontaneous regression. This model of anti tumor response in the natural host of a primary tumor has been extensively studied these last ten years. The quantitative data which have accumulated allow one to test some hypotheses concerning the mechanism of tumor rejection in this system. Our analysis supports the existence of an inter-effector cooperative anti-tumor reaction.

Transformed cell production: The MSV-M is a type C, RNA virus which is associated with a helper virus in MSV producer cells. When inoculated intramuscularly or subcutaneously a transforming activity develops in situ and cell replication of the transformed clone begins [30,31]. It seems that these two mechanisms of transformed cell production coexist during all the tumor evolution.

Immunodepressed mice and some mice strains which do not reject the tumor constitute appropriate systems in order to study the transformed cell production separetely, i.e. without interference of the immune system. The growth curves observed in these systems follow an exponential law despite the fact that transformed cells are also produced by the virus induced transformation and not only by cell replication. Thus a phenomenological rate constant λ may be determined which characterizes the whole production process. Remarkably, in all cases a similar value is found: $\lambda = 0.56$ day^{-1}.

Transformed cell destruction: The developing tumors progress very rapidly at the site of inoculation and are lethal for newborn mice (in agreement with Burnet's predictions). On the contrary in almost all inbred lines, adult animals completely reject the tumor after 2-4 weeks. A small percentage of the animals shows spontaneous recurrence.

Histologically the tumors show an infiltration by inflammatory cells which increases in the course of tumor rejection. The inflammatory infiltrate is rapidly predominated by T lymphocytes and macrophages. The extensive immunological studies of this system have demonstrated the following: (i) the existence of humoral immunity both against the transformed cells and the virus. (ii) a T-cell mediated cytotoxic reaction against the transformed cells. (iii) the existence of macrophages-mediated cytotoxic and cytostatic effects. (iv) a cytotoxic activity of other effector cells as natural killer cells.

According to different authors T lymphocytes are the principal effector cell of this system. Our model permits one to test this hypothesis.

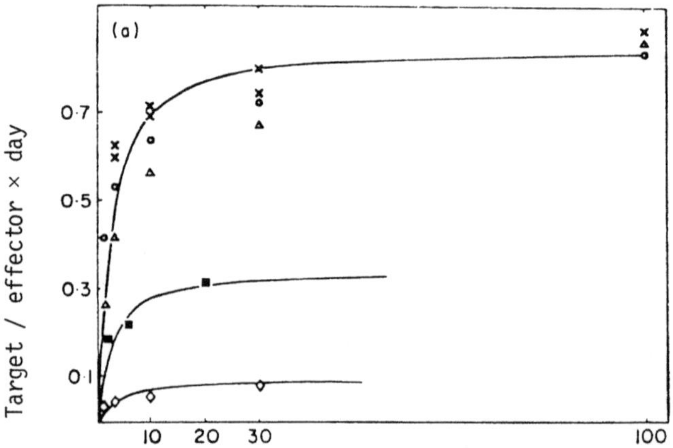

Fig. 7 Cytotoxic titration curves of anti-MSV-M lymphocytes for days: 4 (■), 7 (O), 9 (Δ), 12 (×) and 15 (◇).

T lymphocytes contribution in MSV-M tumor rejection: In the case of MSV-M tumors, the cytotoxic titration curves with intratumoral lymphocytes obtained by Plata and Sordat permit us to calculate the numerical values of k_1N and k_2 for each post inoculation day [32]. We can see from Fig. 7 and Table 2 that these kinetic constants reach a constant, maximum value between days 7 and 12, a period during which the tumor shows maximum size and regression becomes evident. This result agrees with our assumption that the immune response passes through quasi-stationary states (of several days here) over time intervals which are much bigger than the typical replication time.

Table 2: in vitro values of the cytotoxic kinetic constants of intra(MSV-M)tumoral T lymphocytes

post-inoculation day	k_1N (day^{-1})	k_2 (day^{-1})	E_c (effector/target)
4	0.17	0.34	3.3
7	0.425	0.85	1.32
9	0.425	0.85	1.32
12	0.425	0.85	1.32
15	0.045	0.09	12.4

If the kinetic values discussed here are representative of the in vivo ones, one may assert: (i) since $\theta>1$ (see Table 2), the rejection phenomenon which follows the infiltration of the tumor by the T-cells is not associated with a bistability phenomenon. (ii) from the condition which must be fulfilled for rejection to become possible ($\beta>1$), knowing the values of k_1 and k_2, we can determine at each post-inoculation day the intratumoral levels of T lymphocytes which insure tumor rejection (E_c in Table 2; see also Fig. 8).

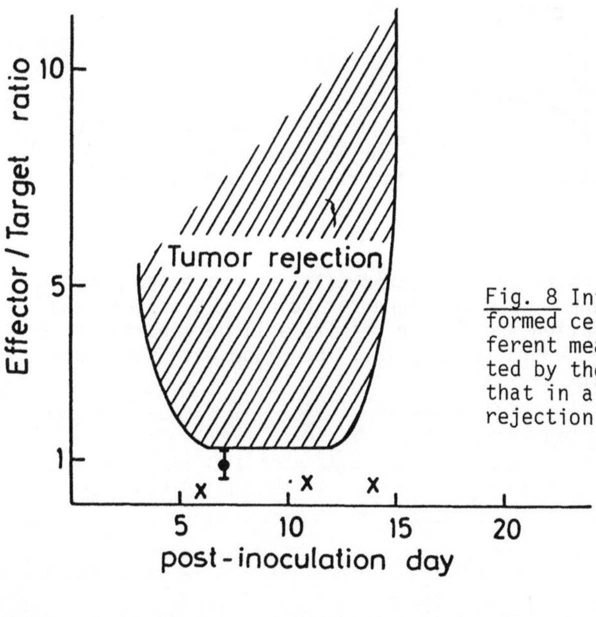

Fig. 8 Intratumoral T lymphocyte/transformed cell ratios corresponding to different measurements have been represented by the × and the ●. It can be seen that in all cases they fall outside the rejection zone indicated by hatchings.

It is clear that the experimental determination of the intra-tumoral T lymphocyte / transformed cell ratio is crucial for the evaluation of the contribution of the T lymphocytes in the tumor rejection. The techniques of disaggregation and identification of tumor constituent cells have many drawbacks. Their results must therefore be analyzed with caution. Nevertheless it seems clear that the values which can be calculated from the experiments of several authors fall within the non-regression zone (cf. Fig. 8). Accordingly the T-cell mediated reaction seems unable by itself to account for the tumoral rejection phenomenon.

The importance of other anti-tumoral reactions: Among non-T mechanisms of tumor rejection, two categories of processes may be retained: (i) cytotoxic processes mediated by antibodies, macrophages,natural killer cells, B lymphocytes, etc... (ii) cytostatic processes mediated by neutralizing antibodies against the virus (a process which inhibits the rate of cell transformation) and by macrophages against transformed cells (a process which inhibits the rate of cell replication).

The non-T cytotoxic process seem not to be more powerful than T cytotoxicity and only play an additive role. On the contrary, the cytostatic processes diminish the value of λ allowing the cytotoxic rate to become larger than the tumor cell production rate. This cooperative role of cytostatic processes may be essential in order to assure an efficient tumor rejection.

5. Modelisation of tumor progression or rejection

Within the framework of this theory the growth of the tumor must be viewed as a plane wave solution of the system of equations (3) connecting either an unstable and stable stationary state ($\theta>1$) or two stable stationary states ($\theta<1$, $1<\beta<\beta^+$, see Fig. 10). Such solutions have translational symmetry and depend on the simple combination of variables $\tau = r + ct$, in which c is the rate of propagation of the wave. We shall consider the cases $\theta>1$ and $\theta<1$ successively.

$\theta>1$: All possible wave like solutions correspond to growing tumors and connect the unstable x-nul state to a stable stationary state. Numerous authors have investigated a similar problem in the case of Fisher equation [33-37]. Fisher equation however has the advantage to be a scalar equation while here we must deal with a system of equations. The results concerning the stability of the wave solutions 38

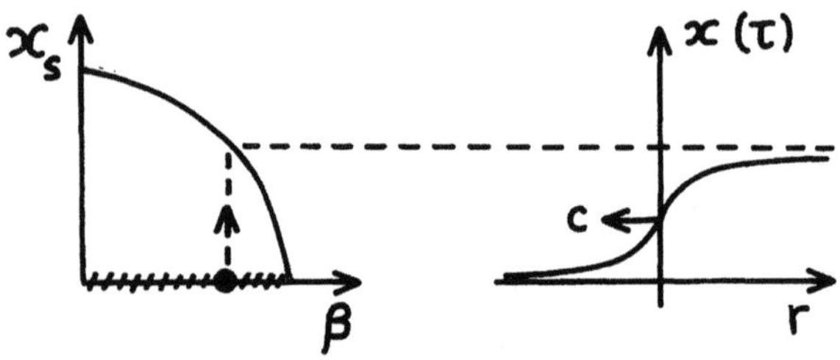

Fig. 9 Sketch of the relation between the homogeneous steady state properties and the wave solutions when $\theta>1$.

can therefore not be extended here. The results sketched below have been obtained using a method based on bifurcation theory and developed principally by Kopell and Howard [39]. The details of the derivation and futher results will be reported elsewhere 40 . Here we focus on some general properties. The plane wave solutions of (3), satisfying the boundary conditions

$$x(-\infty) = 0 \qquad x(\infty) = \frac{1 - \theta + ((1+\theta)^2 - 4\theta\beta)^{1/2}}{2\theta}$$

$$y(-\infty) = 1 \qquad y(\infty) = \frac{1}{1 + x(\infty)}$$

$$z(-\infty) = 0 \qquad z(\infty) = \frac{x(\infty)}{1 + x(\infty)} \tag{4}$$

have been determined by means of the perturbation expansion

$$x(\tau) = x_1(\tau) + x_2(\tau)\varepsilon^2 + \dots$$
$$y(\tau) = y_1(\tau) + y_2(\tau)\varepsilon^2 + \dots$$
$$z(\tau) = z_1(\tau) + z_2(\tau)\varepsilon^2 + \dots \qquad (5)$$

is a small quantity defined by the relation $(1 - \beta) = (\theta - 1)\varepsilon$

For β tending to 1, the wave solution for x is approximated by

$$x = \frac{1 - \beta}{\theta - 1} \times \frac{1}{1 + \exp - (1-\beta)\tau/c} + O((1-\beta)^2) \qquad (6)$$

in which c must obey the condition

$$c^2 \geqq 4a^2(1 - \beta) \qquad (7)$$

in order that a physically acceptable solution exists.

<u>$\theta<1$</u>: We look for wave solutions connecting the two stable stationary states in the domain $1<\beta<\beta^+$. Only if the perturbation expansion is performed so that the velocity and all other parameters have precisely matching values, it is possible to find the unique stable wave solution which exists under these conditions. It reads:

$$x = \frac{1}{1 + \exp c_0\mu x'(1-\theta)} \times (1-\theta) + O((1-\theta)^2) \qquad (8)$$

with

$$c_0^2 = \frac{1}{2}\mu\,(1 - \sqrt{1 - 4\beta_0}) \qquad (9)$$

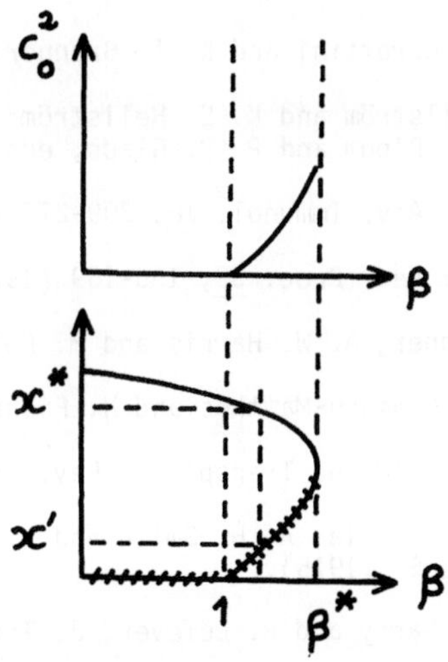

These relations show that for $\beta = 1$, the velocity is zero as well for the class of solutions which correspond for $\beta<1$ to progressing tumors as for the new class of solutions which branch off there and which correspond to regressing tumors. The velocity of the wave of regression increases with β up to a maximum value which is determined by the diffusion coefficient of the effector cells. This points to the fact that the influence of some of the metabolites secreted by the T-cells during the immune response may when $\theta<1$ favor the persistence of the tumor. This applies in particular to the effect of lymphokines which tends to inhibit the mobility of macrophages. It would be very interesting to determine wether there is a relation with the phenomenon of immune facilitation found in some systems.

<u>References</u>

1. Theoretical Immunology. (G. I. Bell, A. S. Perelson and G. Pimbley, eds., New York, 1978).
2. A. S. Perelson: Lectures on Mathematics in the Life Sciences <u>11</u>, 109-163 (1979).

3. R. Lefever and R. P. Garay: in reference 1 .

4. L. Thomas: In: Cellular and Humoral Aspects of the Hypersensitve State. (H. S. Lawrence, ed., Hoeber, New York, 1959)

5. F. H. Burnet: Br. Med. Bull. 20, 154-158 (1964).

6. F. M. Burnet: Lancet 1, 1171-1174 (1967).

7. F. M. Burnet: Transplant. Rev. 7, 3-25 (1971).

8. J. Rygaard and C. O. Povlsen: Transplant. Rev. 28, 43-61 (1976).

9. B. N. Ames: Environ. Health Perspectives Exp. 6, 115-118 (1973).

10. M. Calvin: Naturwissenschaften 62, 405-413 (1975).

11. E. J. Foley: Cancer Res. 13, 835-842 (1953).

12. R. T. Prehn and J. Main: J. Nat. Cancer Inst. 18, 769-778 (1957).

13. K. Habel: Proc. Soc. Exp. Biol. N. Y. 106, 722-725 (1961).

14. L. A. Zilber: Adv. Cancer Res. 5, 291-329 (1958).

18. G. Haugton and A. C. Moore: Transplant. Rev. 28, 76-97 (1976).

15. R. W. Baldwin and M. Moore: Nature 220, 287-289 (1968).

19. K. T. Brunner, J. Mavel, J. C. Cerottini and B. Chapuis: Immunology 14, 181-196 (1968).

20. J. C. Cerottini and K. T. Brunner: Adv. Immunol. 18, 67-132 (1974).

21. I. Hellström and K. E. Hellström: In: In vitro Methods in Cell-Mediated Immunity (B. R. Bloom and P. R. Glads, eds., Academic, New York, 1971).

22. ibid : Adv. Immunol. 18, 209-277 (1974).

23. ibid : Fed. Proc. 32, 156-159 (1973).

24. H. Wagner, A. W. Harris and M. Feldmann: Cell. Immunol. 4, 39-50 (1972).

25. M. L. Lohmann-Matthes and H. Fisher: Transplant. Rev. 17, 150-171 (1973).

26. R. W. Baldwin: Transplant. Rev. 28, 62-74 (1976).

27. O. J. Plescia, A. H. Smith and K. Grinwich: Proc. Nat. Acad. Sci. (USA) 72, 1848-1851 (1975).

28. R. P. Garay and R. Lefever: J. Theor. Biol. 73, 417-438 (1978).

R. Lefever and R. P. Garay: In: Developments in Cell Biology, volume 2 (A.- J. Valleron and P. D. M. Macdonald, eds., Elsevier, Amsterdam,1978).

29. T. Erneux, R. Lefever and H. Chi: in preparation.

30. M. Stanton, L. Law and R. Ting: J. Nat. Cancer Instit. 40, 1113-1129 (1969).

31. J. Papadimitriou, D. Mc Cully and P. Simons: J. Nat. Cancer Instit. 53, 829-835 (1974).

32. F. Plata and B. Sordat: Int. J. Cancer 19, 205-211 (1977).

33. R. A. Fisher: Ann. of Eugenics 7, 355-369 (1937).

34. K. P. Hadeler and F. Rothe: J. Math. Biol. 2,251-263 (1975).

35. D. H. Sattinger: Topics in Stability and Bifurcation Theory. Lect. Notes in Math. 309, Springer (1973).

36. J. D. Murray: Lectures on Non Linear Differential Equation Models in Biology. Clarendon Press, Oxford (1977).

37. P. C. Fife: Mathematical Aspects of Reacting and Diffusing Systems. Lect. Notes in Biomathematics 28, Springer, Heidelberg (1979).

38. D. Aronson and H. Weinberger: Adv. in Math. 30, 33 (1978).

39. N. Kopell and L. N. Howard: Adv. in Math. 18, 306-358 (1975).

Part VIII

Models of Psychological and Social Behavior

Bifurcations in Cognitive Networks: A Paradigm of Self-organization via Desynchronization

J.S. Nicolis

Department of Electrical Engineering, University of Patras
Patras, Greece

ABSTRACT

The concept of "synchronization" is employed as an index of good organization in usual ("preallocation") communication networks; it should be relaxed though and enlarged in cognitive networks (which are hierarchical structures) in order to incorporate dynamical deliberations taking place in "dynamic allocation" or packet - switching networks, finite state automata as well as biological, ecological, economic and social systems.

Here the need to deal with decision making algorithms which involve hierarchical control of competitive processes becomes imperative. In such systems "self" - organization can occur as a result of bifurcation processes triggered by learning.

These processes may lead via destabilization of the old control strategy to the abrupt emergence of a metalanguage or a new decision making algorithm among the members - "users" of the system at the expense perhaps of temporal or spatial coherence among the interacting partners. The result of such a "catastrophe" is a richer perceptive and behavioral repertoire of the system and an amelioration of his adaptability to "noisy" environments. In that sense the system "Soviet Union" is more synchronized but the system "United States" is more organized. In short the way we envisage the role of organization in a "cognitive network" of interacting variables (organisms) - performing under conditions of uncertainty and conflict has not so much to do with the congruence between the sequences of behavioral mode turnover of the individual subsystems constituting the networks: it has rather to do with the solution of a dual objective optimization problem compromizing defacto conflicting factors such as homeostasis (e.g. persistence at a given state) and adequate crosscorrelations with the partner - subsystem.

The whole idea is fully developed in the present paper in a simple network consisting of two interacting hierarchical subsystems.

Each subsystem - organism posesses in turn two irreducible hierarchical levels playing the role of "store - forward" and "traffic" processes or "behavior" and "experience" respectively.

The evolution of the system is pursued in a unifying contentless formalism of inductively played nonnegotiable games between the hierarchical levels of each partner - subsystem.

Fig. 1 The general layout of the communication process between two hierarchical systems

A detailed isomorphic model concerning nonresolvable ambiguity of communication dynamics (envisaged as bidirectional information transfer) between two hierarchical structures, is worked out.

The scope is to design feed forward control algorithms leading to limited resolution of the intrasystemic conflict via inter systemic communication.

I. INTRODUCTION.

Modern digital communication networks are characterized by an Asynchronous Multiple Access modus operandi : These systems are "asynchronous" in the sense that individual users may employ them independently of the other users with no system coordination of frequency band occupancy or timing [1] . In such systems the predominant form of fluctuations comes from multiple - user interference. Coding techniques and store-forward stategies have been suggested to cope with the problem (see e.g [1] , [2] , [3], [4] and references therein).

In this paper we discuss the issue of the self - organization of a cognitive network from a completely different dynamical point of

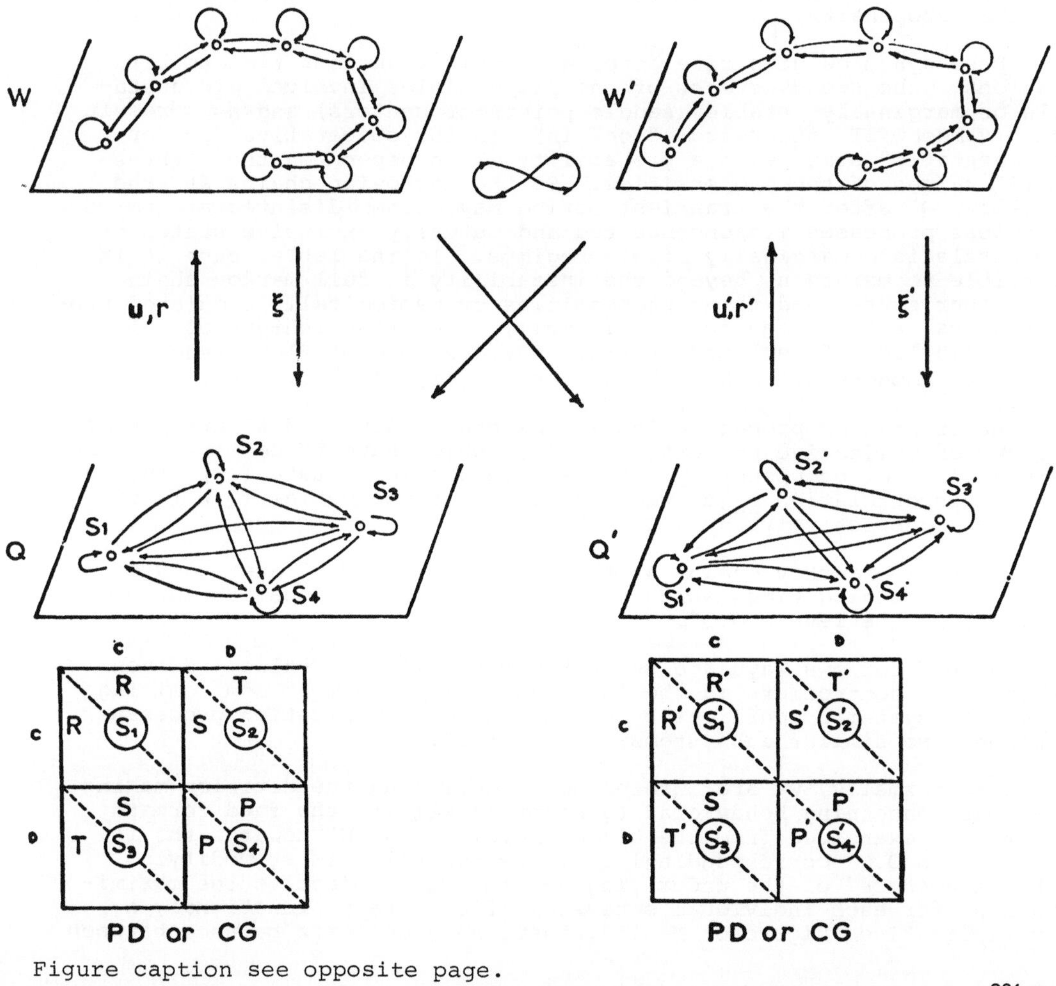

Figure caption see opposite page.

view: the "ecological" one. We accordingly introduce the basic concepts of competition and conflict which permeate the interaction between the individual "users" and couch the model in the formalism of nonnegotible games - played iteratively. [5], [6].

More specifically we consider our network consisting of two subsystems which interact as shown in Fig. 1 . Each "partner" possesses two irreducible hierarchical levels Q,W and Q´, W´ respectively.

The dynamical processes at each level (Q,Q´) , (W,W´) are described correspondingly as Markov and semi - Markov chains. The transition probabilities at the lower levels Q,Q´ stem from the parameters of underlying nonnegotiable games which emulate all activity below Q,Q´ and are played inductively with conditional mixed strategies or "propensities".

The Markovian kinetics at the levels Q,Q´ do not evolve however under fixed propensities: Since the games are played in iteration, learning takes place which means that the time evolution of the propensities is governed by a system of two first order non - linear differential equations; the time derivatives of the propensities are proportional to the gradient of the expected pay off (s) with respect to that propensity.

These dynamics give rise after a certain transient time to bifurcations: the steady states of the propensities involved are unstable or marginally stable (saddle points or centers) and as a result the games will afterwards "lock in" in the cooperative (cc) or "homeostatic" state with a probability which depends on the "threshold" values of the propensities. So the markovian chains at the levels Q,Q´ after the transient period may either disintegrate to memoryless processes i.e. unconnected and mutually exclusive states or, to settle into marginally stable regimes. In the latter case it is possible to maintain beyond the instability a full markov chain with four states and fixed propensities or periodically modulated propensities, a "metalanguage" as it were; the establishment of this chain (in lieu of isolated states) provides some of "long range coherence" amongst all available states $S_i (i = 1,.......,4)$.

The transition probabilities at the higher levels W,W´ are parametrized on collective properties at Q,Q´ which have to do with (a) the probability of occupancy u of the "synergetic" state (cc) and (b) the crosscorrelation r between sequences of the chains (Q,W´) and (Q´, W) respectively.

The semi - Markov chains at the higher levels W,W´ are endowed with mean holding times which are inversely proportional to the redundancy of the levels Q,Q´.

In what follows ways are suggested for designing value - driven stochastic controllers in the individuals which aim at compromizing the intrasystemic conflicts by establishing communication processes between two similar subsystems - "networks".

More formally, we are interested in designing the decision making process concerning behavioral turnover as well as the feed forward controls exercised from the higher levels W and W´ towards the lower levels Q,Q´ correspondingly. These controls aim at modifying the parameters of the underlying game(s) with objective the maximization for each individual subsystem of a certain functional, a multiplicative "Figure of Merit", ensuring a delicate balance between the two conflicting drives: homeostatic tendency versus good cros-correlations with the communicating partner. The underlying games alternate between prisoner´s dilemma (PD) and the game of "chicken" (CG). The changeover between the games (PD) and (CG) takes place within each individual when his figure of merit with the previous game seems to be reaching local plateaus i.e when stationarity is established.

II. THE DYNAMICS AT THE BASE LEVELS Q,Q´ AND THE UNDERLYING GAME(S)

We now try to deduce the dynamical activities at the base "experience" levels Q,Q´ as off - shoot of an underlying game. We will emulate these conflicts by nonconstant sum games which lead (when deductively played) to paradoxical outcomes; namely by the prisoner´s dilemma (PD) game and the chicken game (CG) [8].

In the present context we refrain from identifying the two agents playing the game in each individual subsystem; The essential feature of the game is that the paradoxical outcome (DD) leads to "behavioral paralysis".

A more generalized game (CG) comes up when the states (DC) and (CD) are local equilibria because (DD) is too expensive to afford. In such a paradigm the retaliation of the "betrayed" agent is predicted with probability $a \neq 1$. The "intrapersonal" conflict in each partner is going to swing between (PD) and (CG) i.e. between uninhibited mutual defection and "blackmail".

Let us now calculate for each paradigm (PD), (CG) the P_{ij} elements of the transitional probabilities of the Markov chain, which comes up as a result of inductively playing the above games with time varying "propensities". Specifically, for the (PD) game having four states S_1, S_2, S_3, S_4 (Fig.2), following Rapoport [5] we define the propensities:

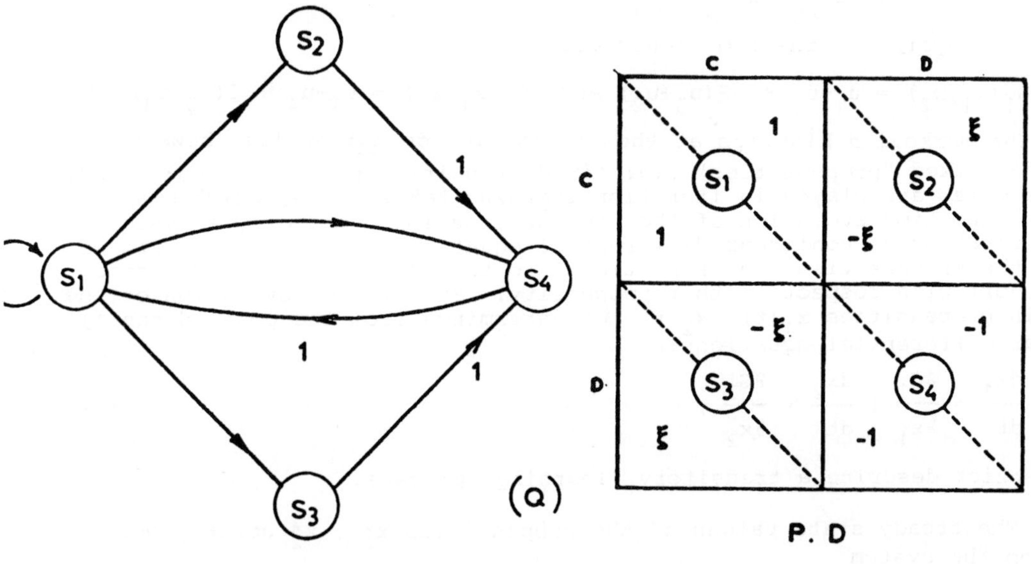

Fig. 2 The Prisoner's Dilemma played inductively and the resulting Markov chain.

$$P(C_1/S_1) = x_1, \quad P(C_1/S_2) = y_1 = 0, \quad P(C_1/S_3) = z_1 = 0$$

$$P(C_1/S_4) = w_1=1, \quad P(C_2/S_1) = x_2 \quad P(C_2/S_2) = y_2 = 0 \tag{1}$$

$$P(C_2/S_3) = z_2=0, \quad P(C_2/S_4) = w_2 = 1$$

thereby implying an immediate retaliation from the part of the "betrayed" partner and an immediate return to the state S_1 after the paradoxical outcome S_4. The sixteen transitional probabilities defining the Markov chain at Q are therefore deduced as follows:

$$P_{11} = x_1x_2 \quad P_{12} = x_1(1-x_2) \quad P_{13} = x_2(1-x_1)$$

$$P_{14} = (1-x_1)(1-x_2) \quad P_{21} = y_1y_2 = 0 \quad P_{22} = y_1(1-y_2) = 0$$

$$P_{23} = y_2(1-y_1) = 0 \quad P_{24} = (1-y_1)(1-y_2) =1 \quad P_{31} = z_1z_2 = 0$$

$$P_{32} = z_1(1-z_2) = 0 \quad P_{33} = z_2(1-z_1) = 0 \quad P_{34} = (1-z_1)(1-z_2) = 1$$

$$P_{41} = w_1w_2 = 1 \quad P_{42} = w_1(1-w_2) = 0 \quad P_{43} = w_2(1-w_1) = 0$$

$$P_{44} = (1-w_1)(1-w_2) = 0 \tag{2}$$

This is an aperiodic Markov chain with asymptotic values of the probability of occupancy of the states $S_i, u_i, i \in (1,2,3,4)$ which are calculated from the relations

$u_i = \sum_{j=1}^{4} u_j P_{ji}$ and $\sum_{i=1}^{4} u_i = 1$ It comes out that

$$u_1 = \frac{1}{\Sigma}, \quad u_2 = \frac{x_1(1-x_2)}{\Sigma}, \quad u_3 = \frac{x_2(1-x_1)}{\Sigma} \quad \text{and } u_4 = \frac{1-x_1x_2}{\Sigma} \quad \text{where,}$$

$$\Sigma = 2+x_1+x_2-3x_1x_2. \tag{3}$$

The payoffs of the two agents are:

$$G_1(x_1,x_2) = u_1-u_4 + \xi(u_3-u_2) \quad \text{and} \quad G_2(x_1,x_2) = u_1-u_4 + \xi(u_2-u_3) \tag{4}$$

The markovian kinetics at the levels Q,Q´ do not evolve however under fixed "propensities" e.g. fixed transition probabilities: Since the games are played in iteration learning takes place, which means that the time evolution of the propensities is governed by a system of first order nonlinear differential equations; the time derivatives of the propensities are proportional to the gradient of the expected pay-off with respect to that propensity. So the time evolution of the main propensities $x_1(t)$, $x_2(t)$ is determined from the coupled non linear differential equations:

$$\frac{dx_1}{dt} = \frac{\partial G_1}{\partial x_1}, \quad \frac{dx_2}{dt} = \frac{\partial G_2}{\partial x_2}, \tag{5}$$

Which describe a transitory "learning" process.

The steady state values of the propensities x_1^*, x_2^* coming out from the system

$$\frac{dx_1}{dt} = \frac{(3\xi+1)x_2^2 + 2(1-\xi)x_2-2\xi}{(2+x_1+x_2-3x_1x_2)^2} = 0$$

$$\frac{dx_2}{dt} = \frac{(3\xi+1)x_1^2 + 2(1-\xi)x_1-2\xi}{(2+x_1+x_2-3x_1x_2)^2} = 0 \tag{6}$$

are equal to

$$x_1^* = x_2^* = x^* = \frac{\xi-1+\sqrt{1+7\xi^2}}{3\xi + 1}$$; they exist only for $1<\xi\leq 3$.

The value x^* represents a "theshold" above which "locking in" at the (cc) regime is ultimately reached; it is an unstable steady state (a saddle point) as can be easily deduced by investigating conditions of stability.

We next turn to a version of (PD); We present a generalized case where the probability of retaliation of the "betrayed" partner is now $\alpha \neq 1$ where $0\leq\alpha \leq 1$. So we have $y_1 = z_1 = y_2 = z_2 = \alpha$ and $w_1=w_2=1$.

For the new Markov chain at $Q,(Q')$ we will have:

$$\begin{array}{llll} P_{11} = x_1x_2 & P_{21} = \alpha^2 & P_{31} = \alpha^2 & P_{41} = 1 \\ P_{12} = x_1(1-x_2) & P_{22} = \alpha(1-\alpha) & P_{32} = \alpha(1-\alpha) & P_{42} = 0 \\ P_{13} = x_2(1-x_1) & P_{23} = \alpha(1-\alpha) & P_{33} = \alpha(1-\alpha) & P_{43} = 0 \\ P_{14} = (1-x_1)(1-x_2) & P_{24} = (1-\alpha)^2 & P_{34} = (1-\alpha)^2 & P_{44} = 0 \end{array} \tag{7}$$

The asymptotic values of the probabilites of occupancy of the available states S_i are deduced as follows:

$$u_1 = \frac{2\alpha-2\alpha^2-1}{\Sigma_0} \qquad u_2 = \frac{(\alpha-\alpha^2-1)x_1-\alpha(1-\alpha)x_2+x_1x_2}{\Sigma_0} \tag{8}$$

$$u_3 = \frac{-\alpha(1-\alpha)x_1+(\alpha-\alpha^2-1)x_2+x_1x_2}{\Sigma_0} \qquad u_4 = \frac{\alpha^2x_1+\alpha^2x_2+(1-2\alpha)x_1x_2+2\alpha-2\alpha^2-1}{\Sigma_0}$$

Where $\Sigma_0 = (3-2\alpha)x_1x_2 + (\alpha^2-1)(x_1+x_2)-2(2\alpha^2-2\alpha+1)$

By putting $\alpha = 0$ we recover the previous expessions for (PD). The steady states of x_1, x_2 follow as

$$x_1^*=x_2^*=x^* = \frac{-(\alpha^2-1)\xi-(1-2\alpha) + \sqrt{(\alpha^4-8\alpha^3+ 18\alpha^2-16\alpha+7)\xi^2+(2\alpha^2-4\alpha+1)}}{(3-2\alpha)\xi+1} \tag{9}$$

This steady state is unstable. Again,

$$G_1(x_1,x_2) = u_1-u_4 + \xi(u_3-u_2),\; G_2(x_1,x_2) = u_1-u_4+\xi(u_2-u_3) \tag{10}$$

We will consider in what follows the case $\alpha = \frac{1}{2}$; More over, we will modify the normalized matrix of the game by putting both payoffs in the state S_4, $P = -2\xi$ so that the states DC and CD(S_2,S_3) become local equilibria since S_4(DD) state is too expensive to afford. ("Chi-

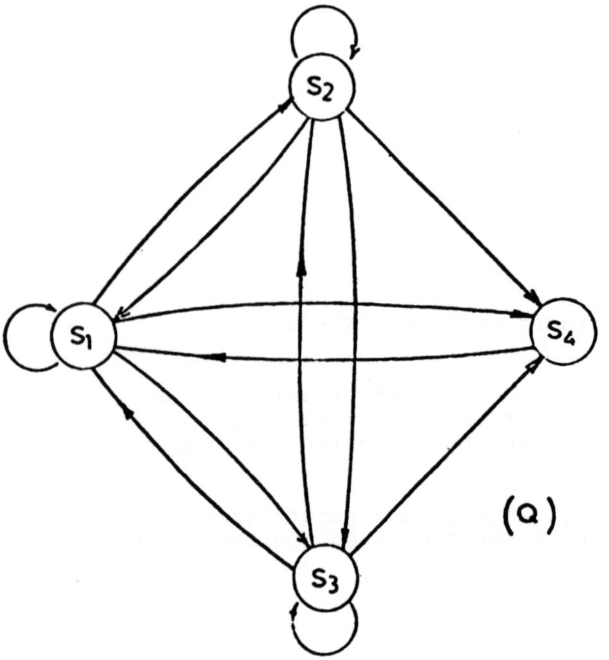

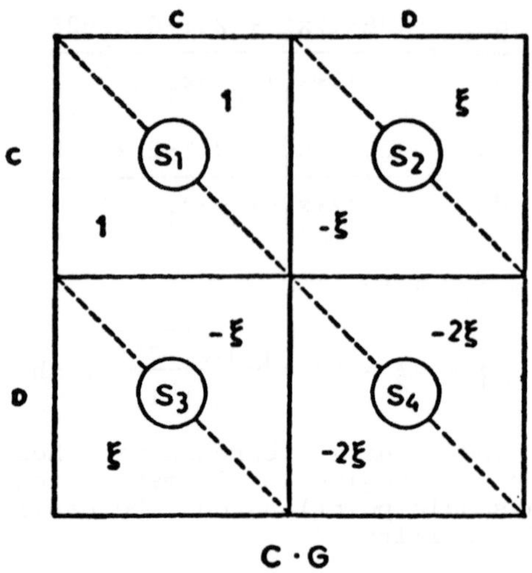

Fig. 3 The Chicken's game and the resulting Markov chain.

cken"-game). After some trivial algebra along the same lines as above concerning the new expression for G_1 and G_2

$$G_1 = u_1 - 2\xi u_4 + \xi(u_3 - u_2) \qquad G_2 = u_1 - 2\xi u_4 + \xi(u_2 - u_3) \tag{11}$$

the system of non-linear differential equations takes now the form:

$$\frac{dx_1}{dt} = \frac{32\xi x_2^2 + 4(4-11\xi)x_2 + 12\xi - 6}{\{4 - 8x_1x_2 + 3(x_1+x_2)\}^2}$$

$$\frac{dx_2}{dt} = \frac{32\xi x_1^2 + 4(4-11\xi)x_1 + 12\xi - 6}{\{4 - 8x_1x_2 + 3(\dot{x}_1+x_2)\}^2} \tag{12}$$

We get for the steady state of the propensities x_1, x_2 the expression

$$x^* = \frac{(\frac{11}{4}\xi - 1) \pm \sqrt{(1 - \frac{5}{4}\xi)^2}}{4\xi} \;\begin{cases} \nearrow 3/8 \\ \searrow \frac{2\xi-1}{2\xi} \end{cases} \tag{13}$$

which turns again to be unstable or marginally stable. More specifically the states $(\frac{3}{8}, \frac{3}{8})$ and $(1 - \frac{1}{2\xi}, 1 - \frac{1}{2\xi})$ are saddle points and the states $(\frac{3}{8}, 1 - \frac{1}{1\xi}), (1 - \frac{1}{2\xi}, \frac{3}{8})$ are centers.

We note in conclusion that the continuous non-linear dynamics governing the time evolution of the propensities $x_1(t)$ and $x_2(t)$ give rise (after a transient time) to discontinuous or switching phenomena both for the PD($1<\xi\leq 3$) and CG($1<\xi<\infty$). More specifically in the case of PD, for $\xi>3$, the temptation for unilateral defection is so great that no

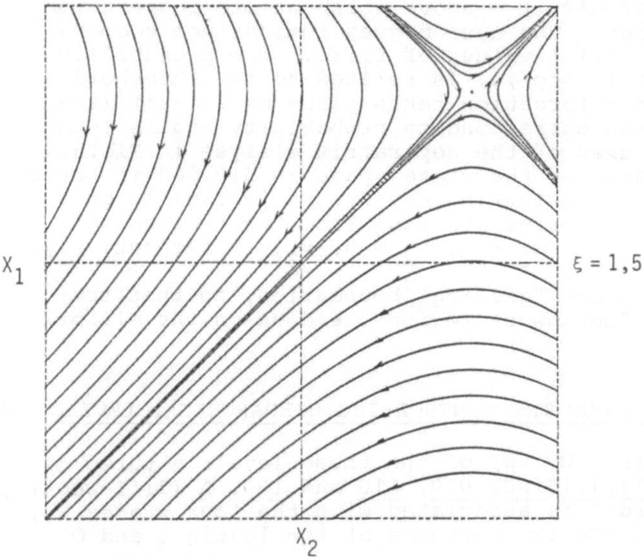

◄ Fig. 4 Phase-space description of the Prisoner's Dilemma

Fig. 5a-c Phase-space description of the Chicken game. ▼

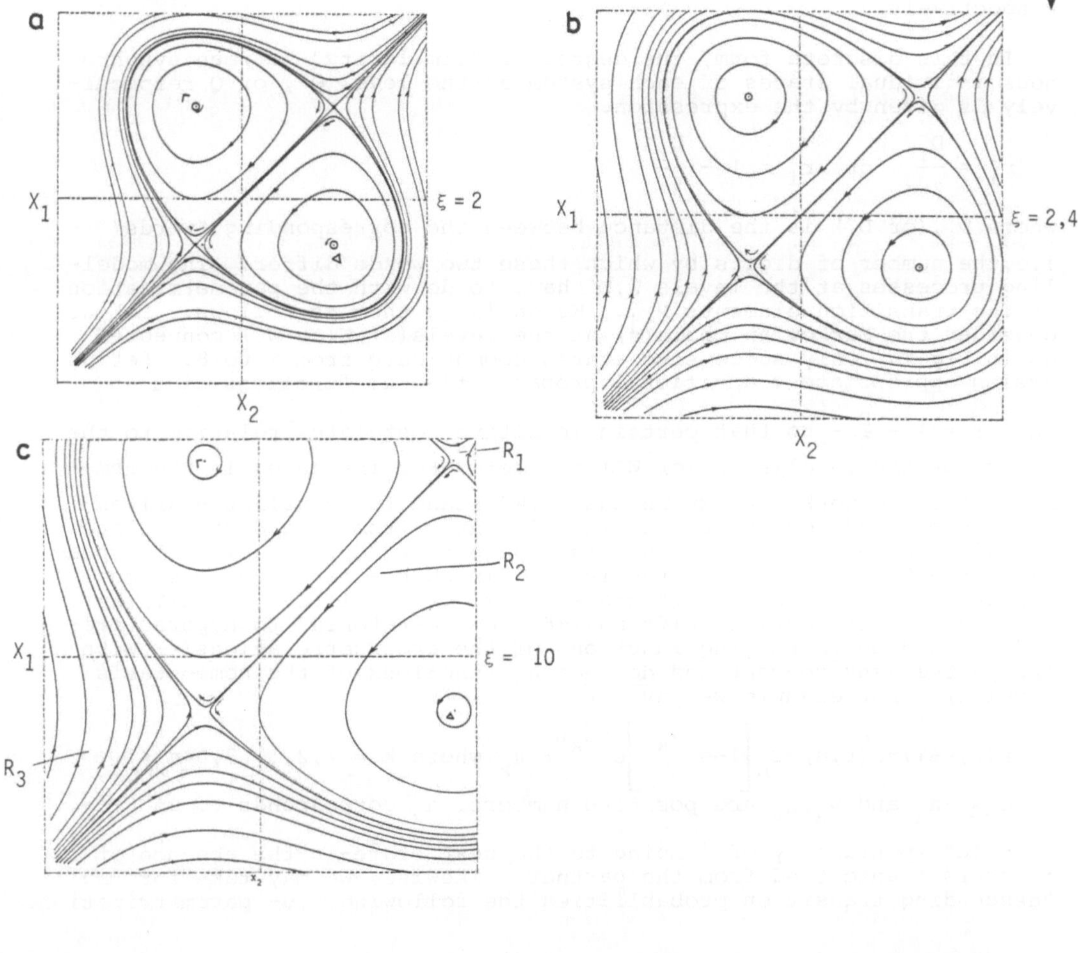

"threshold of trust" ($x^*<1$ or $u^*\leq 1$) - above which stable cooperation takes place-can be established: The system goes towards non cooperation (DD) whatever are the initial values of x_1, x_2. The probability of "locking-in" (cooperation) in (cc), u, is related to the threshold probability x^*, beyond which bifurcation takes place as $2(1-x^*)^2$ (see Fig.4). In the case of CG the corresponding probability equals $1-\pi ab$ for $\xi<2,4$ where a,b are the axes of the separatrix ellipse in Fig.5. For $\xi\geq 2,4$ (Figs 5β, c) the area of the state space is divided in three regions: The probability for locking-in the (cc) state rapidly goes to zero as $\xi\to\infty$ since now trajectories starting from the "southwest" region hit the boundaries $x_1 = 0$, $x_2 = 1$, $x_2 = 0$, $x_1 = 1$ respectively.

In the present paper we assume "gliding" boundaries, so that trajectories starting from the "southwest region" end up on the elliptic separatrix (Fig.5c).

III A SEMI-MARKOV CHAIN MODEL FOR THE CONTROLLING HIERARCHICAL LEVELS W,W´

We associate with each state W_i, W_i' of the these levels a three digit "word" i.e. 000, 001, 011, 111, 100, 010, 110 and 101. A (different) subset of four of these "words" is associated with the four states S_1, S_2, S_3, S_4, and S_1', S_2', S_3', S_4' of the two partners at the levels Q and Q´ respectively.

In this discrete form, the degree of "similarity" between synchronous individual states of each system at the level Q´, or Q respectively is given by the expression.

$$r_i' = \frac{D_i'}{3} \quad \text{or,} \quad r_i = 1 - \frac{D_i}{3}$$

where D_i, (or D_i') is the distance between the corresponding "words" i.e. the number of digits by which these two words differ. The modelling processes at the levels W,W´ have to do with the parametrization of the transition elements P_{ij}. We envisage them as follows: Let us consider the Markovian chain(s) at the levels(s) W(or W´) connected as in Fig 1. Only successive states communicate from 1 to 8. Let us assing appropriate transitional probabilities as functions of u and $r = 1 - \frac{D}{3}$ - so that certain intuitive postulates relative to the nature of the level(s) W (or W´) are met. For instance, in the communication network context in case W,W´ stand for conflict - driven "store-forward" levels, the string of states from 1 to 8 may emulate a scale ranging from "idleness" to "congestion" or, behaviorally speaking from Catatonia to Hyperactivity. In such a paradigm it seems appropriate that "ascending" transitional probabilities $P(k,k+1)$ ($k=1, \ldots, 7$) (i.e. sequential shifting 1→8 from "Catatonia" to Hyperactivity") should be increasing functions of the crosscorrelations(r) with the partner-environment and decreasing functions of the homeostasis level(u). For example we may put

$$P(k,k+1) = P'(r,u) = C_\varkappa\left[1-e^{-\nu_\varkappa r}\right]e^{-\xi_\varkappa u} + \pi_\varkappa \quad \text{where } k = 1,2,\ldots,7,\ 0\leq\pi_\varkappa\leq 1\ ,$$

$C_\varkappa \leq 1-\pi_\varkappa$ and $\nu_\varkappa, \xi_\varkappa$ are positive numbers. $\pi_\varkappa$ corresponds to an "intrinsic" spontaneity of jumping to the next state-in the absence of triggers transmitted from the partner. Likewise, we may take for the "descending transition probabilities the following r,u- parametrization.

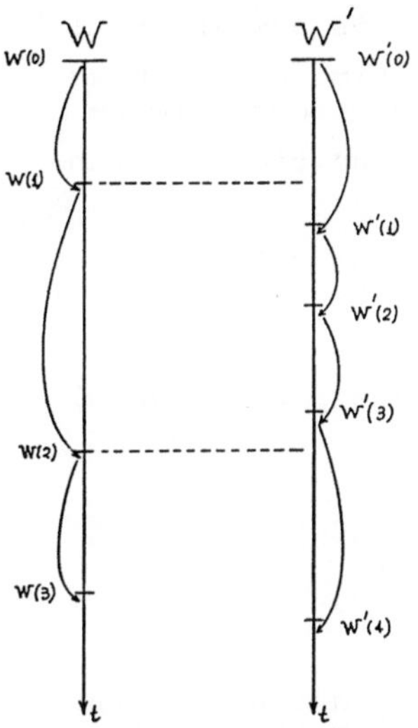

Fig. 6 Computer simulation: State transitions at the levels W and W'.

$P(k,k-1) = P''(r,u) = d_\varkappa \left[1-e^{-\mu_\varkappa u}\right] e^{-\lambda_\varkappa r} + g_\varkappa$ (where $g_\varkappa$ has a similar meaning as $\pi_\varkappa$ above) $k = 2,3,\dots,8, 0 \leq g_\varkappa \leq 1$, $0 \leq d_\varkappa \leq 1-q_\varkappa$ and $\mu_\varkappa, \lambda_\varkappa$ are positive numbers.

Finally the probabilities of remaining in the same state are calculated as:

$$P(k,k) = 1-P(k,k+1) - P(k,k-1) = \quad (k = 2,\dots,7)$$

$$= 1 - \pi_\varkappa - g_\varkappa - C_\varkappa \left[1 - e^{-\nu_\varkappa r}\right] e^{-\xi_\varkappa u} - d_\varkappa \left[1-e^{-\mu_\varkappa u}\right] e^{-\lambda_\varkappa r}$$

$$P(1,1) = 1-P(1,2) \text{ and } P(8,8) = 1-P(8,7)$$

Similar things hold for the deliberations on the level W'. We now introduce a second kind of parametrization on the transition elements P_{ij} at the levels W and W' as follows: When the process at the level W enters a state i, we know that it determines the next state j to which it will move according to the state's i conditional probabilities $P_{ij}(r,u)$ - as described above.

However, after j has been selected, but before making this transition from state i to state j, the process, we imagine, "holds" or "gets stuck" for a time τ_{ij} in state i. The holding times are positive, integer - valued random variables each governed by probability mass function $h_{ij}(\tau)$-called the "holding time density function".

The probability that the system at W will spend τ time units in state i if we do not know its successor state is

$$\Lambda_i(\tau) = \sum_{j=1}^{N} P_{ij} h_{ij}(\tau) \quad \text{where N is the number of all successor states.}$$

We call τ the "waiting" time in state i and $\Lambda_i(\tau)$ the waiting time probability mass function.

Let $P_{ij}(\tau)$ be the probability that a process which is now in state i and which will make a transition out of state i at time τ will make that transition to state j. Thus the $P_{ij}(\tau)$ are transitional probabilities conditioned on holding time; we call them "Conditional" transitional probabilities.

The waiting time distribution $\Lambda_i(\tau)$ and the set of conditional transition probabilities $P_{ij}(\tau)$ provide a complete alternate definition of the semi-Markov process.

So if the process is originally described (as in the present case) in terms of P_{ij} and $h_{ij}(\tau)$ we compute

$$P_{ij}(\tau) = \frac{P_{ij}h_{ij}(\tau)}{\Lambda_i(\tau)} = \frac{P_{ij}h_{ij}(\tau)}{\sum_{j=1}^{N} P_{ij}h_{ij}(\tau)}$$

The holding time-concept has to do operationally speaking with the "jerkyness" of the "clocking" mechanism activating the Markov chain(s) involved i.e. the mechanism responsible for the turnover of behavioral modes (states). We postulate that this cloking mechanism (some sort of master pacemaker) becomes inactivated within time intervals (holding times) which increase with the degree of disorganization at the levels Q and Q′ for the systems concerned respectively.

The redundancy at the level Q is defined as $R_Q = 1 - \frac{H_Q}{H_{max}}$ where

$H_Q = \frac{1}{\tau}\sum_{\nu=1}^{\tau} H_{i\nu}$ is the time average entropy or uncertainty per state, at the level Q during the previous holding time,

$H_i = -\sum_{k=1}^{4} P_{ik}\log_2 P_{ik}$ is the uncertainty of the state i (i=1,2,3,4) and $H_{max} = \log_2 4$; i_ν stands for the index of the state at the particular moment ν.

Let us now define the holding time probability mass function as a geometrical distribution $h_{ij}(\tau) = n_{ij}(1-n_{ij})^{\tau-1}$ where $0 \leq n_{ij} \leq 1$ stands for the conditional probability of staying in the state i at the higher level W exactly one unit time before commuting, given that the transition is from state i to state j. The mean value of the above distribution equals $\frac{1}{n_{ij}}$ and its variance equals

$$\frac{1-n_{ij}}{n^2_{ij}}$$

We now calculate easily the elements $P_{ij}(\tau)$ of the Markovian dynamics at the level W (similar things hold for the level W′). We find reasonable to postulate - as far as the holding time dependence is concerned - that $P_{ij}(\tau)$ should be a decreasing function of τ as one moves from the Idle state to the Congested state and increasing function of τ as one moves from Congestion to Idleness. This means that the longer the system at W stays in an idle state the more probable is to keep staying at this regime and the longer the system stays in a congested state the more probable is to switch to a less congested or more idle state. It is easy to check that the selected h_{ij} fulfill the above constaints.

To demonstrate this we write:

$$n_i(i+1) = n_d \quad \text{for} \quad i = 1,2,\ldots\ldots,7$$

$$n_i(i-1) = n_\mu \quad \text{for} \quad i = 2,3,\ldots\ldots,8 \qquad \text{and}$$

$$n_{ii} = n_0 \quad \text{for} \quad i = 1,2,\ldots\ldots,8$$

assuming $n_\mu < n_0 < n_d$

So the relationship between the corresponding conditional holding times will be

$$\frac{1}{n_\mu} > \frac{1}{n_0} > \frac{1}{n_d}$$

The holding time dependence on the Redundancy of the lower hierarchical level R_Q is introduced in our example as follows:

$$n_\mu = \frac{1+R_Q}{36}, \quad n_0 = \frac{1+R_Q}{30}, \quad n_d = \frac{1+R_Q}{24}$$

We write now,

$$h_{i(i+1)}(\tau) = n_d(1-n_d)^{\tau-1} \quad \text{for } i = 1,2,\ldots\ldots,7$$

$$h_{i(i-1)}(\tau) = n_\mu(1-n_\mu)^{\tau-1} \quad \text{for } i = 2,3,\ldots\ldots,8$$

$$h_{ii}(\tau) = n_0(1-n_0)^{\tau-1} \quad \text{for } i = 1,2,\ldots\ldots,8$$

$$h_{ij}(\tau) = 0 \quad \text{for } (j-i)(j-i-1)(j-i+1)\neq 0$$

The expressions for the holding time conditioned transitional probability elements follow easily.

Some results of the computer simulation concerning the behavioral mode turnover W,W´ for given values of the figure of merit J=u.r are shown on Figs. 7-10; for further details see Ref-[7],[8].

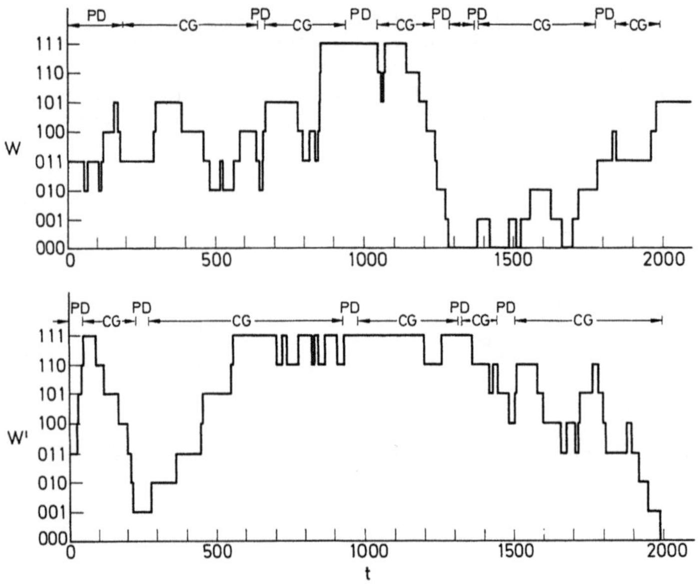

Fig. 7

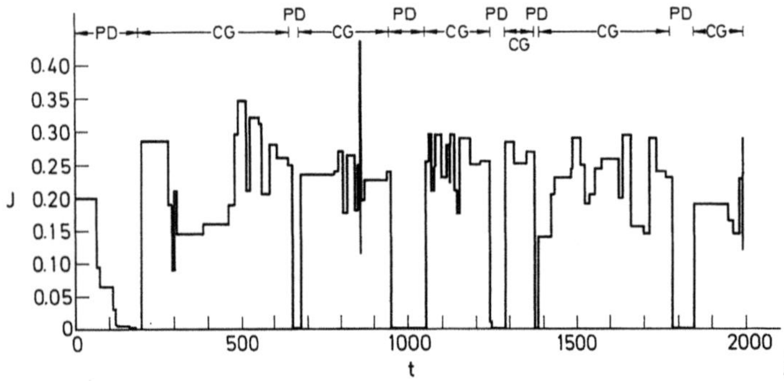

Fig. 8.a

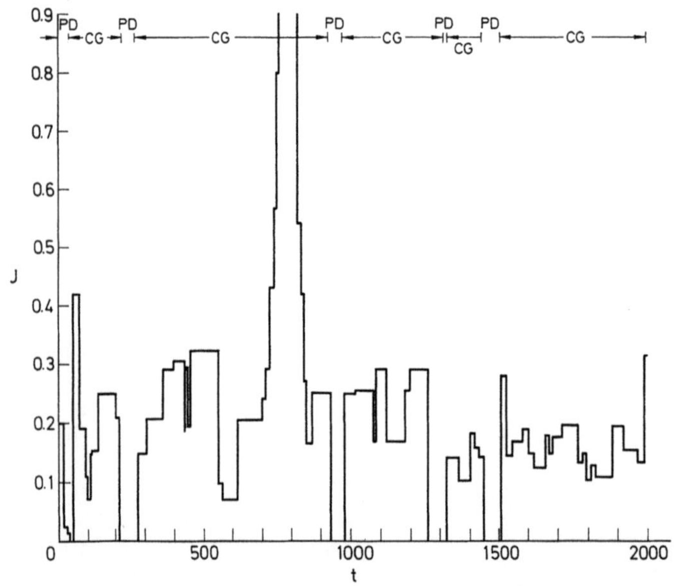

Fig. 8.b

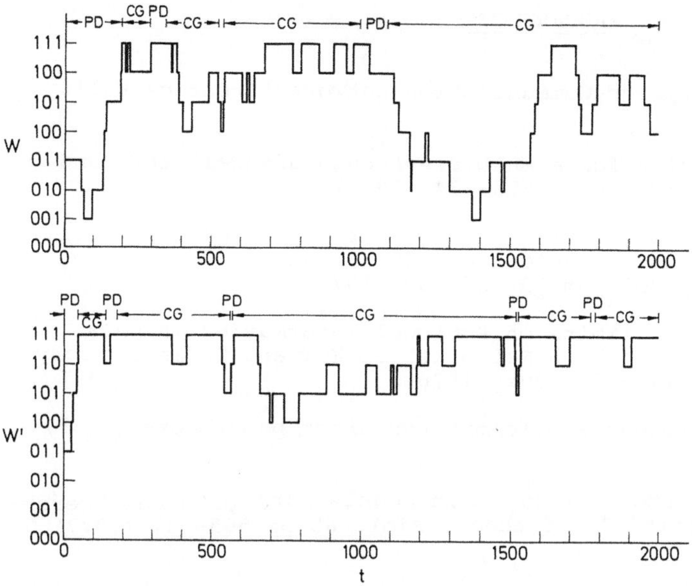

Fig. 9

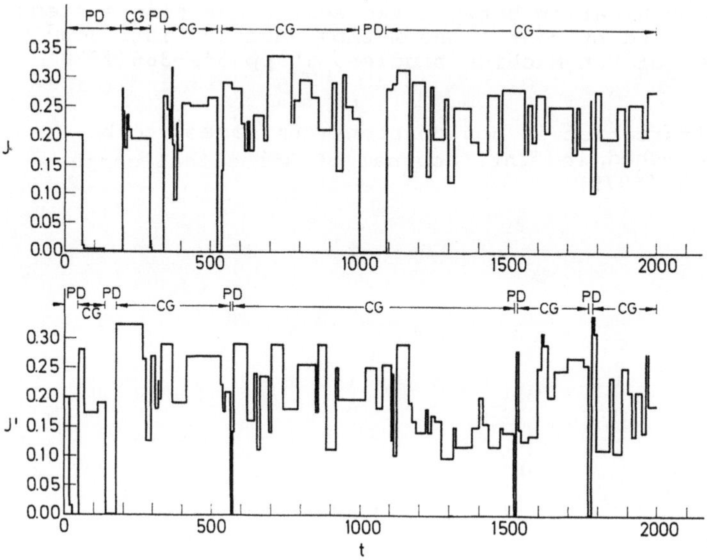

Fig. 10

Fig. 7 Sequence of states at the higher levels W and W', with transition probability parameters at these levels: $\pi_k = 0.1$, $g_k = 0.2$, $c_k = d_k = 0.8$, $\pi'_k = 0.2$, $q'_k = 0.1$, $c'_k = d'_k = 0.8$, and $\lambda_k = \mu_k = \nu_k = \xi_k = \lambda'_k = \mu'_k = \nu'_k = \xi'_k = 2$, for $k = 1,2,\ldots,8$. The initial value of the parameter ξ in the underlying games are $\xi(0) = \xi'(0) = 2$; the figure of merit is given as initial value $J(0) = J'(0) = 0.2$; each partner switches games if $J(n) < 10^{-4}$.

Fig. 8 Evolution of the instantaneous figure of merrit derived from the computer simulation with parameter values as in Fig. 7.

Fig. 9 Sequence of states at the higher levels W and W', with transition probability parameters at these levels: $\pi_k = 0.1$, $g_k = 0.2$, $c_k = d_k = 0.8$, $\pi'_k = 0.2$, $g'_k = 0.1$, $c'_k = d'_k = 0.8$, and $\lambda_k = \mu_k = \nu_k = \xi_k = \lambda'_k = \mu'_k = \nu'_k = \xi'_k = 2$, for $k = 1,2,\ldots,8$. The initial value of the parameter ξ in the underlying games are $\xi(0) = 1.1$ and $\xi'(0) = 1.8$; the figure of merrit is given as initial value $J(0) = 0.2$; switches games if $J(n) < 10^{-4}$.

Fig. 10 Evolution of the instantaneous figure of merrit derived from the computer simulation with parameter values as in Fig. 9.

REFERENCES

1. M SCHWARTZ: "Computer Communication networks"Prentice Hall (1977)

2. E.J. WELDON: "Coding for a Multiple-Access channel",Inf. and Control 36 pp 256-274 (1979)

3. COHEN A.I. HELLER J.A. VITERBI A.J "A new Coding Technique for Asynchronous Multiple Access Communication" IEEE Trans on Comm. Technology vol com-19 No 5 Oct 1971.

4. GALLAGER I.G. "Basic limits on Protocol Information in Data Communication Networks" IEEE Trans On Info.theory vol IT - 22, July 1976.

5. RAPOPORT A. and CHAMMAH A. "Prisoner´s Dilemma" Ann Arbor paperbacks (1965).

6. NICOLIS J.S PROTONOTARIOS E.N VOULODEMOU I. "Control Markov chain models for Biological hierarchies" J. of theor. Biol. 68 pp 563-581 (1977)

7. NICOLIS J.S. PROTONOTARIOS E.N THEOLOGOU M. "Communication between two self organizing systems modelled by controlled Markov chains" Int. Journal of Man-Machine studies, 10 pp 343-366(1978)

8. NICOLIS J.S PROTONOTARIOS E.N "Bifurcation in non constant sum games" to be published in "Int. Journal of Biomedical computing (1979)

Dynamics of Interacting Groups in Society with Application to the Migration of Population

W. Weidlich and G. Haag

Institut für Theoretische Physik der Universität Stuttgart
D-7000 Stuttgart 8, Fed. Rep. of Germany

1. Introduction

Before developing a general frame for the quantitative description of dynamical processes in society, let us make some remarks about the structure of science into which our theory has to be embedded.

Deductive science consists of a hierarchical sequence of levels: The more fundamental levels contain the detailed "microscopic" experimental information and theoretical concepts about the system, while the phenomenological levels contain the "macroscopic" collective gross structures of the system. The properties of the macroscopic level have to be derived from the microscopic level.

It is the great discovery of statistical physics and nowadays of synergetics, that for a large class of systems in physics, chemistry, biology and sociology the relation between the microscopic units and the collective cooperative features of these systems can be understood rather independently of the specific nature of the elementary units. This is the reason for the interdisciplinary universality of the concepts of synergetics.

In order to include quantitative sociology into this conceptional frame, we have 1. to introduce appropriate "microscopic" and "macroscopic" levels of description, 2. to establish the general relation between both levels, and 3. to fill the frame with content by chosing appropriate models for concrete sociological situations.

2. Fundamental Concepts

It is trivial, that the society consists of men and women and that these human beings should be considered as the "elementary units" of our system. However, we may decompose society into subpopulations P_α which are internally homogeneous with respect to the background and traditions or other selected properties of their members.

Dynamical processes in society on a *microscopic level* may now be described as follows: There exists a number of quasi independent *aspects* of life for each member of society. Examples are: religion, politics, education, habitation, occupation, economic standard, consumer habits, family, sports etc. Let us select n aspects and consider them to span an n dimensional vectorspace, called the *aspectspace* A. Every individual now may take a certain *attitude* out of a discrete or continuous set of possible attitudes with respect to each of the n aspects. We quantify these attitudes by assigning a vector component v_{ai} to an attitude i with respect to an aspect a (for instance, if "a" is the political aspect, attitude "i" may be one of the opinions represented by political parties). Hence, the total attitude of an individual to *all* aspects is characterized by his *attitude vector* $\underline{v}_i \in A$, which is an element of the aspect space A.

Microscopic changes occur in the state of society, if a single individual belonging to subpopulation P_α changes his attitude, say, from $\underline{v}_i$ to $\underline{v}_j$. The fundamental dynamical quantity governing these changes is the transition probability per time unit w^α_{ji}, for a member of subgroup P_α to change his attitude from $\underline{v}_i$ to $\underline{v}_j$. Formally, we put $w^\alpha_{ii} \equiv 0$. Here, we assume for simplicity a discrete set $\{\underline{v}_i\}$ of attitudes.

Later, we will have to make model assumptions for w^α_{ji} in concrete cases. The level, on which the form of w^α_{ji} would find its theoretical explanation, is a still more microscopic and fundamental one and may be called *"political psychology"* i.e. the theory of individual decisions under the constraints of a given society. In our context, w^α_{ij} is treated as a given quantity.

Let us now go over to dynamical processes in society on a *Macroscopic Level*. We introduce the *state configuration* $\underline{m}(t) = \{m_{\alpha i}(t)\}$, where $m_{\alpha i}(t)$ is the number of members of subpopulation P_α taking attitude $\underline{v}_i$ at time t. We assume $m_{\alpha i}(t) \gg 1$ and mention, that $m_{\alpha i}(t)$ is a stochastic variable because of the attitude-fluctuations of the individuals. The state configuration characterizes the society on a global, collective level, which is, however, determined by the individual processes.

The microscopic and macroscopic level are coupled, in particular, because the individual transition probability in general will depend on $\underline{m}$. Hence, $w^\alpha_{ji} = w^\alpha_{ji}(\underline{m})$. This means, that individual decisions are made in dependence of the existing global configuration of society.

Because of the probabilistic description of the fluctuations of $\underline{m}(t)$, we may now either introduce the probability $p\{\underline{m};t\}$ to find configuration $\underline{m}$ at time t, or the mean value $\underline{\tilde{m}}(t) = \{\tilde{m}_{\alpha i}(t)\}$ of the configuration $\underline{m}(t)$. In the first case we include fluctuations and take an ensemble stand point; in the second case we only obtain the mean behaviour. We will derive equations of motion for $p\{\underline{m};t\}$ and for $\underline{\tilde{m}}(t)$ in the next section.

3. Fundamental Equations of Motion

We find the *fundamental master equation* for the probability $p\{\underline{m};t\}$ by taking into account, that $p\{\underline{m};t\}$ may *decrease* in time by transitions from configuration $\underline{m}$ into a neighbouring $\underline{m}'$, and may *increase* in time by transitions from neighbouring configurations, e.g. $\underline{m}' = \{\ldots m_{\alpha j}-1;\ldots m_{\alpha i}+1\ldots\}$ into $\underline{m}$. These transitions are induced by the individual changes of attitude with transition probability $w^\alpha_{ji}(\underline{m})$. This simple probability rate consideration yields:

$$\frac{d\,p\{\underline{m};t\}}{d\,t} = -\sum_{\alpha i j}^{gss} m_{\alpha j}\, w^\alpha_{ij}(\underline{m})\, p\{\underline{m};t\} \tag{3.1}$$

$$+ \sum_{\alpha i j}^{gss} (m_{\alpha j}+1)\, w^\alpha_{ij}(m_{\alpha j}-1;m_{\alpha i}+1)\, p\{m_{\alpha j}-1;m_{\alpha i}+1;t\}$$

Here, we have assumed g interacting populations P_α and s different attitudes. In the last term of the r.h.s. of (3.1) we have only written explicitly the components of the neighbouring configuration $\underline{m}'$ *differing* from $\underline{m}$ in the argument of w^α_{ij} and p.

The definition of *mean values* $\tilde{m}_{\alpha i}$ reads

$$\tilde{m}_{\alpha i} = \sum_{\underline{m}} m_{\alpha i}\, p\{\underline{m};t\} \tag{3.2}$$

From definition (3.2) and (3.1) there follow exact equations for the time development of $\tilde{m}_{\alpha i}$:

$$\frac{d\,\tilde{m}_{\alpha i}}{d\,t} = \sum_{\underline{m}} \Big\{ \sum_{j}^{s} (m_{\alpha j}+1)\, w_{ij}^{\alpha}(m_{\alpha i}-1;m_{\alpha j}+1)\, p\{m_{\alpha i}-1;m_{\alpha j}+1;t\} \qquad (3.3)$$

$$- \sum_{j}^{s} (m_{\alpha i}+1)\, w_{ji}^{\alpha}(m_{\alpha i}+1;m_{\alpha j}-1)\, p\{m_{\alpha i}+1;m_{\alpha j}-1;t\} \Big\}$$

which only yield $\tilde{m}_{\alpha i}(t)$, if $p\{\underline{m};t\}$ is known by solving (3.1). If, however, $p\{\underline{m};t\}$ can be considered as a probability distribution sharply peaked around $\underline{m} = \underline{\tilde{m}}$, we obtain from (3.3) the approximate mean value equation

$$\frac{d\,\tilde{m}_{\alpha i}}{d\,t} = \sum_{j}^{s} \tilde{m}_{\alpha j}\, w_{ij}^{\alpha}(\underline{\tilde{m}}) - \sum_{j}^{s} \tilde{m}_{\alpha i}\, w_{ji}^{\alpha}(\underline{\tilde{m}}) \qquad (3.4)$$

This is a closed coupled set of nonlinear ordinary differential equations for the $\tilde{m}_{\alpha i}$.

The Eqs. (3.1)...(3.4) may be applied to a large variety of sociological situations. According to the interpretation of "aspects", "attitudes" and the corresponding transition probabilities, they may for instance describe the dynamics of the development of political opinions, of consumer habits, or, as we will consider now, the migration behaviour of mixed populations in a town.

4. The Model for Migration of Populations in a Town

Let us now interpret "attitude i" (i=1,2,...s) as "to live in part i of a town consisting of s parts". Correspondingly, $w_{ij}^{\alpha}(\underline{m})$ is the probability per unit time for a member of subpopulation P_α to move from part j to part i of the town. This transition probability may depend on all $m_{\alpha k}$ (α=1...g; k=1...s), i.e. the number of people of P_α living in part k.

We specialize this model to the simplest non trivial case of two populations P_μ, P_ν and two parts i=1,2 of the town. Further, we only treat the mean value equations (3.4).

Neglecting immigration or emigration into or from the town, respectively, we have the conservation laws:

$$m_{\mu 1} + m_{\mu 2} = 2\bar{m} = \text{const} \qquad (4.1)$$

$$m_{\nu 1} + m_{\nu 2} = 2\bar{n} = \text{const}$$

and may introduce the relevant variables

$$m = m_{\mu 1} - \bar{m} = \bar{m} - m_{\mu 2} \qquad (4.2)$$

$$n = m_{\nu 1} - \bar{n} = \bar{n} - m_{\nu 2}$$

The Eqs. (3.4) for the mean values $\tilde{m}$, $\tilde{n}$ of m, n, then read

$$\frac{d\tilde{m}}{dt} = \{ (\bar{m}-\tilde{m})\, w^{\mu}_{12}(\tilde{m},\tilde{n}) - (\bar{m}+\tilde{m})\, w^{\mu}_{21}(\tilde{m},\tilde{n}) \} \tag{4.3}$$

$$\frac{d\tilde{n}}{dt} = \{ (\bar{n}-\tilde{n})\, w^{\nu}_{12}(\tilde{m},\tilde{n}) - (\bar{n}+\tilde{n})\, w^{\nu}_{21}(\tilde{m},\tilde{n}) \}$$

In order to make the model explicit and applicable, we have to make an ansatz for the transition probabilities realistically describing the behaviour of people belonging to the subgroups, and flexible enough to comprise several possibilities. A sufficiently flexible ansatz for $w^{\alpha}_{ij}(m,n)$ is given by

$$w^{\mu}_{12}(m,n) = \gamma \exp\{u(m,n)\} \tag{4.4}$$

$$w^{\mu}_{21}(m,n) = \gamma \exp\{-u(m,n)\}$$

$$w^{\nu}_{21}(m,n) = \gamma \exp\{v(m,n)\}$$

$$w^{\nu}_{21}(m,n) = \gamma \exp\{-v(m,n)\}$$

with

$$u(m,n) = (\delta_1 + \tilde{\alpha}_1 m + \tilde{\beta}_1 n) \tag{4.5}$$

$$v(m,n) = (\delta_2 + \tilde{\beta}_2 m + \tilde{\alpha}_2 n)$$

The interpretation of parameters $\tilde{\alpha}_j, \tilde{\beta}_j, \delta_j$ follows from the meaning of (4.4):

As $\delta_1 > 0$ leads to favouring of part 1 before part 2 by population P_μ, we denote δ_1 (and analogously δ_2) as *natural preference parameter*.

As $\tilde{\alpha}_1 > 0$ leads to a clustering trend of population P_μ in the same part of the town, (trend to "live together"), we denote $\tilde{\alpha}_1$(and $\tilde{\alpha}_2$) as *internal sympathy parameter*.

As $\tilde{\beta}_1 > 0$ means, that population P_μ prefers to live together with P_ν in the same part of town, (analogously $\tilde{\beta}_2 > 0$ after exchange $P_\mu \leftrightarrow P_\nu$), we denote $\tilde{\beta}_1$ as *external sympathy parameter*.

The change of sign transforms "sympathy" into "antipathy". In our ansatz we have so far neglected saturation effects which, however, could be included easily.

Introducing the quasi continuous variables

$$\tilde{x} = \frac{\tilde{m}}{\bar{m}} \; ; \quad \tilde{y} = \frac{\tilde{n}}{\bar{n}} \; ; \quad -1 \leq \tilde{x},\tilde{y} \leq +1 \; ; \quad \tau = 2\gamma t \; ; \tag{4.6}$$

$$\tilde{u} = (\delta_1 + \alpha_1 \tilde{x} + \beta_1 \tilde{y}) \; ; \quad \tilde{v} = (\delta_2 + \beta_2 \tilde{x} + \alpha_2 \tilde{y}) \qquad \text{with} \tag{4.7}$$

$$\alpha_1 = \bar{m}\tilde{\alpha}_1 \; ; \quad \alpha_2 = \bar{n}\tilde{\alpha}_2 \; ; \quad \beta_1 = \bar{n}\tilde{\beta}_1 \; ; \beta_2 = \bar{m}\tilde{\beta}_2$$

and inserting (4.4)...(4.7) in (4.3), we obtain the explicit mean value equations:

$$\frac{d\tilde{x}}{dt} = \{sh(\tilde{u}) - \tilde{x}\, ch(\tilde{u})\} \equiv \{th(\tilde{u}) - x\}\, ch(\tilde{u}) \tag{4.8}$$

$$\frac{d\tilde{y}}{dt} = \{sh(\tilde{v}) - \tilde{y}\, ch(\tilde{v})\} \equiv \{th(\tilde{v}) - \tilde{y}\}\, ch(\tilde{v})$$

For $\delta_1 = \delta_2 = 0$, i.e. without natural preference, the equations (4.8) are invariant under the transformations:

(i) $\tilde{x} \to \tilde{x}' = -\tilde{x}$; $\tilde{y} \to \tilde{y}' = -\tilde{y}$

(ii) $\tilde{y} \to \tilde{y}' = -\tilde{y}$; $\beta_1 \to \beta_1' = -\beta_1$; $\beta_2 \to \beta_2' = -\beta_2$

where (i) means exchange (part 1) ↔ (part 2) and (ii) means exchange (external sympathy) ↔ (external antipathy) and simultaneously (mutual clustering trend) ↔ (mutual separation trend).

Of high relevance for the migration process are the stationary (singular) points $P_j(x_j,y_j)$ of the eq. (4.8), defined by $\frac{d\tilde{x}}{dt} = 0$ and $\frac{d\tilde{y}}{dt} = 0$, or

$$F_1(x_j,y_j) = \{th(u(x_j,y_j)) - x_j\} = 0 \tag{4.9}$$

$$F_2(x_j,y_j) = \{th(v(x_j,y_j)) - y_j\} = 0$$

The decisive question, whether the migration process comes to rest or not at a singular point P_j in configuration space, depends on the stability of this point. We can answer this question by investigating the equation of motion (4.8) linearizing it around $P_j(x_j,y_j)$.

With

$$\xi(\tau) = \tilde{x}(\tau) - x_j \ ; \quad \eta(\tau) = \tilde{y}(\tau) - y_j \tag{4.10}$$

one obtains from (4.8):

$$\frac{d\xi}{d\tau} = -\sigma_{1j}\xi + \beta_{1j}\eta \tag{4.11}$$

$$\frac{d\eta}{d\tau} = \beta_{2j}\xi - \sigma_{2j}\eta$$

where

$$u_j = u(x_j,y_j) \ ; \quad v_j = v(x_j,y_j) \ ; \tag{4.12}$$

$$\alpha_{1j} = \frac{\alpha_1}{ch(u_j)} \ ; \quad \alpha_{2j} = \frac{\alpha_2}{ch(v_j)} \ ;$$

$$\beta_{1j} = \frac{\beta_1}{ch(u_j)} \ ; \quad \beta_{2j} = \frac{\beta_2}{ch(v_j)} \ ;$$

$$\sigma_{1j} = \{ch(u_j) - \alpha_{1j}\} \ ; \quad \sigma_{2j} = \{ch(v_j) - \alpha_{2j}\} \ ;$$

The ansatz

$$\xi(\tau) = \xi_o\exp(\lambda_j\tau) \ ; \quad \eta(\tau) = \eta_o\exp(\lambda_j\tau) \tag{4.13}$$

then solves the linearized eq. (4.11) with eigenvalues

$$\lambda_{j\pm} = -\frac{(\sigma_{1j}+\sigma_{2j})}{2} \pm \frac{1}{2}\{(\sigma_{1j}+\sigma_{2j})^2 + 4\,\varepsilon_j\}^{1/2} \tag{4.14}$$

where

$$\varepsilon_j \equiv \{\beta_{1j}\beta_{2j} - \sigma_{1j}\sigma_{2j}\}$$

Because of (4.12) and (4.7), (4.9) all eigenvalues are functions of the fundamental parameters $(\delta_1,\delta_2;\alpha_1,\alpha_2;\beta_1,\beta_2)$.

The detailed analysis shows, that - corresponding to different choices of parameters $\alpha_1,\alpha_2,\beta_1,\beta_2$ - there may exist 1 or 3 or 5 or 9 singular points (x_j,y_j) as solutions of (4.9) in the (x,y)-configuration space, and that the following kinds of singular points $P_j(x_j,y_j)$ may occur in different situations:

(a) P_j = stable focus, with $\lambda_{j\pm}$ real

$\lambda_{j+} < 0 \;;\; \lambda_j < 0$

conditions: $(\sigma_{1j}+\sigma_{2j}) > 0 \;;\; -\frac{(\sigma_{1j}+\sigma_{2j})}{2} < \varepsilon_j < 0.$

(b) P_j = stable focus, with $\lambda_{j\pm}$ conj. complex

$Re(\lambda_{j+}) < 0 \;;\; Re(\lambda_j) < 0.$

conditions: $(\sigma_{1j}+\sigma_{2j}) > 0 \;;\; -\infty < \varepsilon_j < -\frac{(\sigma_{1j}+\sigma_{2j})}{2}$

(c) P_j = unstable focus, with $\lambda_{j\pm}$ real

$\lambda_{j+}\cdot\lambda_j < 0$

conditions: $\varepsilon_j \equiv \{\beta_{1j}\beta_{2j}-\sigma_{1j}\sigma_{2j}\} > 0$

(d) P_j = unstable focus, with $\lambda_{j\pm}$ real

$\lambda_{j+} > 0 \,;\, \lambda_{j-} > 0$

conditions: $(\sigma_{1j}+\sigma_{2j}) < 0 \;;\; -\frac{|\sigma_{1j}+\sigma_{2j}|}{2} < \varepsilon_j < 0$

(e) P_j = unstable focus, with $\lambda_{j\pm}$ conj. complex

$Re(\lambda_{j+}) > 0 \;;\; Re(\lambda_{j-}) > 0$

conditions: $(\sigma_{1j}+\sigma_{2j}) < 0 \;;\; -\infty < \varepsilon_j < -\frac{|\sigma_{1j}+\sigma_{2j}|}{2}$

There exists a case of particular interest for conditions:

$$(\alpha_1+\alpha_2) > 2 \;;\; \beta_1\cdot\beta_2 < 0 \tag{4.15}$$

Here, the origin $P_0(0,0)$ is the only singular point, and $P_0(0,0)$ is an unstable focus of kind (e). On the other hand, the flux lines of the differential equation (4.8) point inwards the domain D $(-1 \le \tilde{x} \le +1 \;;\; -1 \le \tilde{y} \le +1)$ of the variables at the boundary because of

$$\frac{d\tilde{x}}{d\tau} < 0 \,,\ \text{for } \tilde{x} = 1 \;;\; \frac{d\tilde{x}}{d\tau} > 0 \,,\ \text{for } \tilde{x} = -1 \tag{4.16}$$

$$\frac{d\tilde{y}}{d\tau} < 0 \,,\ \text{for } \tilde{y} = 1 \;;\; \frac{d\tilde{y}}{d\tau} > 0 \,,\ \text{for } \tilde{y} = -1$$

According to the Poincaré-Bendixon Theorem we therefore expect for this case a solution of (4.8) with a limit cycle in D around the origin.

Instead of doing further analytical investigations, it is more enlightening to look at the flux lines found by computer solution of (4.8) for several characteristic cases exhibited in Fig. 1 to Fig. 6. Afterwards we give the sociological interpretation of the processes described by the model.

Fig. 1

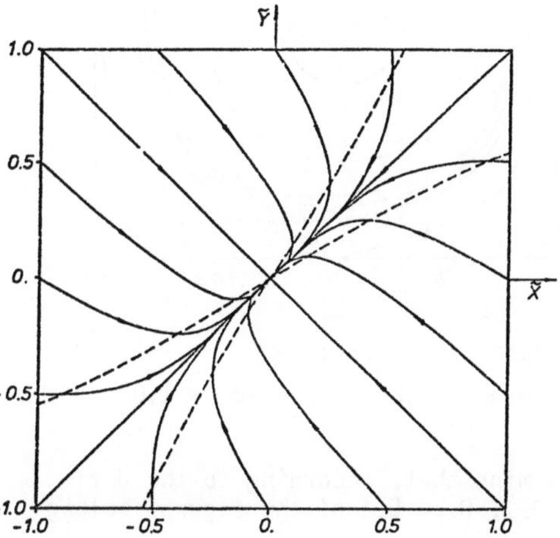

Parameters: $\delta_1 = \delta_2 = 0$;
$\alpha_1 = \alpha_2 = 0{,}2$; $\beta_1 = \beta_2 = 0{,}5$

Fig. 2

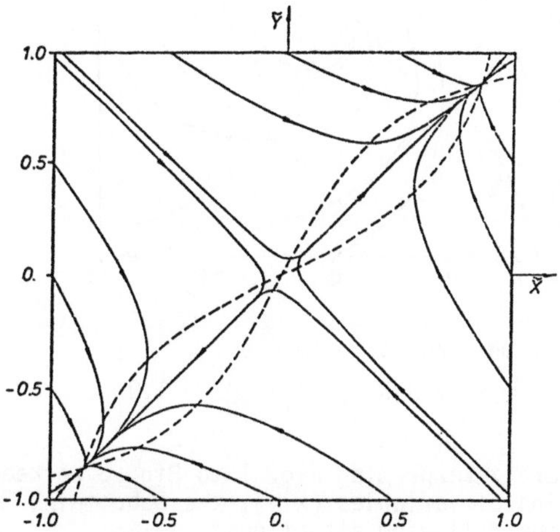

Parameters: $\delta_1 = \delta_2 = 0$;
$\alpha_1 = \alpha_2 = 0{,}5$; $\beta_1 = \beta_2 = 1{,}0$

Fig. 3

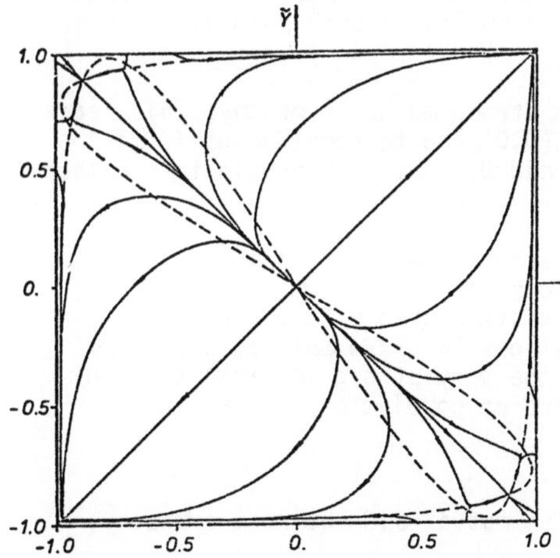

Parameters: $\delta_1 = \delta_2 = 0$
$\alpha_1 = \alpha_2 = 2{,}6$; $\beta_1 = \beta_2 = 1{,}0$

Fig. 4

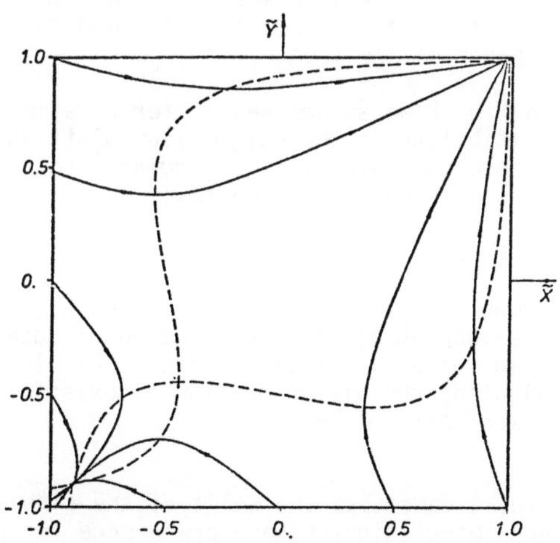

Parameters: $\delta_1 = \delta_2 = 0{,}5$
$\alpha_1 = \alpha_2 = 1{,}2$; $\beta_1 = \beta_2 = 1{,}0$

Fig. 5

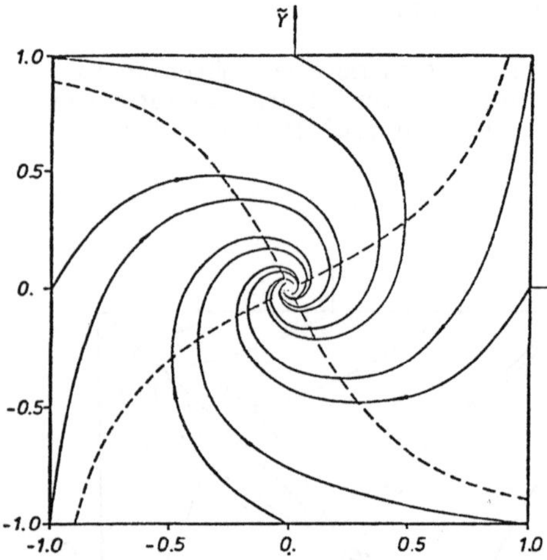

Parameters: $\delta_1 = \delta_2 = 0$

$\alpha_1 = \alpha_2 = 0{,}5$; $\beta_1 = -\beta_2 = 1{,}0$

Fig. 6

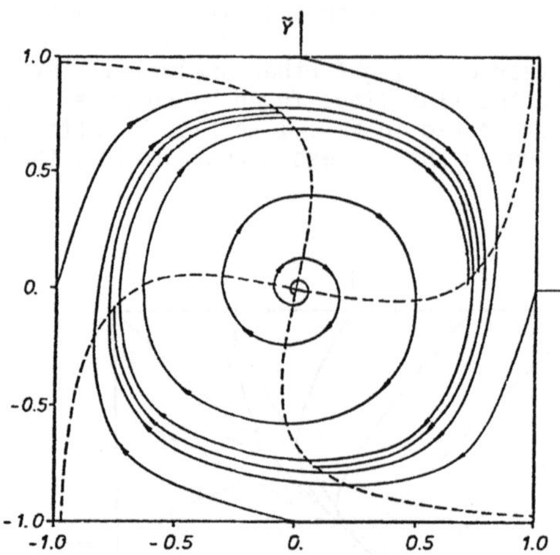

Parameters: $\delta_1 = \delta_2 = 0$

$\alpha_1 = \alpha_2 = 1{,}2$; $\beta_1 = -\beta_2 = 1{,}0$

Before interpreting Fig. 1 to Fig. 6 we keep in mind that, according to the definition of variables (x,y), the subdomains D_{++}, D_{-+}, D_{--}, D_{+-} of the figures belong to the following situations:

$D_{++}(x>0;y>0)$: P_μ clusters in part 1 ; P_ν clusters in part 1

$D_{-+}(x<0;y>0)$: P_μ clusters in part 2 ; P_ν clusters in part 1

$D_{--}(x<0;y<0)$: P_μ clusters in part 2 ; P_ν clusters in part 2

$D_{+-}(x>0;y<0)$: P_μ clusters in part 1 ; P_ν clusters in part 2

In Fig. 1 we assume weak internal sympathy and weak mutual sympathy. This leads to stability of $P_0(0,0)$, i.e. the homogeneous distribution of both populations over both parts of the town.

In Fig. 2 we assume weak internal sympathy but strong mutual sympathy. This leads to instability of the homogeneous distribution $P_0(0,0)$ and to spontaneous formation of stable concentrations (singular points in D_{++} and D_{--}) of both populations either in part 1 or part 2 of the town.

In Fig. 3 we assume extremely strong internal sympathy and strong mutual sympathy. Beyond the stable concentrations of **both** populations either in part 1 or part 2 of the town (singular points in D_{++} and D_{--}) there exists a stable focus in D_{-+} and D_{+-} corresponding to concentration of both populations in *different* parts of the town in spite of mutual sympathy. This is due to the very strong internal sympathy prohibiting the disintegration of existing clusters of populations P_μ, P_ν in *different* part of town.

Fig. 4 describes the shift of the situation with mutual and internal sympathy, if both populations have a preference for part 1.

In Fig. 5 we assume weak internal sympathy but strong assymmetric mutual sympathy: P_μ "likes" P_ν, but P_ν "dislikes" P_μ. Nevertheless, the fluxlines approach the stable focus $P_0(0,0)$, i.e. the homogeneous distribution of populations P_μ, P_ν.

In Fig. 6 we assume strong internal sympathy and the same strong assymmetric mutual sympathy as in Fig. 5. This is the limit cycle case discussed above leading to a permanent afflicting process which nevertheless might be realistic: Starting from both populations P_μ, P_ν concentrated in part 1 ($P \in D_{++}$), P_ν tends to evade into part 2 and to cluster there, leaving behind P_μ in part 1 ($P \in D_{+-}$). But, as P_μ "likes" P_ν, population P_μ will follow sooner or later and cluster in part 2, too ($P \in D_{--}$). The process is now repeated and leads to the limit cycle. Although oversimplified, this case of the model may describe the sequential erosion of parts of a town by migration of populations of different social standards living under some mutual asymmetric tension.

According to transformation (i,i), a reflection of Fig. 1,2,3,5,6 on the x-axis leads to the description of cases with parameters ($-\beta_1$, $-\beta_2$) instead of (β_1,β_2). In particular, the stable situations in Fig. 1...3 where both populations live in the *same* part of the town are transformed into stable situations where they live in *different* parts of the town.

Conclusion

The application of our general formalism for a quantitative description of process in society to migration problems shows that a variety of structure formations in the development of a town may be explained at least semi quantitatively already by the simplest version of our model.

Part IX

Mathematical Concepts and Methods

Structural Instability in Systems Modelling

T. Poston

Département de Physique Théorique, Université de Genève
CH-1211 Genève 4, Switzerland

Abstract

The predictions of many scientific models are highly sensitive to small perturbations of the equations. Topological study of these can clarify and improve robustness of the models, sometimes revealing new phenomena implicitly associated with them.

"Choices aimed to simplify the calculations may complicate the problem" ([7], p. 302). This corollary of Murphy's Law applies equally to models of cooperative phenomena, to the partial differential equations studied in non-equilibrium thermodynamics, to any modelling approach down to directly writing down a low-dimensional ODE.

"The problem" here is to understand what the system modelled will really do, if it corresponds best, not to the model chosen, but to a small perturbation of it. One hopes, evidently, for almost the behaviour predicted by the first model. But if a laser was supposed to have zero external magnetic field, the analysis is a poor guide to what a real laser in a small field will do [7], and the properties of the 'Brusselator' model chemical reaction change radically if the assumption is relaxed that one reagent diffuses with infinite speed [8]. Frequently, however, such instability can be detected by examining the topology of the predicted behaviour from the viewpoint of catastrophe theory, and the model embedded in a larger one that is demonstrably robust.

Of course, no model is robust in an absolute sense. For example, if a model's conclusions rest on the assumption that a particular curve is differentiable, one might object that this is infinitely improbable : Wiener measure on a space of continuous curves makes those differentiable nowhere into the typical kind, all other curves forming a set of measure zero. Reasonable perturbations must always be chosen within an appropriate class, and 'appropriate' must always be defined as much on scientific as on mathematical grounds. The world cannot be deduced by pure thought.

However, it is sometimes possible to supplement the scientific tradition (well exemplified at this meeting) of asking "what happens if you add a term of" some very specific type. Frequently, indeed, adding such a term changes the result, and a long and highly refined process of trial and error is needed before the experimental and theoretical community reaches consensus that the model contains everything necessary. A priori, on this approach, there is no guarantee that the process will converge.

By contrast, in some cases one can analyse a model and say : This is stable; no small perturbation that is suitably differentiable can produce a perturbed model not equivalent to the original. 'Small' of course has a precise (topological) meaning, and 'equivalent' means that one can be given as a reparametrization of the other. The statement is strengthened when strong conditions (e.g. of differentiability) can be proved about the reparametrization, and when wide classes of perturba-

---

Research supported by the Swiss National Science Foundation

tion are allowed. Numerical 'smallness' conditions satisfying to the experimentalist may be hard to derive (just as singular perturbation theory often guarantees solutions only for perturbations much smaller than experiment and numerical work support), but the kinds of perturbation admitted are usually a rather wide class.

An excellent kind of stability result is represented by Prof. Arnold's contribution to this meeting, showing that certain smooth models correspond well to the statistical results of the same models enlarged by the treatment of suitable noise - though Dr. Horsthemke showed us that noise can have more interesting consequences. Here, however, I will consider largely the differentiable models which often form the 'skeleton' of something richer, before one adds the effect of oscillation, fluctuation or quantum behaviour (an activity Berry has called "sewing the quantum flesh on the classical bones"). I would expect to find a fair set of examples of my points in any issue of any good journal in a mathematical science, but it is clearly natural to take them from this meeting[1]). Where the resulting comments may appear critical, they are not intended to be hostile : it is precisely when work is good that it becomes interesting to comment on it topologically.

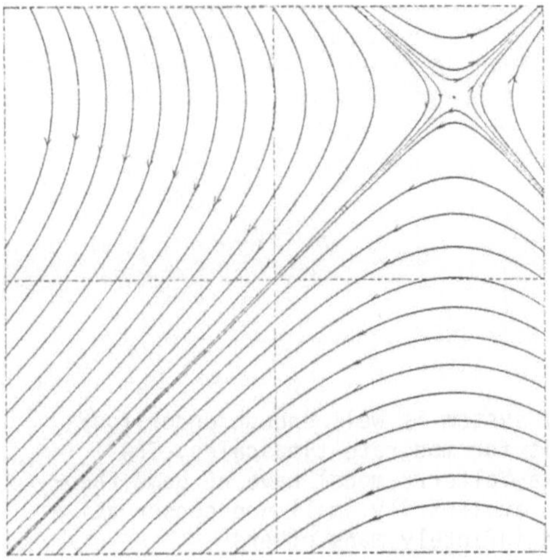

Fig. 1: Flow for the evolution of the 'Prisoner's Dilemma' game (J.S. Nicolis, these proceedings).

Good examples of both stability and instability appear in the contribution of Prof. Nicolis. Fig. 1 reproduces his flow for the evolution of the 'Prisoner's Dilemma' game. It is straightforward and elementary in Dynamical Systems theory (see e.g. 1,4,6) to show that this is structurally stable. Any small perturbation of the ODE for the flow will give a phase portrait that can be transformed (by a topological transformation near the identity) to precisely Fig. 1 again. *A fortiori*, small perturbations of the game model giving rise to the ODE (which may not be able to realize arbitrary perturbation terms at the ODE level) will leave Fig. 1 topologically unchanged - so long as their smallness is inherited by their effect on the ODE terms. This makes the associated predictions notably robust. By contrast, consider the analogous Fig. 2, for the 'Chicken' game. This has a number of unstable features, such as the flow-lines connecting saddles and the fact that (like the Lotka-Volterra predator and prey model) it admits the construction of a conserved quantity. The flow is not area-preserving (as is easily checked from its equation), and there is no *a priori* reason for such a conserved quantity to exist. (The situation is different in Hamiltonian mechanics - classical or quantum - which are conservative by general principle. It is different even for the pendulum. This is always damped in practice,

1) In consequence, I must omit detailed references, not always knowing final titles, possible co-authors, etc.

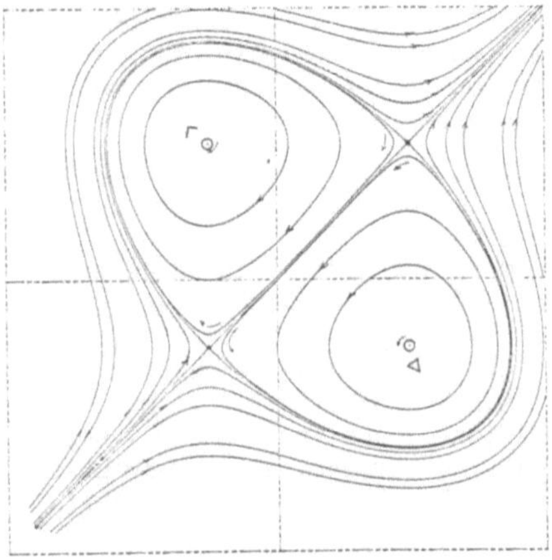

◄ Fig. 2: (From J.S. Nicolis, these proceedings.) Flow for the evolution of the 'Chicken's' game.

Fig. 3: Three flow types arbitrarily close to Fig. 2; (A) with everything flowing to the 'cooperative' top right corner, (B) two point attractors to which many initial data evolve, (C) two limit cycles.

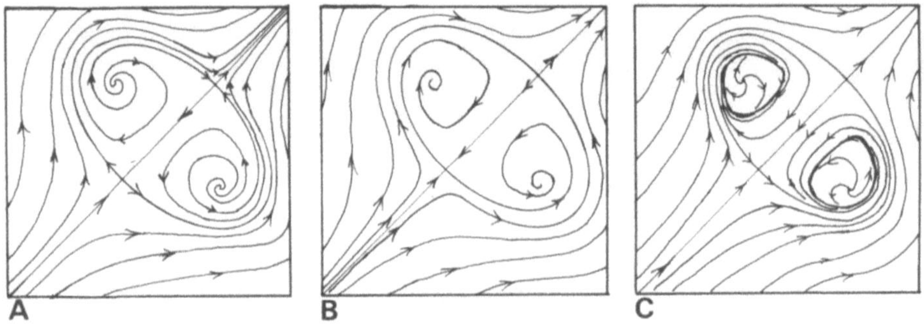

but the conserved quantity -energy- of the ideal system is well enough understood that we can make an experiment nearly conserve it for moderate timescales. The quantities constructible for Fig. 2 and the Lotka-Volterra model have no game-theoretic or biological meaning). The addition of an arbitrarily small non-conservative term can produce any of the forms in Fig. 3, and infintely many others.

Note that each of the perturbed flows in Fig. 3 preserves the symmetry of the original about the diagonal, and consequently the diagonal saddle-connection. Now, this symmetry is a consequence of the standard formulation of game theory. It means, not that the players are identical in all respects, but that they assess values and chances similarly in the game. Such an assumption is clearly doubtful in most applications, such as to wars. (The U.S.A. and the Vietnamese querrillas were applying radically different schemes of analysis, for instance). It is doubtful here, where the players are taken as two parts of the brain with different evolutionary depth; indeed, if their assessments were always identical the point of distinguishing them would be unclear!

The assumption of symmetry is harmless in Fig. 1, where symmetry-breaking perturbations do not significantly change the picture. We may suppose the model applies well to similar players, at least. But, even without an extension of game theory to differing players, we can see that the symmetry in Figs. 2 and 3 is unrobust.

Fig. 4 shows what general asymmetric perturbations could do (note that Fig. 4-C is arbitrarily close to 3-C, which is thus itself unstable in the absence of symmetry constraints). Any asymmetric generalization of game theory is likely to generate at least some of these forms, for arbitrarily small asymmetry : hence the symmetrical predictions (even if conservatism could be guaranteed, which would

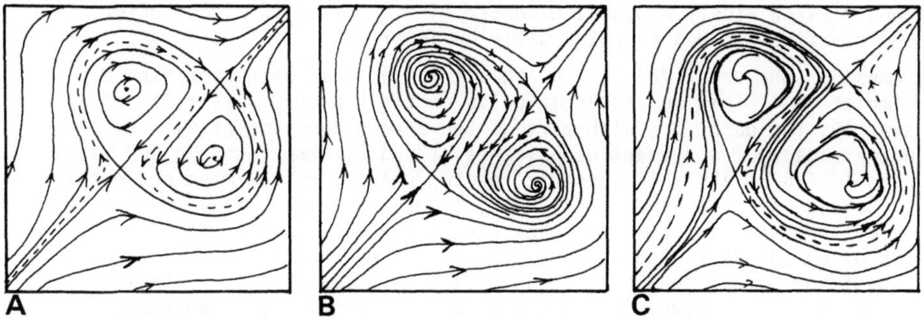

Fig. 4 : Asymmetric perturbations of Fig. 2 ; (A) conservative, (B) nonconservative with one point attractor, (C) arbitrarily close to Fig. 3-C , with a new orbit type (dotted) in common with A.

make them topologically stable among symmetric possibilities) are experimentally nonrobust unless the experimenter can guarantee symmetry to a very good approximation. In some contexts (s)he can, but certainly not here.

Consider now Prof. Sigmund's discussion of the hypercycle, or loop of chemical species catalysing each others' production, from the structural stability viewpoint.

The first assumption introduced, that the total concentration Σx_i of the species $x_1,\ldots,x_n$ is held constant, can be seen to be stable, under certain conditions. Indeed, the entire equations, with the fixed concentration normalized to 1, have

$$\frac{d}{dt}(\Sigma x_i) = \sum_{i=1}^{n} G_i(1 - \Sigma x_i) \tag{1}$$

so that if all the G_i are positive Σx_i must exponentially approach 1 when it does not start there. (Consequently, the dynamics will be dominated by those in the simplex S_n , where $\Sigma x_i = 1$ and all $x_i \geq 0$). In the model $G_i = k_i x_i x_{i-1}$ with $k_i > 0$, so that the positivity of G_i follows - though I shall return to this point.

Secondly, we have the proof that for $n < 4$ chemical species there is a globally attracting equilibrium with all concentrations present, representing a time-invariant hypercycle. For $n > 4$ the corresponding point repels, and numerical evidence suggests the new attractor is a limit cycle. (The change is intuitive if one considers the hypercycle as a feedback loop. With a few species, variation in x_i is rapidly damped by its effects on x_{i+1} through to x_{i-1} . With many, there is enough delay for the system to 'hunt' like the temperature of an overcorrected bathroom shower). For $n = 4$, the point is globally attractive, i.e. asymptotically stable, but structurally unstable. The linearized equations there have two pure imaginary eigenvalues, and only the higher-order terms show (by a Liapunov argument) that it is attractive. Consequently an arbitrarily small perturbing linear term could destabilize it, creating a small limit cycle by a Hopf bifurcation.

This of course does not detract from the general analysis, since for $n > 4$ the point is repelling, stably, in any case. But it is an interesting example of an unstable conclusion, not least because sensitivity to perturbation is closely associated with sensitivity to noise. Assuming exact equations for the deterministic part, an ordinary point attractor will usually merely blur a little with noise (though see Dr. Horsthemke's contribution). The weakness of restoring forces around a structurally unstable attractor like the one above means that a noise-driven

$(x_1,...,x_n)$ would lurch widely around the centre like a possessed witch doctor. Around such points fluctuations cannot usually be ignored.

Thirdly, consider the proof for $n > 4$ that there is some attractor strictly within the simplex S_n. This depends critically on the boundary $\{x_i = 0\}$ of S_n coinciding exactly, for each i, with part of the zero-production-of-x_i set $\{\dot{x}_i = 0\}$. Geometrically, such exact coincidence of two hypersurfaces in $\mathbb{R}^n$ is extremely unstable (as unlikely 'in general' as two general curves meeting, not in $\mathbb{R}^3$ or $\mathbb{R}^n$, but in infinite dimensions). It follows, certainly, from the hypotheses : but its mathematical instability leads one to question whether these hypotheses can be supported as exact. And in fact, the addition of an arbitrarily small term to represent some additional process involving x_i destroys the coincidence. If a little of each x_i is always being produced, the theorem is strengthened. If any x_i is suffering destruction not included in the equations, the proof too is destroyed. Indeed, since we heard that for large n the numerical limit cycles came very close to the boundary, the conclusion is probably false if small amounts of x_i are independently destroyed. The instability at the corners, where highly multiple zeros of the ODE can be perturbed with arbitrarily small terms into almost anything (including chaos), casts further doubt on the robustness of the model. (These degenerate zeros come from the assumption that all reaction rates between x_i and x_{i-1} are exactly zero).

Finally, return to the assumed restriction to the simplex. This is stable, as above, as long as the G_i are all positive. But they are not bounded away from 0. Indeed, the failure of

$$\tilde{G}_i = k_i x_i x_{i-1} + \text{small terms}$$

to be positive exactly inside the boundary $\{x_i = 0\}$ is just the possibility discussed in the previous paragraph. Consequently, we see that the control mechanism maintaining Σx_i constant is itself unstable and liable to breakdown for any x_i small - insofar as it is modelling by the equations given. I would expect any flow reactor whose control system stably achieves Σx_i = constant to correspond rather to equations with the G_i strictly positive on all of S_n, boundary included. This would also make the hypercycle theorem stably true - and, easier to prove.

So much for individual stable and unstable equations. My own concerns are more with families of equations, changing - in particular, bifurcating - as an external parameter α is varied. Real change must include unstable equations for some parameter values (if for every α the equations were topologically equivalent to those for nearby α, no change of type could occur at any α), but one can study the stability with which such points are encountered. Analogously, a single point in Europe is usually in one country, and only this is stable - a frontier point can be perturbed off the frontier. Individual frontier points are 'non-generic'. Journeys from Geneva to Bonn must include frontier points - usually isolated and finite in number. They need not, and usually do not, include points where three countries meet; and any journey that does is easily perturbed to one meeting only double frontier points. In the mathematical sciences, this comes down to the principle that "more equations than unknowns means no solutions" and, conversely, that if we have a solution to n equations in n unknowns then we will generally still have one, close by, if the equations are perturbed a little. The formalization of these ideas, most complete in the case of systems with behaviour dominated by some real valued function (energy, utility, phase, action, stream function...), has been christened 'catastrophe theory' by Zeeman [10] in development of the work of Thom [9] and many others. It turns out as much more potent than the informal thinking it refines, as group representation theory in comparison to pre-19th Century symmetry arguments. For an introduction to its techniques, see [7].

Here, however, I am concerned less with methods of finding what is typical (or, technically, 'generic') than with the common practice in science of writing down systems that are untypical - like Geneva to Bonn via a triple frontier point - and consequently unstable. Again, this is in no critical spirit. These systems are often interesting, important ones. But their instability is now worth pointing out, because new techniques often let us enlarge the models to stable ones that include those first written down as special cases, and give new insights about them.

I will limit myself here to one specific unstable bifurcation phenomenon, which occurs in almost exactly half of the talks at this meeting - in a few instances, with some or all of the nearby behaviour described.

It is convenient to begin with the case of the explicit potential

$$V(x) = \frac{\gamma}{2\lambda}\left(1 - \frac{2\nu}{\lambda}\right) \ln \tan(x) - \frac{\mu-\gamma}{\gamma} \ln \cos(x) + \tfrac{1}{2}\cos^2(x) \tag{2}$$

quoted by Prof. Gardiner, to illustrate both the power and the limitations of the techniques here discussed. For

$$\frac{\gamma}{2\lambda}\left(1 - \frac{2\nu}{\lambda}\right) \neq 0 ,$$

$V(x)$ is defined only for $x > 0$, while for $\lambda = 2\nu$ it is defined for all x. Moreover even on $\{x > 0\}$, the potential for $\lambda - 2\nu$ nonzero but arbitrarily small is infinitely far (by most tests) from that with $\lambda = 2\nu$. The techniques of catastrophe theory are defined for families of 'nearby' functions defined on a common domain and near both in values and in suitably many derivatives, which are required to exist. This restriction is sometimes physical, sometimes not : it applies here to the family of potentials for varying μ,γ with $\lambda = 2\nu$, so I consider this limited family

$$V^{\mu,\gamma}(x) = \tfrac{1}{2}\cos^2(x) - \frac{\mu-\gamma}{\gamma} \ln \cos(x) . \tag{3}$$

$$= \left[\frac{1}{6}x^4 + \left(\tfrac{1}{2} - \frac{\mu}{\gamma}\right)x^2 + \frac{7\mu-\gamma}{12\gamma}\right] + O(6) . \tag{4}$$

Now, elementary calculus will show ([7], p. 57) that a regular change of variable near 0 will change $(x^4/6 + O(6))$ to the form $\tilde{x}^4/6$, independently of whether the whole expansion converges. But it essentially requires the Malgrange Preparation Theorem, first proved in 1964 [5], to show that the <u>family</u> $V^{\mu,\gamma}$ can be reduced by a regular (μ,γ)-dependent change of x-variable, to the form

$$V^{\tilde{\mu},\tilde{\gamma}}(\tilde{x}) = \frac{1}{6}\tilde{x}^4 + \left(\tfrac{1}{2} - \frac{\tilde{\mu}}{\tilde{\gamma}}\right)\tilde{x}^2 + \frac{7\tilde{\mu}-\tilde{\gamma}}{12\tilde{\gamma}} \tag{5}$$

<u>exactly</u> around $x = 0$, $\gamma = 2\mu$. This is highly nontrivial, but true, independently of whether the whole Taylor series for V converges. The stable and unstable equilibria of $V^{\mu,\gamma}$ are then given exactly by

$$\frac{2}{3}\tilde{x}^3 + \left(1 - \frac{2\tilde{\mu}}{\tilde{\gamma}}\right)\tilde{x} = 0 \tag{6}$$

Dopping ~s again in a linear change of variable

$$x = \left(\frac{3}{2}\right)^{1/3}\tilde{x} , \quad a = \left(\frac{2}{3}\right)^{1/3}\left(\frac{2\tilde{\mu}}{\tilde{\gamma}} - 1\right)$$

we have the standard equation

$$x^3 - ax = 0 \tag{7}$$

whose solutions are shown in Fig. 5; a familiar picture. As the reduction above illustrates, most examples of

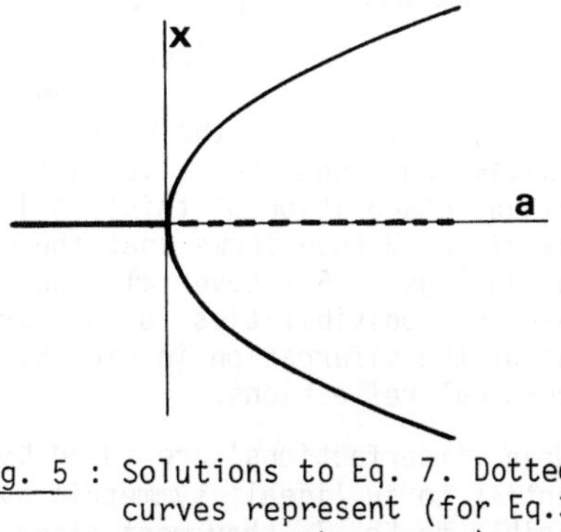

<u>Fig. 5</u> : Solutions to Eq. 7. Dotted curves represent (for Eq.3 and many other cases) unstable equilibria.

this diagram (arising even in infinite dimensions) can be reduced to Eq. 7 as an exact form.

However, unless some physical constraint guarantees symmetry in x (so that the potentials V must be even functions of x, the gradients $\partial V/\partial x$ odd), Eq. 7 is unstable. For arbitrarily small $p > 0$, the equation

$$x^3 - ax + px^2 = 0 \qquad (8)$$

has the solutions shown in Fig. 6. This again is unstable : if q is small relative to p, the solutions of

$$x^3 - ax + px^2 + q = 0 \qquad (9)$$

are as shown in Fig. 7. Note that for $q > 0$, the equilibrium value of x cannot evolve smoothly as a is varied. (Whether it jumps at a determined value, or displays hysteresis, depends on whether the physics admits metastable equilibria). Notice that the evolution of the maxima and minima of Prof. Haug's Ginsburg-Landau potentials for various pumping intensities in a semi-conductor are described by Fig. 7A for $a < 0$.

The process of adding terms can now stop. Any small perturbation of $\partial V/\partial x = 0$ for V as in Eq. 3 can be reduced to the form of Eq. 9 by a suitable transformation : indeed any parametrized family

$$\frac{\partial V^{\mu,\gamma}_{\alpha,\beta,\ldots,\omega}}{\partial x} = 0 \qquad (10)$$

of equations with $V^{\mu,\gamma}_{0,0,\ldots,0} = V^{\mu,\gamma}$ of Eq. 3 can be reduced locally to the form of Eq. 9 with p and q functions of $(\alpha,\beta,\ldots,\omega)$. (See the work of Golubitsky and Schaeffer [1,2,8] for proof and elaboration of this). A little study of Eq. 9 then shows that the diagrams in Figs. 5,6,7 cover all the topological possibilities for perturbations of the bifurcation in Fig. 5, up to vertical reflections.

When 'imperfections' are added by a scientist to an 'ideal' symmetric system reducible to Eq. 7, they most often take a form locally equivalent to

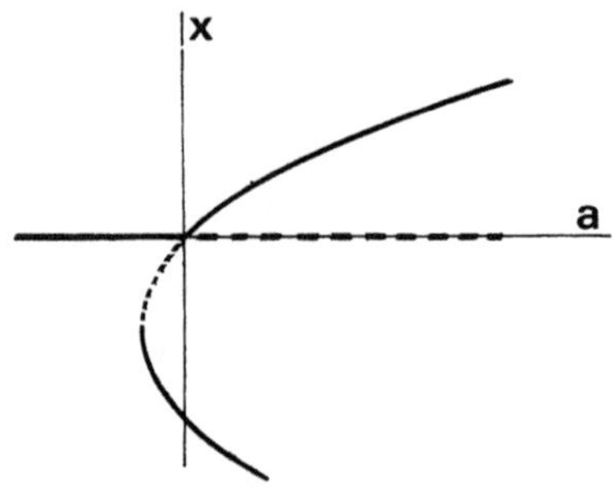

Fig. 6 : Solutions to Eq. 8

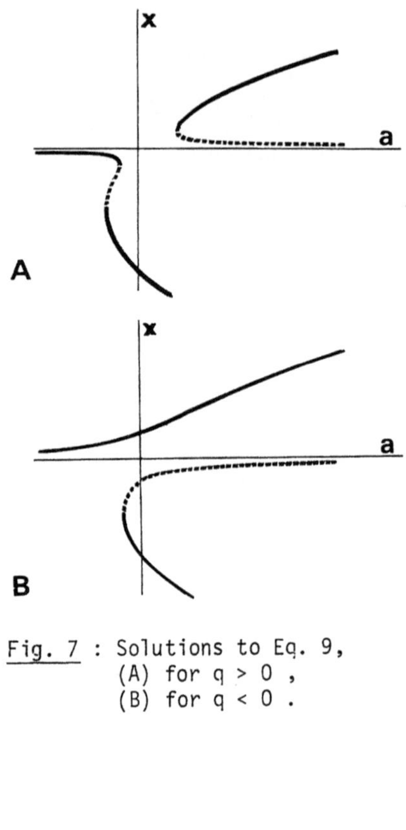

Fig. 7 : Solutions to Eq. 9, (A) for $q > 0$, (B) for $q < 0$.

$$x^3 - ax + r(1+\rho x^2) = 0 \qquad (11)$$

giving the diagrams of Fig. 8. For example, in Prof. Brun's account at this meeting of his work on rasers (radio-frequency masers), we have the undotted parts of Fig. 8, representing stable and metastable states. There, r corresponds to the strength of an asymmetric external field, with ρ a constant given by the expansion of the field effect. If a different asymmetric effect is added (and many are usually physically possible), its expansion will generally give a term $s(1 + \sigma x^2)$ with $\sigma \neq \rho$ to add to (11). Setting

$$p = (\rho r + \sigma s) \ , \quad q = r + s \qquad (12)$$

we recover Eq. 9. Thus we see that near $r + s = 0$, i.e. near where the two asymmetric terms cancel to first order, we may expect to see hysteresis effects and sudden jumps in raser output.

Very generally, Figs. 6 and 7 live arbitrarily close to Fig. 5 and can usually be realized both theoretically - by the addition of physically meaningful asymmetric terms - and experimentally, if symmetry is not deeply built in to the nature of the problem that has been complicated by the simplifying assumption of symmetry. (For such an 'unfolded' analysis of the Dicke laser hamiltonian, and many other physical models, see [7]). Furthermore, no other behaviour is nearby in the same strong sense.

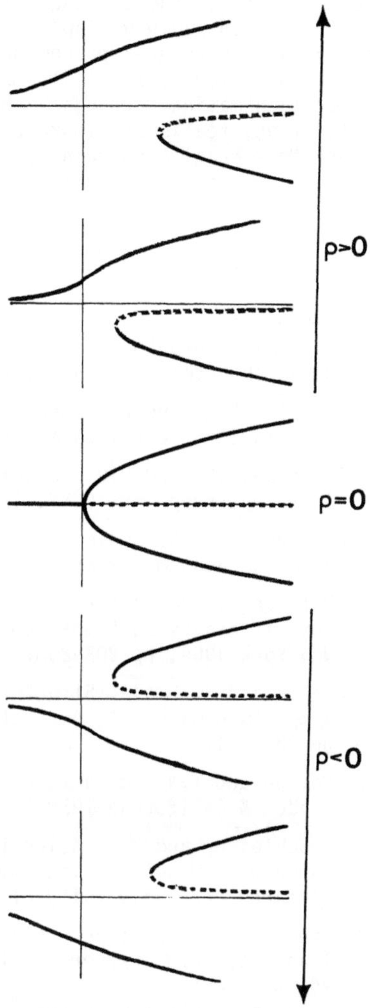

Fig. 8: Solutions to Eq.11 for various ρ.

General considerations of this type would apply equally to - for instance - Prof. Kuramoto's discussion of the stability of chemical plane waves. By his symmetry assumptions, the straight solution is compelled to lose stability at a bifurcation like Fig. 5. One can predict with confidence that chemically meaningful asymmetry terms, however small, can produce smooth onset of nonstraightness of the wave (by $\rho \neq 0$ in Fig. 8) or jump to a suddenly increased sinuosity (by Fig. 7A). Nothing else, locally, can happen if only small terms may be added. (The progress to further bifurcations, leading finally to chaos, described by Prof. Kuramoto, is of course independent of these remarks).

These techniques apply equally to more complex bifurcations than Fig. 5, as a wealth of examples shows. Where the problem is dominated by a potential or other real valued function, see in particular [7]; for more general problems, the methods of Golubitsky and Schaeffer become particularly helpful. The general utility of this approach is not dependent on the existence of a potential, as is often supposed - indeed, the universality of Eq.9 relative to Eq. 7 requires, for technical reasons, that we ignore the fact that it comes from a potential. (Eq. 5 cannot be 'unfolded' in the same exhaustive way, if we insist on keeping the 'bifurcation variable' $\frac{1}{2} - \mu/\gamma$ distinct from 'perturbation variables' like p and q).

The relevance of these methods to science depends, finally, only on the relevance of differentiable variables. Any science using these in an essentially nonlinear way will encounter phenomena like those above (for an intuitive explanation of their omnipresence, see [7]). Moreover the mathematics will be able both to query - on mathematical grounds - the robustness of particular models, and to suggest, often, realisable improvements where these queries cannot be settled by reference to special constraints. This will remain true, by the fundamental character of topology, even if physics again changes as radically as it has since topology was created by Poincaré.

Bibliography

1. D.R.J. Chillingworth, Differential Topology with a View to Applications, Research Notes in Math. 9, Pitman, London 1976.
2. M.A. Golubitsky and D.G. Schaeffer, A Theory for Imperfect Bifurcation via Singularity Theory, Comm. Pure & Appl. Math., XXXII (1979), 21-98.
3. M.A. Golubitsky and D.G. Schaeffer, Imperfect Bifurcation in the Presence of Symmetry, preprint, CUNY, 1978.
4. M.W. Hirsch and S. Smale, Differential Equations, Dynamical Systems and Linear Algebra, Academic Press, New York 1974.
5. B. Malgrange, The preparation theorem for differentiable functions, in Differential Analysis, Bombay Colloquium, Oxford University Press, Oxford and New York 1964, pp 203-208.
6. J. Palis and S. Smale, Structural Stability Theorems, in Global Analysis, Proc. Symp. Pure Math. 14, Am. Math. Soc., Providence, Rhode Island, 1970, pp 223-231.
7. T. Poston and I.N. Stewart, Catastrophe Theory and its Applications, Pitman, London & California 1978.
8. D.G. Schaeffer and M.A. Golubitsky, Bifurcation Analysis near a Double Eigenvalue of a Model Chemical Reaction, MRC Technical Summary Report 1859, Math. Research Ctr, University of Wisconsin 1978.
9. R. Thom, Stabilité Structurelle et Morphogénèse, Benjamin, New York, 1972. Translated D.H. Fowler as Structural Stability and Morphogenesis, Benjamin-Addison Wesley 1975.
10. E.C. Zeeman, Catastrophe Theory-Selected Papers 1972-1977, Addison-Wesley, 1977.

Stationary and Time Dependent Solutions of Master Equations in Several Variables

J.Wm. Turner

Faculté des Sciences, Université Libre de Bruxelles, Campus Plaine C.P. 231
B-1050 Bruxelles, Belgium

1. Introduction

It is a well known fact that exact solutions of the Master Equation

$$\frac{dP(\{X_i\},t)}{dt} = \sum_\rho w(\{X_i - r_{i\rho}\}\to\{X_i\})P(\{X_i - r_{i\rho}\},t)$$

$$-\sum_\rho w(\{X_i\}\to\{X_i + r_{i\rho}\})P(\{X_i\},t)$$

which describes the time evolution of the probability function $P(\{X_i\},t)$ in terms of the various gain and loss processes,are few and far between.Already in the case of one variable,the absence of detailed balance precludes any straightforward general solution, even for the stationary solution. (1,2)

Our purpose here is to study the stationary and time-dependent solutions of a Master Equation in more than one variable, describing a given chemical kinetic scheme in the case where the stationary solution of the corresponding deterministic equations becomes unstable and a stable limit cycle solution appears.

2. Stationary and time dependent solutions.

For the sake of illustration (see remarks at the end of this section) we shall consider the so-called Brusselator scheme (2)

$$A \to X$$
$$B + X \to Y + D$$
$$2X + Y \to 3X$$
$$X \to E$$

where the concentrations of A and B are kept fixed. The Master Equation for the stationary solution reads

$$0 = A\left(P_s(X-1,Y;\not{t})-P_s(X,Y;\not{t})\right)+ \frac{B}{\Omega}\left((X+1)P_s(X+1,Y-1;\not{t})-XP_s(X,Y;\not{t})\right)$$
$$+\frac{1}{\Omega^2}\left((X-1)(X-2)(Y+1)P_s(X-1,Y+1;\not{t})-X(X-1)YP_s(X,Y;\not{t})\right)$$
$$+(X+1)P_s(X+1,Y;\not{t})-XP_s(X,Y;\not{t}). \qquad (1)$$

Put $X=\Omega\xi$, $Y=\Omega\eta$, $A=\Omega\alpha$, $B=\Omega\beta$ and $P_s=\exp\Omega\Phi^s(\xi,\eta)$ to obtain

$$0 = \alpha\left(\exp\{\Omega\Phi^s(\xi-\Omega^{-1},\eta;\not{t})-\Omega\Phi^s(\xi,\eta;\not{t})\}-1\right)$$
$$+\beta(\Omega\xi+1).\exp\{\Omega\Phi^s(\xi+\Omega^{-1},\eta-\Omega^{-1};\not{t})-\Omega\Phi^s(\xi,\eta;\not{t})\} -\beta\Omega\xi$$
$$+\Omega^{-2}(\Omega\xi-1)(\Omega\xi-2)(\Omega\eta+1).\exp\{\Omega\Phi^s(\xi-\Omega^{-1},\eta+\Omega^{-1};\not{t})-\Omega\Phi^s(\xi,\eta;\not{t})\}$$
$$-\Omega^{-2}\Omega\xi(\Omega\xi-1)\Omega\eta$$
$$+(\Omega\xi+1).\exp\{\Omega\Phi^s(\xi+\Omega^{-1},\eta;\not{t})-\Omega\Phi^s(\xi,\eta;\not{t})\} -\Omega\xi \qquad (2)$$

where Ω represents the size of the system.

We now assume Φ to have a Taylor expansion around (ξ,η), i.e.

$$\Phi^s(\xi-\Omega^{-1},\eta)=\Phi^s(\xi,\eta) -\Omega^{-1}.\partial_\xi\Phi^s(\xi,\eta)+ O(\Omega^{-2}) \;\; ; \;\; \ldots$$

and Φ itself to be expandable in powers of Ω^{-1}, namely

$$\Phi^s(\xi,\eta)= \Phi_0^s(\xi,\eta) + \Phi_1^s(\xi,\eta).\Omega^{-1} + \Phi_2^s(\xi,\eta)\;.\Omega^{-2} + \ldots$$

The dominant term of (2) is found to be

$$0 = \alpha\left(\exp\{-\partial_\xi\Phi_0^s\}-1\right) +\beta\xi\left(\exp\{\partial_\xi\Phi_0^s-\partial_\eta\Phi_0^s\}-1\right)+\xi^2\eta\left(\exp\{-\partial_\xi\Phi_0^s+\partial_\eta\Phi_0^s\}-1\right)$$
$$+\xi\left(\exp\{ \partial_\xi\Phi_0^s\}-1\right). \qquad (3)$$

We now seek a solution of (3) which is expandable around the stationary solution (u,v) of the deterministic equations

$$0= \alpha-(\beta+1)u+u^2v$$
$$0= \beta u-u^2v. \qquad (4)$$

Inserting

$$\Phi_0^s(\xi,\eta)= \sum_r\sum_s a_{rs}(\xi-u)^r(\eta-v)^s \quad (r,s\geq 0)$$

into (3) and identifying same powers of $\xi-u$ and $\eta-v$ leads to sets of equations which, but for the set concerning a_{20}, a_{11}, a_{02}, prove to be linear. Furthermore the set of equations for the second order terms can easily be transformed into a linear set by noticing that to this order the description is equivalent to a Gaussian approximation. As there exists a linear set of equations connecting the variances $\langle(\delta\xi)^2\rangle$, $\langle(\delta\eta)^2\rangle$ and the covariance $\langle(\delta\xi\;\delta\eta)\rangle$ in this case, the appropriate transformation is obtained by identifying the quadratic form

$$a_{20}(\xi-u)^2+2a_{11}(\xi-u)(\eta-v)+a_{02}(\eta-v)^2$$

with the standard quadratic form as it appears in the bivariate normal distribution, viz.

$$\frac{1}{2}(1-\rho^2)^{-1}\left((\xi-u)^2\sigma_\xi^{-2}-2\rho(\xi-u)(\eta-v)\sigma_\xi^{-1}\sigma_\eta^{-1}+(\eta-v)^2\sigma_\eta^{-2}\right)$$

where $\sigma_\xi^2=\langle(\delta\xi)^2\rangle$, $\sigma_\eta^2=\langle(\delta\eta)^2\rangle$ and $\rho=\langle(\delta\xi\ \delta\eta)\rangle\sigma_\xi^{-1}\sigma_\eta^{-1}$.

Despite the non-linear aspect of (3), the calculation of the coefficients a_{rs} is thus seen to reduce to a linear problem. It is well known that if β lies under a critical value $\beta_c=1+\alpha^2$, the stationary solution (u,v) is stable. Beyond this value, the stationary solution becomes unstable and a stable limit cycle solution appears. Correspon--dingly, when $\beta<\beta_c$, Φ_0^s has a local maximum at (u,v); as β increases and goes through β_c, Φ_0^s gradually flattens out until when $\beta>\beta_c$, Φ_0^s has a local minimum at (u,v). In this latter case, Φ_0^s takes on a crater-like aspect: the bottom of the crater is projected onto (u_{st},v_{st}) and the lip of the crater is projected onto the deterministic limit cycle in the (u,v) plane. From a practical point of view however, the convergence of the series solution for Φ_0^s is poor but for values of β in a very small neighbourhood of β_c.

The case $\beta>\beta_c$ is better studied by considering the time-dependent version of (2). Replacing the left hand side of (2) by $\partial_t\Phi(\xi,\eta;t)$, and making similar assumptions as to the expansions of $\Phi(\xi,\eta;t)$, we obtain an equation similar to (3), where Φ_0^s is replaced by $\Phi_0(\xi,\eta;t)$ and the L.H.S. by $\partial_t\Phi_0(\xi,\eta;t)$. We now seek a solution in the neighbourhood of the deterministic limit cycle: let $(u(t),v(t))$ be the periodic solutions of the deterministic equations

$$\dot u=\alpha-(\beta+1)u+u^2v$$
$$\dot v=\beta u-u^2v.$$

As before, insert

$$\Phi_0(\xi,\eta;t)=\sum_r\sum_s a_{rs}(t)(\xi-u(t))^r(\eta-v(t))^s$$

and identify same powers of $(\xi-u(t))$ and $(\eta-v(t))$. To lowest order, i.e. $r+s=2$, a set of three closed non-linear equations is obtained, the solution of which is somewhat more involved than in the stationary case. Nevertheless the following two important results can be shown:

The quadratic form

$$Q\equiv a_{20}(t)(\xi-u(t))^2+2a_{11}(t)(\xi-u(t))(\eta-v(t))+a_{02}(t)(\eta-v(t))^2$$

has at all times its principal axes coinciding with the tangent and normal to the limit cycle at (u(t),v(t)). Furthermore the curvature of Q along the tangent is zero.

Although the details given so far concern a specific reaction, they can be shown to hold for any scheme having a stable limit cycle:one of the principal axes of the quadratic form Q always lies along the tangent to the limit cycle.The remaining axes lie thus in the normal subspaces to the limit cycle at each of its points.There is however no apparent connection between these axes and the geometric properties of the limit cycle.For instance in the case of three variables, the two remaining axes of Q do not in general coincide with the binormal and the principal normal to the cycle.

The connection between the stationary solution $\Phi^s(\xi,\eta)$ and the time dependent solution $\Phi(\xi,\eta;t)$ is obtained by noticing that the stationary solution P_s of the Master Equation can be seen as a time average over a period of the limit cycle of P(t),the time dependent solution. It follows then from the theory of asymptotic integrals that

$$\Phi_0^s(\{X_i\}) = \sup_{t\varepsilon(0,T)} \Phi_0(\{X_i\},t)$$

$$\Phi_1^s(\{X_i\}) = \Phi_1(\{X_i\},t^\dagger\{X_k\})$$

$t^\dagger\{X_k\}$ being the value of t for which $\Phi_0(\{X_i\},t)$ reaches its greatest value.

Φ_0^s can thus be seen as the *envelope* of $\Phi_0(t)$ as it moves in time. For instance in the case of two variables, the dome shaped $\Phi_0(t)$ will have, as it moves along the neighbourhood of thelimit cycle, an envelope Φ_0^s which can easily be imagined to be crater shaped.

3. Time and Space averages

Let u(t),v(t) be the limit cycle solutions (of period T) of the deterministic equations of a given chemical kinetic scheme in two variables, and let f(u,v) be a sufficiently smooth function.

Consider the time average of f

$$\overline{f} = \lim_{\tau\to\infty} \frac{1}{\tau} \int dt\ f(u(t),v(t)) = \frac{1}{T} \int dt\ f(u(t),v(t))$$

the second integral being taken over a period T of the limit cycle. Consider on the other hand the space average of f

$$\langle f\rangle = \iint du\, dv\ f(u,v)\ P_s(u,v)$$

where P_s is the stationary solution of the Master Equation.

To evaluate this latter integral asymptotically, note that the dominant contribution comes from the neighbourhood of the limit cycle, and that by rotating the axes till they coincide with the normal and tangent directions (ν,s) to the limit cycle, one recovers a one dimensional integral

$$\langle f \rangle = \int_{\Gamma} ds \; \rho(s) \; f(u(s),v(s))$$

where the curvilinear integral is taken along the limit cycle Γ and

$$\rho(s) \sim (\exp \Phi_1^S(u(s),v(s))) . |\partial^2_{\nu\nu}\Phi_0^S|^{-\frac{1}{2}}$$

It follows as a consequence of the equations for Φ_0^S and Φ_1^S that this density $\rho(s)$ (and not it should be pointed out simply the weight exp Φ_1^S of the delta function in the thermodynamic limit) is just the reciprocal of the tangential speed $v(s)$ of the deterministic point along the limit cycle. Consequently

$$\bar{f} = \langle f \rangle$$

obtained in this case in the absence of a hamiltonian formalism.

Details and more explicit calculations will be appearing in a forthcoming publication.

References

An outline and summary of the various approximations methods available for solving the Master Equation as well as further references can be found in

1. Haken,H., Synergetics,An Introduction (Springer-Verlag) Chapter 4, §§5-11.
2. Nicolis,G. & Prigogine, I., Self-Organization in Nonequilibrium Systems (Wiley) Chapter 10.

Poissonian Techniques for Chemical Master Equations

C.W. Gardiner

I. Institut für Theoretische Physik der Universität Stuttgart
D-7000 Stuttgart 8, Fed. Rep. of Germany and
Physics Department, University of Waikato
Hamilton, New Zealand

1. Introduction

I shall present here a review and updating of the work on Poisson Representations and, in particular, give the theoretical justification for the use of "negative noise", which we originally introduced without rigorous justification a few years ago.

The aim of the Poisson Representation is to reduce the chemical master equations, introduced and used by many authors (1-9) to Fokker-Planck and Stochastic Differential equations. Professor Arnold has given in his article in this volume, certain limits under which the master equation converges to a Fokker-Planck equation. Our work here can be viewed as a different way of achieving the same aim. The poisson representation transforms a much more restricted class of master equations exactly into Fokker-Planck equations.

The Fokker Planck equations which arise are equivalent to stochastic differential equations, which, however, are not the same as conventional chemical Langevin equations, which have also been widely used, and which are the processes to which an ppropriate master equation can be shown to converge, as professor Arnold has demonstrated. I shall show that the Fokker-Planck equations which result from the Poisson representation have one very significant property; namely, that the diffusion constant need not be positive semidefinite. There are two principal ways of resolving the consequential problems, both of which involve defining stochastic processes in the complex plane. One of these methods gives Fokker-Planck equations defined on contours, and thus involves analytic functions. The other involves stochastic processes in independent real and imaginary parts of a complex variable. This can be shown to be equivalent to a complex langevin equation, in which the coefficient of the fluctuating force is obtained simply by taking the square root of the (non positive semidefinite) diffusion coefficient. We thus justify a procedure which we have been using without proof up to now. The advantages of these methods are i) By complex variable methods we may develop asymptotic representations, etc. in a particularly transparent manner ii) The Langevin equations are easy to manipulate, and are particularly adapted to perturbation theory. Furthermore, the ability to manipulate and change variables (using the Itô calculus) brings to bear on chemical master equations the powerful techniques developed for stochastic differential equations.

The results given in this paper have been derived and developed in a series of papers (refs. 14, 15, 16, 17, 18, 19), and have been obtained in collaboration with S. Chaturvedi, D.F. Walls, P. D. Drummond and the late I.S. Matheson.

2. Definition of the Poisson Representation

We consider P(x) to be the probability that a system contains x particles (x = 0, 1,...), and assume an expansion in poisson distributions thus

$$P(x) = \int d\mu(\alpha) \quad f(\alpha) \left[\frac{\alpha^x e^{-\alpha}}{x!}\right] \tag{1}$$

We note the following points concerning equation (1).

I) f(α) is a quasiprobability, which may take on a variety of values, (positive, negative, complex), depending on the measure μ(α). However we readily derive the following moment formulae: if

$$\langle x^r \rangle_f = \sum_x x(x-1) \; .... \; (x-r+1)P(x) \tag{2}$$

and $$\langle \alpha^r \rangle = \int d\mu(\alpha) \; f(\alpha)\alpha^r \tag{3}$$

then $$\langle x^r \rangle_f = \langle \alpha^r \rangle \tag{4}$$

In particular, putting r = 0 shows that f(α) is normalised to 1.

II) Measure dμ(α) is adaptable to the circumstances. We have found the following choices to be useful

a) dμ(α) dα : α has range on the real line

b) dμ(α) dα : α varies over a contour in the complex plane

c) dμ(α) $d^2\alpha$: α varies over the whole complex plane.

III) Existence of the Poisson Representation

We have shown[18,19] that, under certain circumstances, representations of types (a) and (b) do not always exist, but we can always find a representation of type (c) in which, furthermore, f(α) is non-negative. Thus, in the last case, f(α) can be regarded as a genuine probability distribution.

IV) Uniqueness and Inversion

Representations of types (a) and (b) when they exist are normally unique[18,19], but representations of type (c), always exist, and are generally not unique. This non - uniqueness is a result of the analyticity of the Poisson function

$$\frac{e^{-\alpha} \; \alpha^x}{x!} \; .$$

For example let $$f(\alpha) = [2\pi\sigma^2]^{-1} \exp\left[-[\alpha-\alpha_0]^2/2\sigma^2\right] \tag{5}$$

Notice that if $g(\alpha)$ is any analytic function of α, we can expand

$$g(\alpha) = g(\alpha_0) + \sum_{n=1} \frac{d^n g(\alpha_0)}{d\,\alpha_0^{\,n}} \frac{(\alpha-\alpha_0)^n}{n!} , \tag{6}$$

and thus

$$\int [2\pi\sigma^2]^{-1} \; d^2\alpha \; \exp\,[-[\alpha-\alpha_0]^2/2\sigma^2]\; g(\alpha) = g(\alpha_0) \tag{7}$$

since the terms in (6) with $n \geq 1$ vanish when integrated as in (7). Noting next that $\frac{e^{-\alpha}\alpha^x}{x!}$ is an analytic function of α, we see that by choosing $f(\alpha)$ according to (5) we obtain, for any value of σ^2,

$$P(x) = \int d^2\alpha f(\alpha) \frac{e^{-\alpha}\alpha^x}{x!} = \frac{e^{-\alpha_0}\alpha_0^{\,x}}{x!} \tag{8}$$

In practice it does not matter that $f(\alpha)$ is not unique. In what follows, all that will be needed is the existence of at least one $f(\alpha)$.

V) Two Time Correlations (Ref. 16)

We very often deal with a markovian system, for which we define time dependent joint and conditional probabilities. The 2 time correlation function is defined by

$$\langle x(t),\, x(t')\rangle = \langle x(t)\; x(t')\rangle - \langle x(t)\rangle\langle x(t')\rangle \tag{9}$$

where, in the above

$$\langle x(t)\rangle = \sum_{x,x'} x\, P(x,t;\, x',\, t') \tag{10}$$

$$\langle x(t')\rangle = \sum_{x,x'} x' P(x,t;\, x',t') \tag{11}$$

and

$$\langle x(t)x(t')\rangle = \sum_{x,x'} x\; x'\; P(x,t;\, x',t'). \tag{12}$$

In a markovian system, $P(x,t;x',t')$, which is the joint probability that the system had values x and x' at times t and t', factorizes into a conditional probability and an unconditional probability,

$$P(x,t;x',t') = P(x,t \mid x',t')\; P(x',t'). \tag{13}$$

Under these conditions we have shown that the Poisson representation can give a very nice formula for the 2 time correlation function. Let

$$\langle \alpha(t) \mid [\,\alpha',t']\rangle = \int f(\alpha,t \mid [\alpha',t']\,)\,\alpha\, d\mu(\alpha), \tag{14}$$

where $f(\alpha,t| [\alpha',t'])$ is the conditional quasiprobability, of having α at time t, given that its value was α' at time t', thus

$$f(\alpha,t|[\alpha',t']) = \delta(\alpha-\alpha') \tag{15}$$

Then with some manipulation, one can show that

$$\langle x(t),\ x(t')\rangle = \langle\alpha(t),\ \alpha(t')\rangle + \left\langle\alpha' \frac{\partial}{\partial\alpha'}\langle\alpha(t)|[\alpha',t']\rangle\right\rangle \tag{16}$$

The first term in (16) is defined by a similar procedure to that used for eqs. (9), (10), (11), (12) - the second term is a response function, namely it is the average over the unconditional distribution at time t' of the response of the conditional mean (14) to a change in its initial condition α' at time t'. In reference 16 we treat this result quite fully, and show that it is a generalisation of a result of Bernard + Callen[20] to nonequilibrium situations.

3. Application to Chemical Master Equations

For chemical reactions with only bimolecular steps, the poisson representation converts the chemical master equation for P(x) into a Fokker-Planck equation exactly[14]. For example, let us consider the Master equation for Schlögl's reaction

$$A + X \underset{k_4}{\overset{k_2}{\rightleftharpoons}} 2X \tag{18a}$$

$$B + X \underset{k_2}{\overset{k_1}{\rightleftharpoons}} C \tag{18b}$$

for which the appropriate chemical master equation is[14]

$$\frac{dP(x,t)}{dt} = k_1 B(x+1)P(x+1,t) + k_3 C\, P(x-1,t) + k_2 A(x-1)\, P(x-1,t) + k_4(x+1)xP\,(x+1,t) - k_1 Bx + k_3 C + k_2 Ax + k_4 x(x-1)\, P(x,t) \tag{19}$$

Then by using a poisson representation of type (a) or (b), and appropriately integrating by parts, one obtains the Fokker-Planck equation

$$\frac{\partial f(\alpha,t)}{\partial t} = -\frac{\partial}{\partial\alpha}\left\{\left[\kappa_3 V + (\kappa_2 - \kappa_1)\alpha - \kappa_4 V^{-1}\alpha^2\right] f(\alpha,t) + \frac{1}{2}\frac{\partial}{\partial\alpha}\left[2(\kappa_2\alpha - k_4 V^{-1}\alpha^2) f(\alpha,t)\right]\right\} \tag{20}$$

where $k_3 C = \kappa_3 V$, $k_2 A = \kappa_2$, $k_1 B = \kappa_1$, $k_4 = \kappa_4 V^{-1}$,

where we have explicitly indicated the dependence of the parameters on the system volume V.

However, it is not necessary to derive the Fokker-Planck equation separately for every different reaction. We have derived rules for writing the appropriate Fokker-Planck equation for an arbitrary set of coupled chemical reactions, as long as these contain only uni- and bimolecular steps.

Rules for writing Poisson representation Fokker-Planck and Langevin equations. Directly from Chemical Reactions.

n component system with s different reactions (components X_i

$$\sum_{i=1}^{n} N_i^p X_i \underset{k_p^B}{\overset{k_p^F}{\rightleftharpoons}} \sum_{i=1}^{n} M_i^p X_i \qquad (p = 1,2,\ldots s)$$

Define

$$\text{I} \quad J_p(\alpha) = \prod_{i=1}^{n} k_p^F \alpha_i^{N_i^p} - \prod_{i=1}^{n} k_p^B \alpha_i^{M_i^p}$$

$$\text{II} \quad A_i^p = M_i^p - N_i^p$$

$$\text{III} \quad B_{ij}[J(\alpha)] = \sum_{p=1}^{s} J_p(\alpha) \Big\{ \delta_{ij}\big[M_i^p(M_i^p-1)-N_i^p(N_i^p-1)\big] + (1-\delta_{ij})\big[M_i^p M_j^p - N_i^p N_j^p\big]\Big\}$$

Fokker Planck Equation (if only bi+unimolecular steps)

$$\frac{\partial f}{\partial t} = -\sum_{i=1}^{n} \frac{\partial}{\partial \alpha_i} \Big[\sum_p A_i^p J_p(\alpha) f(\alpha,t)\Big] + \frac{1}{2}\sum_{ij=1}^{n} \frac{\partial^2}{\partial\alpha_i \partial\alpha_j} \Big[B_{ij}\,[J(\alpha)]\, f(\alpha,t)\Big]$$

Corresponding Itô stochastic differential equation

$$\frac{d\alpha_i}{dt} = \sum_p A_i^p J_p(\alpha) + \sum_j \Big\{B\,[J(\alpha)]\Big\}_{ij}^{1/2} \xi_j(t)$$

where $\langle \xi_j(t) \rangle = 0$

$$\langle \xi_i(t)\, \xi_j(t') \rangle = \delta_{ij}\delta(t-t')$$

4. Fokker-Planck and Langevin Equations in the Complex Plane.

A: Example: Consider Schlögl's reaction, given in equations (18), (19) and (20). A steady state solution of the Fokker-Planck equation (20) is given, up to a normalisation factor by

$$f_s(\alpha) = e^{\alpha}(\kappa_2 V - \kappa_4 \alpha)^{V(\kappa_1/\kappa_4 - \kappa_3/\kappa_2) - 1} \alpha^{(\kappa_3 V/\kappa_2 - 1)} \tag{21}$$

This is a solution obtained by setting equal to zero the term inside the curly bracket of eq. (20). If one uses the poisson representation and derives (20) directly (using integration by parts) it is not difficult to show that eq. (21) gives a steady state solution of (19) provide (21) satisfies

I $\quad f_s(0) = 0$

II $\quad (\kappa_2\alpha - \kappa_4 V^{-1}\alpha^2) f_s(\alpha) = 0$ when $\kappa_4\alpha = \kappa_2 V$ (22)

III $\quad$ The range of α is $0, \kappa_2 V/\kappa_4$

Condition I) requires $\kappa_3/\kappa_2 > 1/V$ for its validity. We shall assume this in what follows. We now distinguish three sitiations depending on the value of $\delta = \kappa_1/\kappa_4 - \kappa_3/\kappa_2$. The quantity δ gives a measure of the direction in which the reaction system (18) is proceeding when a steady state exists. If $\delta > 0$, we find that, when X has its steady state value, reaction (18a) is producing X, while reaction (18b) consumes X. When $\delta = 0$, both reactions balance separately - thus we have chemical equilibrium. When $\delta < 0$, reaction (18a) consumes X, while reaction (18b) produces X.

(I) $\delta > 0$

According to equation (22) this is the condition for $f(\alpha)$ to be a valid quasi probability on the real interval $[0, \kappa_2 V/\kappa_4]$. In this range, the diffusion coefficient $D(\alpha) = 2(\kappa_2 - \kappa_4\alpha^2/V) \geq 0$.
The deterministic mean of α, given by

$$\alpha = \frac{V}{2\kappa_4}\left[(\kappa_2 - \kappa_1) + \sqrt{(\kappa_2 - \kappa_1)^2 + 4\kappa_3\kappa_4}\,\right] \tag{23}$$

lies within the interval $[0, \kappa_2 V/\kappa_4]$. We are therefore dealing with the case of a genuine Fokker-Planck equation and $f_s(\alpha)$ is a function vanishing at both ends of the interval, and peaked near the deterministic steady state.

(II) $\delta = 0$

Since both reactions now balance separately, we expect a poissonian steady state. We note that $f_s(\alpha)$ in this case has a pole at $\kappa_2 V = \kappa_4\alpha$, and we choose the range of α to be a contour in the complex plane enclosing this pole. Since this is a closed contour, there are no boundary terms arising from partial integration, and $P_s(x)$ given by choosing this type of poisson representation clearly satisfies the steady state master equation. Using now the calculus of residues, we see that

$$P_s(x) = \frac{e^{-\alpha_0}\alpha_0^{\,x}}{x!} \tag{24}$$

with $\alpha_0 = \kappa_2 V/\kappa_4$

(III) $\delta < 0$

When $\delta<0$ we meet some very interesting features. The steady state solution eq. (21) now no longer satifies the condition (22). However, if the range of α is chosen to be a contour C in the complex plane (Fig. 1)

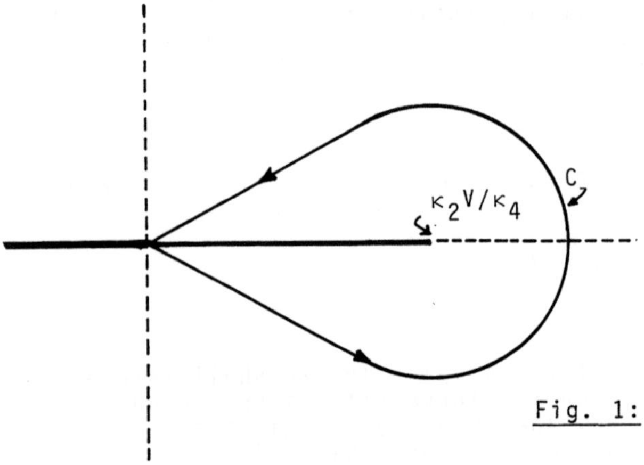

Fig. 1: Contour C used when $\delta>0$.

it is not difficult to show that $P_s(x)$ constructed as

$$P_s(x) = \int_C d\alpha f_s(\alpha) \frac{e^{-\alpha}\alpha^x}{x!} \tag{25}$$

is a solution of the master equation. The deterministic steady state now occurs at a point on the real axis to the right of the singularity at $\alpha = \kappa_2 V/\kappa_4$, and asymptotic evaluations of means, moments, etc. may be obtained by choosing C to pass through the saddle point that occurs there. In doing so, one finds that the variance of α, defined as

$$\mathrm{var}\,[\alpha] = \langle\alpha^2\rangle \quad \langle\alpha\rangle^2 \tag{26}$$

is negative, so that

$$\mathrm{var}\,[x] = \langle x^2\rangle - \langle x\rangle^2 = \langle\alpha^2\rangle - \langle\alpha\rangle^2 + \langle\alpha\rangle < \langle\alpha\rangle \tag{27}$$

This means that the steady state is narrower than the poissonian. Finally, it should be noted that all three cases can be obtained from the contour C. In the case that $\delta = 0$, the cut from the singularity at $\alpha = \kappa_2 V/\kappa_4$ to $-\infty$ vanishes, and C may be distorted to a simple contour round the pole, while if $\delta > 0$, the singularity at $\alpha = \kappa_2 V/\kappa_4$ is now integrable, so the contour may be collapsed onto the cut, and the integral evaluated as a discontinuity integral over the range $[0. \kappa_2 V/\kappa_4]$. (When δ is a positive integer, this argument requires modification).

B: Langevin Equations with complex noise

In the situation in which $\delta < 0$ we can formally write a langevin equation equivalent to the Fokker-Planck equation as

$$\frac{d\alpha}{dt} = \kappa_3 V + (\kappa_2-\kappa_1)\alpha-\kappa_4 V^{-1}\alpha^2+ \sqrt{2(\kappa_2\alpha-\kappa_4 V^{-1}\alpha^2)}\,\xi(t) \tag{28}$$

This procedure is not justifiable in general, since the coefficient of the fluctuating force can become imaginary if $\kappa_4\alpha > \kappa_2 V$. In the situation in which δ is positive, we note that the noise term vanishes at $\alpha = 0$ and $\alpha = V\kappa_2/\kappa_4$, and in these cases, the drift term is such as to return α to the range $[0, V\kappa_2/\kappa_4]$ whenever it approaches the end points. Thus for $\delta > 0$, the Langevin equation (28) is valid, since α cannot leave the range in which the noise is real.

If $\delta < 0$, this no longer holds, and we cannot use the Langevin equation (28) without further justification.

We resolve this dilemma by using a type C poisson representation

$$p(x) = \int d^2\alpha f(\alpha)\ \frac{e^{-\alpha}\alpha^x}{x!}$$

If we derive a Fokker Planck equation for $f(\alpha)$, an intermediate step will be

$$\frac{dp}{dt} = \int d^2\alpha f(\alpha)\ \left[a(\alpha)\frac{\partial}{\partial\alpha} + \frac{1}{2}[b(\alpha)]^2\frac{\partial^2}{\partial\alpha^2}\right]\frac{e^{-\alpha}\alpha^x}{x!} \tag{29}$$

Notice, however, that $e^{-\alpha}\alpha^x/x!$ is analytic in $\alpha = \alpha_x + i\alpha_y$, so we may make the substitutions

$$\frac{\partial}{\partial\alpha} \longleftrightarrow \frac{\partial}{\partial\alpha_x} \longleftrightarrow -i\frac{\partial}{\partial\alpha_y} \tag{30}$$

Write now

$$\begin{aligned} a(\alpha_x + i\alpha_y) &= a_R + ia_I \\ b(\alpha_x + i\alpha_y) &= b_R + ib_I \end{aligned} \tag{31}$$

And using (30), replace as follows

$$(a_R + ia_I)\frac{\partial}{\partial\alpha} \longrightarrow a_R\frac{\partial}{\partial\alpha_x} + a_I\frac{\partial}{\partial\alpha_y} \tag{32}$$

$$(b_R^2+2ib_Rb_I - b_I^2)\frac{\partial^2}{\partial\alpha^2} \longrightarrow b_R^2\frac{\partial^2}{\partial\alpha_x^2} + 2b_Rb_I\frac{\partial^2}{\partial\alpha_x\partial\alpha_y} + b_I^2\frac{\partial^2}{\partial\alpha_y^2} \tag{33}$$

so that integrating by parts gives the Fokker-Planck equation

$$\frac{\partial f}{\partial t} = -\frac{\partial}{\partial\alpha_x}a_R f - \frac{\partial}{\partial\alpha_y}a_I f + \frac{1}{2}\left[\frac{\partial^2}{\partial\alpha_x^2}b_R^2 f + 2\frac{\partial^2}{\partial\alpha_x\partial\alpha_y}b_Rb_I f + \frac{\partial^2}{\partial\alpha_y^2}b_I f\right] \tag{34}$$

The noise term is now non negative, so we may make the correspondence between this 2 variable Fokker-Planck equation and a Langevin equation in two variables. Noting that the diffusion matrix can be written

$$D = \begin{pmatrix} b_R^2 & b_R b_I \\ b_I b_R & b_I^2 \end{pmatrix} = \begin{pmatrix} b_R & 0 \\ b_I & 0 \end{pmatrix} \begin{pmatrix} b_R & b_I \\ 0 & 0 \end{pmatrix} \qquad (35)$$

we can write the corrsponding Langevin equations with only one noise source, and these can be conveniently collected together as

$$\frac{d}{dt}(\alpha_x + i\alpha_y) = a_R(\alpha) + ia_I(\alpha) + [b_R(\alpha) + ib_I(\alpha)]\,\xi(t) \qquad (36)$$

This is, of course exactly the same equation as obtained by the simple minded and "illegal" procedure which gives equation (28).

Thus we are led to the conclusion that the "illegal procedure" is in fact valid, and that pure imaginary noise is acceptable.

5. Summary

This paper has presented only a brief outline of how Poisson Representation methods may be used. For a full account, the reader is referred to references 14, 15 and 16. The basic advantages found are those of computational ease and simple physical insight. We note the following points.

I. Complex variable methods using type (a) or (b) representations can give simple asymptotic evaluation methods of mean, variance, etc., in simple one variable system[14].

II. Linear Processes, such as reactions A X , or diffusion in multivariate master equations, contribute only to the drift terms and not the noise terms, which is a considerable simplification over the more conventional chemical master equations. Diffusion can be shown[14] to correspond to a poisson variable $\eta = \alpha/\Delta V$, where ΔV is the size of the cell in the multivariate master equation[6,8,9], which obeys the equation

$$\frac{\partial \eta}{\partial t} = D\,\nabla^2\eta + \kappa_3 + (\kappa_2 - \kappa_1)\eta - \eta^2 + [2(\kappa_2\eta - \eta^2)]^{1/2}\,\xi(r,t) \qquad (35)$$

Here $\langle \xi(r,t), \xi(r',t') = \delta(r-r')\delta(t-t')$, (36)

which differs from the conventional chemical Langevin equation, in which the coefficient of $\xi(r,t)$ is different, and involves gradient operators.

III. Perturbation Theory: if it is desired to evaluate higher order corrections a perturbation theory can easily be developed[14], which corresponds to an expansion in inverse powers of $\sqrt{V}$ of the equation

$$\frac{d\eta}{dt} = a(\eta) + \frac{1}{\sqrt{V}}\,b(\eta)\,\xi(t) \qquad (37)$$

The appropriate perturbation theory is developed extensively in refs. 14 and 15. One should note that we use Itô rules, which mean that, if $\phi(t)$ is an arbitrary function of t,

$$\int_0^t dt'\ \phi(t')\langle\xi(t'),\xi(t)\rangle$$

$$= \int_0^t dt'\phi(t')\delta(t-t')\ (t-t') = 0,\ (t>0) \tag{38}$$

$$= \phi(t) \quad (t<0)$$

(the alternative Stratanovich rules make the result $\phi(t)/2$ in both cases). This is also discussed in ref. 14.

IV. Transformation Theory: The stochastic differential equations may be transformed by the use of nonlinear transformations. Pasquale, Tartalgia + Tombesi have done this (Ref. 21) to the Schlögl model, and have found several interesting properties, analogous with those of Laser instabilities.

V. Analogies with Quantum Optics: In ref. 16, and particularly in ref. 19 we have studied the close connection between the Poisson Representation and the Glauber-Sudarshan P-representation of Quantum optics. We have shown that in Quantum-Optics, we may also find non-positive definite noise, and deal with it in very similar ways, which have found significant useful applications (refs. 18, 22, 23, 24).

References

1) D.A. McQuarrie, J. Appl. Prob. 4, 570 (1967)
2) G. Nicolis + I. Prigogine, Proc. Natl. Acad. Sci. USA, 68, 2102 (1971)
3) M. Malek-Mansour + G. Nicolis, J. Stat. Phys. 13, 197 (1975)
4) K.J. McNeil + D.F. Walls, J. Stat. Phys. 10, 439 (1974)
5) I.S. Matheson, D.F. Walls + C. W. Gardiner, J. Stat. Phys. 12, 21 (1975)
6) C.W. Gardiner, K.J. McNeil, D.F. Walls, I.S. Matheson, J. Stat. Phys. 14, 307 (1976)
7) N.G. van Kampen,"Fluctuations in Continuous Systems" in "Topics in Statistical Mechanics + Biophysics", ed. R. Piciarelli (Am. Ins. Phys., N.Y. 1976)
8) H. Haken, Zeitschrift f. Physik B 20, 413 (1975)
9) H. Haken, Zeitschrift f. Physik B 22, 69 (1975)
10) Joel Keizer, J. Chem. Phys. 63, 398 (1975)
11) A. Nitzan, P. Ortoleva, J. Deutch, J. Ross, J. Chem. Phys. 61, 1056 (1974)
12) C.W. Gardiner, J. Stat. Phys. 15, 451 (1976)
13) S. Grossman, J. Chem. Phys. 65, 2007 (1976)
14) C.W. Gardiner + S. Chaturvedi, J. Stat. Phys. 17, 429 (1977)
15) S. Chaturvedi, C.W. Gardiner, I.S. Matheson, D. F. Walls, J. Stat. Phys. 17, 469 (1977)
16) S. Chaturvedi, C. W. Gardiner, J. Stat. Phys. 18, 501 (1978)
17) C. W. Gardiner, D. F. Walls, J. Phys. A 11, 161 (1977)
18) P.D. Drummond, C.W. Gardiner, J. Phys. A (to appear)
19) C. W. Gardiner, P.D. Drummond, (to appear)
20) W. Bernard, H.B. Callan, Rev. Mod. Phys. 31, 1017 (1959)
21) F. de Pasquale, P. Tartaglia, P. Tombesi, (Rome Preprint) (1979)
22) S. Chaturvedi, P.D. Drummond, D.F. Walls, J. Phys. A10:L187 (1977)
23) P.D. Drummond, K.J. McNeil, D.F. Walls, Opt. Commun. 28:255 (1979)
24) P. D. Drummond, D.Phil. Thesis, Univ. Waikato (unpublished 1979)

Index of Contributors

MIX
Papier aus verantwortungsvollen Quellen
Paper from responsible sources
FSC® C105338

If you have any concerns about our products,
you can contact us on
ProductSafety@springernature.com

In case Publisher is established outside the EU,
the EU authorized representative is:
Springer Nature Customer Service Center GmbH
Europaplatz 3, 69115 Heidelberg, Germany

Printed by Libri Plureos GmbH
in Hamburg, Germany